THE POTOMAC

RIVERS OF AMERICA BOOKS

already published are:

KENNEBEC *by Robert P. Tristram Coffin*

UPPER MISSISSIPPI (New Revised Edition, 1944) *by Walter Havighurst*

SUWANNEE RIVER *by Cecile Hulse Matschat*

POWDER RIVER *by Struthers Burt*

THE JAMES (New Revised Edition, 1945) *by Blair Niles*

THE HUDSON *by Carl Carmer*

THE SACRAMENTO *by Julian Dana*

THE WABASH *by William E. Wilson*

THE ARKANSAS *by Clyde Brion Davis*

THE DELAWARE *by Harry Emerson Wildes*

THE ILLINOIS *by James Gray*

THE KAW *by Floyd Benjamin Streeter*

THE BRANDYWINE *by Henry Seidel Canby*

THE CHARLES *by Arthur Bernon Tourtellot*

THE KENTUCKY *by T. D. Clark*

THE SANGAMON *by Edgar Lee Masters*

THE ALLEGHENY *by Frederick Way, Jr.*

THE WISCONSIN *by August Derleth*
LOWER MISSISSIPPI *by Hodding Carter*
THE ST. LAWRENCE *by Henry Beston*
THE CHICAGO *by Harry Hansen*
TWIN RIVERS: The Raritan and the Passaic, *by Harry Emerson Wildes*
THE HUMBOLDT *by Dale L. Morgan*
THE ST. JOHNS *by Branch Cabell and A. J. Hanna*
RIVERS OF THE EASTERN SHORE *by Hulbert Footner*
THE MISSOURI *by Stanley Vestal*
THE SALINAS *by Anne B. Fisher*
THE SHENANDOAH *by Julia Davis*
THE HOUSATONIC *by Chard Powers Smith*
THE COLORADO *by Frank Waters*
THE TENNESSEE: THE OLD RIVER, Frontier to Secession, *by Donald Davidson*
THE CONNECTICUT *by Walter Hard*
THE EVERGLADES *by Marjory Stoneman Douglas*
THE TENNESSEE: THE NEW RIVER, Civil War to TVA, *by Donald Davidson*
THE CHAGRES *by John Easter Minter*
THE MOHAWK *by Codman Hislop*
THE MACKENZIE *by Leslie Roberts*
THE WINOOSKI *by Ralph Nading Hill*
THE OHIO *by R. E. Banta*
THE POTOMAC *by Frederick Gutheim*

SONGS OF THE RIVERS OF AMERICA, *edited by Carl Carmer*

RIVERS OF AMERICA

EDITED BY

Hervey Allen AND *Carl Carmer*

AS PLANNED AND STARTED BY

Constance Lindsay Skinner

ASSOCIATE EDITOR

JEAN CRAWFORD

ART EDITOR

FAITH BALL

THE POTOMAC

BY

Frederick Gutheim

ILLUSTRATED BY

Mitchell Jamieson

NEW YORK · TORONTO

RINEHART & COMPANY · INC.

Grateful acknowledgment is made to Little, Brown & Co., Boston, Mass., for permission to reprint lines from "First Families, Move Over!" by Ogden Nash, from *The Face Is Familiar*, copyright, 1931, 1933, 1935–1940, by Ogden Nash.

PRINTED IN THE UNITED STATES OF AMERICA

DESIGNED BY STEFAN SALTER

TO MY MOTHER AND FATHER

CONTENTS

MAPS

THE POTOMAC

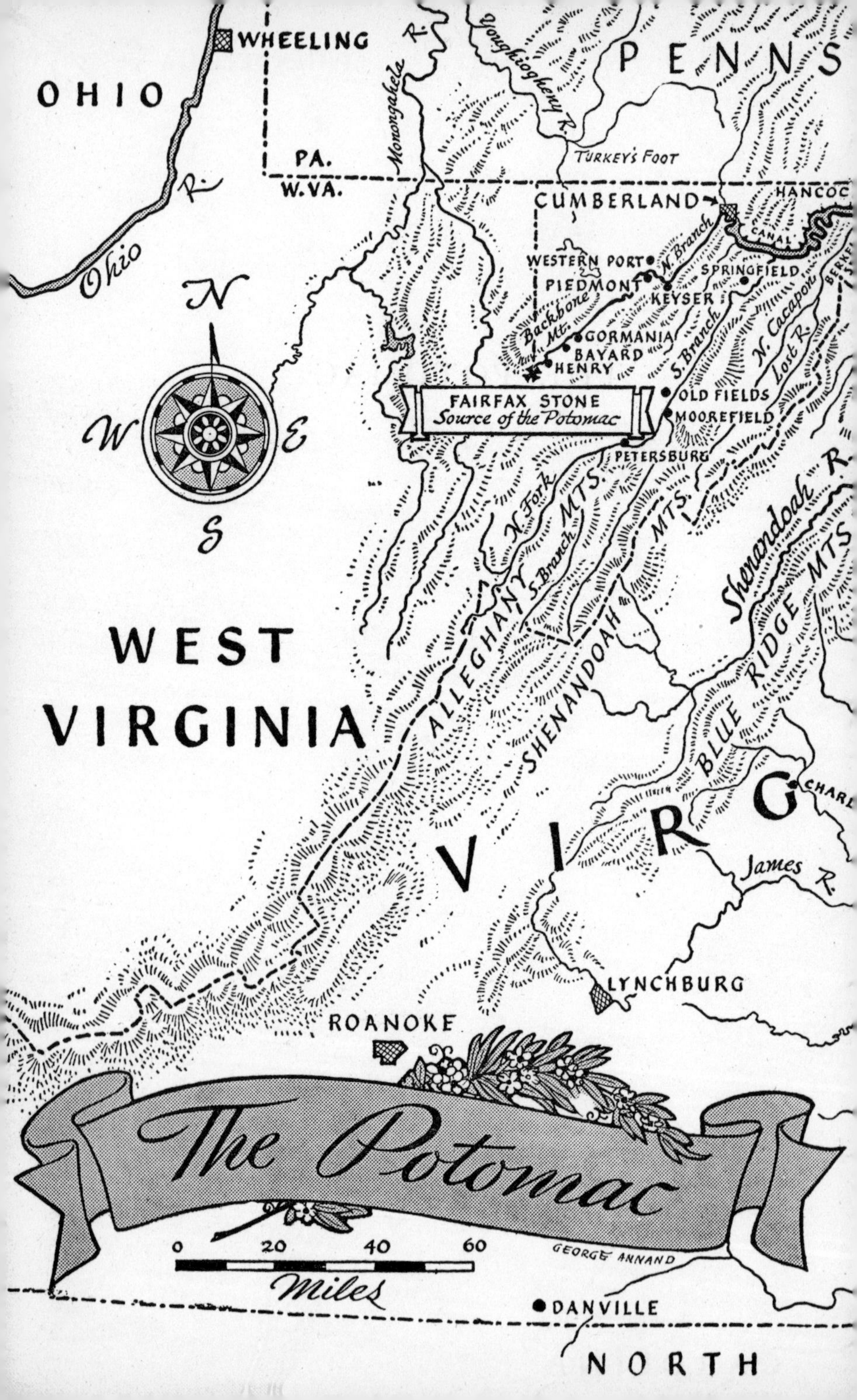

WHEELING
OHIO
Ohio R.
PA.
W.VA.
Monongahela R.
Youghiogheny R.
PENNS
TURKEY'S FOOT
CUMBERLAND
HANCOC
CANAL
WESTERN PORT
N. Branch
PIEDMONT
KEYSER
SPRINGFIELD
Backbone Mt.
GORMANIA
BAYARD
HENRY
S. Branch
N. Cacapon
Lost R.
FAIRFAX STONE
Source of the Potomac
OLD FIELDS
MOOREFIELD
PETERSBURG
N
W
E
S
N. Fork
S. Branch
MTS.
ALLEGHANY
SHENANDOAH
MTS.
Shenandoah R.
BLUE RIDGE MTS.
WEST
VIRGINIA
VIRG
CHARL
James R.
LYNCHBURG
ROANOKE
The Potomac
0 20 40 60
Miles
GEORGE ANNAND
DANVILLE
NORTH

VANIA
Conococheague Cr.
GETTYSBURG
WAYNESBORO
COLUMBIA
YORK
Susquehanna R.
PHILADELPHIA
DELAWARE R.
WILMINGTON
N. J.
HAGERSTOWN
WILLIAMSPORT
Monocacy R.
SHEPHERDSTOWN
FREDERICK
MARYLAND
HAVRE DE GRACE
BALTIMORE
BRUNSWICK
POINT OF ROCKS
HARPERS FERRY
WHITES FERRY
C.&O. Canal
Patuxent R.
LEESBURG
GREAT FALLS
BETHESDA
ANNAPOLIS
Anacostia River
LITTLE FALLS
GEORGETOWN
ALEXANDRIA
WASHINGTON
DELAWARE
Delaware Bay
CAPE MAY
MARYLAND
VIA
PORT TOBACCO
MD
VA.
DEL.
MD.
FREDERICKSBURG
ST. MARYS CITY
Potomac R.
WESTMORELAND
Rappahannock R.
Mattaponi R.
Chesapeake Bay
MD.
VA.
Pamunkey R.
York R.
RICHMOND
Appomattox R.
James R.
JAMESTOWN
CAPE CHARLES
Atlantic Ocean
NEWPORT NEWS
NORFOLK
PORTSMOUTH
CAROLINA

ONE

A POTOMAC PRELUDE

WHEN PRESIDENT JOHN ADAMS FIRST SAW THE NEW CAPITAL OF the United States from the heights of Georgetown, late on a June afternoon in the year 1800, it was a broad panorama of trees, fields, and water. Of the Federal City itself little was yet to be seen. Below him the Potomac spread in all its fullness and grandeur, a tidal waterway nearly half a mile from bank to bank, its majestic stream carrying the largest ships. Toward the south bank Mason's Island filled a good part of the river. On the steeper Georgetown bank, under the tall masts of sailing vessels, rose the sturdy brick and stone warehouses of the Scotch merchants, and above these on higher ground were handsome and substantial residences. Through the heavily wooded slopes that reached to the river's edge roads could be distinguished, and here and there along them scattered groups of houses. Beyond Georgetown was one little settlement, and at a greater distance the yellow frames and new red brick of the houses of Greenleaf's Point stood out against the uniform background of green hills and blue water. Here the widening Potomac, joined by its eastern branch, the Anacostia, flowed on to the Chesapeake, the headlands emphasized here and there by large plantation houses.

President Adams had a cold, calculating New England eye, and however beautiful the prospect that nature presented, that created by the hand of man caused him a few misgivings. He had left the tidy red-brick city of Philadelphia on May 27 in a carriage drawn by four horses, accompanied only by his private secretary, and journeyed a roundabout way through the towns of central Pennsylvania and Maryland. It was an election year, and merely to be seen in those remote places in tie-wig and small-clothes was a capital political advantage. As he approached the Federal City that morning of June 3, the Maryland countryside unrolled before him, a rough and unfinished land, heaving to the west with blue mountain ranges, the forest punctuated occasionally by thin plumes of smoke in the mild morning air. Occasionally the road passed a hut, where the naked wood showed silver as the bark peeled from the logs, or a cluster of cabins around a crossroads tavern or store. In contrast to the lush landscape of the German farmers around Lancaster and York, or the fat prosperity of the thrifty burghers of Fredericktown, what he now saw was a lubberland, a hard and poor country.

Could President Washington have erred when he placed the nation's capital city here on the Potomac? In this ragged, deserted, and unproductive countryside it was hard to see the outlines of the great city Washington had so often promised would arise where a great river, the longest arm of the ocean, reached almost to the mountains beyond which lay the rich lands of the Ohio Valley. Behind the scene that appeared to promise so little to a man who had known the much-traveled, well-manicured roads from Passy to Paris, from Scheveningen to The Hague, or even the sea of clay that connected Bush Hill to Philadelphia, there must be a great faith, a great promise, a design. It was upon George Washington, whose vice-president he had been for eight years, that he must rely once more for his belief in the Federal City, and he rehearsed Washington's theory of the consolidated community arising at the focus of converging highways and waterways.

Washington had explained it again and again. The great coastal road from north to south crossed the Potomac at this point. Colonel George Mason's hand-operated chain ferry plying from the foot of the Frederick road could be seen as evidence of its traffic. To this highest point on the river below

WHERE THE POTOMAC RISES

the falls the largest ships that sailed the ocean could navigate. To reassure a doubting eye, they lay in full view alongside the grain warehouses in Georgetown. Up the river and beyond the falls, which commerce would soon find no barrier because of the new canal, were rich farms; and the river stretched on, a navigable waterway, threading the mountain passes to the very spine of the Appalachians. Here began the road over the mountains which Washington had known for fifty years, which led to the Ohio. Where these great natural routes converged, Washington had convinced many, a great city must inevitably arise.

The theory continued. In the mountains lay iron, coal, forests. In the valleys were farms with fine wheat and fat cattle. As these goods flowed down the valley to the head of navigation at the falls, where every stream was a potential mill stream, a great manufacturing activity would commence to process raw materials into finished goods for trade among the states and with the world. Here bulk cargoes would be broken for shipment to the interior. Here great markets and exchanges would be established. In a diffuse agricultural nation still with few cities, stretching along the narrow coastal plain between the ocean and the mountains, there would be a great center. Here obviously was the proper place to locate the seat of government. Further, in Washington's mind, the rise of manufactures would extinguish slavery in Maryland and Virginia and the greatest step toward homogeneity and national unity would be taken. The interests of the eastern and western states would be united in the trade over the mountains. A nation stronger in fact than even in law would develop, and here would find its natural capital.

All this Washington had told Adams, as he had told Jefferson and Mason and scores of others. This he had told Thomas Johnson, the Marylander who was in charge of the development of the Federal City, and Major Charles Pierre L'Enfant, the French engineer who drew the plan for the capital. Visiting Englishmen around the mahogany table at Mount Vernon had heard it. The widely shared belief had become a commonplace in Georgetown and Alexandria as well as in the new Federal City, the governing motivation in men's acts. The Potomac vision was seen by others, by John Quincy Adams, by Albert Gallatin, and it long animated their view of national policy.

In 1800 it was still a vision, the faith of a man, a vision with obscure roots, hard to materialize. For President Adams, who had never been so far south before, it was enough to plunge him into a thoughtful preoccupation from which he scarcely emerged while he remained in the Federal City. Each new experience there comforted him or raised new doubts.

In Georgetown the presidential party was met by an escort of citizens, led by Francis Deakins, Esq., who accompanied them into the city. A detachment of Marines also greeted them with appropriate salutes. Consonant with the custom of the day, the president listened to an effusive address "expressive of the high respect entertained by the citizens for his character and gratitude for his distinguished services." To this he made a brief reply.

After a night in Georgetown, the president crossed over the Rock Creek bridge and continued into the new city. On his way to Tunnicliffe's Hotel he visited the President's House and the Treasury buildings, still under construction. The following day he was escorted to the Capitol building by its architect, Dr. Thornton, and Tristam Dalton, where he received another address and made a short reply. After the ceremony at the Capitol the president dined with his son's father-in-law, Joshua Johnson. He was to have received guests afterwards, but as the observant Mrs. Thornton recorded in her diary, "The President did not come into the drawing room till tea was ready and went immediately after." She also noted that he regretted being unable to call at the Thorntons' to inspect the plans for the Capitol.

The following day, June 6, a further ceremony was held at dinner in McLaughlin's Tavern in Georgetown, and seventeen toasts were drunk. The President again replied briefly and left early, before sunset.

The next day fell on a Sunday, and was marked by a dinner at the home of Thomas Peter, a leading merchant and banker, who lived next door to the Johnsons, on the street leading from Georgetown to the Federal City. Visitors were invited to come after dinner but, as Mrs. Thornton again noted, "The President went early in the afternoon."

On Monday the President paid his respects to Mrs. Washington at Mount Vernon, and on his return was given a dinner by the citizens of Alexandria. He left early, returning to Georgetown.

The Potomac natives were clearly disappointed. Polite but inquisitive, they were gregarious to the point of waylaying guests at crossroads taverns in the interest of sociability and news. The program to receive the elusive and retiring president had been planned to show their hospitality, but also to satisfy a natural curiosity about strangers. The already cosmopolitan Potomac country easily accustomed itself to dealing with newcomers, and it was prepared to recognize that their habits and values often differed from its own. The natives knew representatives of all sections and all interests, and as citizens of a capital they were now obliged to wait upon and entertain many others who could hardly be ignored. But they expected something in return, and were not used to being disappointed.

It was not part of President Adams's plan to satisfy the natural curiosity of the people who entertained him. His principal business just then was to make sure that his order of May 15 providing for the removal of the government offices from Philadelphia had been carried out, and to inspect the accommodations provided in the city for the government. Beyond that the face he presented to the Potomac people was the one he had earlier revealed to Mrs. Mercy Warren: "You know, madam, that I have no pleasure or amusement which has any charms for me. Balls, assemblies, concerts, cards, horses, dogs, never engaged any part of my attention or concern. Nor am I ever happy in large and promiscuous companies. Business alone, with the intimate, unreserved conversation of a few friends, books, and familiar correspondence, have ever engaged all my time; and I have no pleasure, no ease, in any other way. . . . Thus have I sketched a character for myself, of a morose philosopher and a surly politician." When the painter Copley depicted him in a full-length portrait, standing before the globe, holding a map of

Europe, the artist took the proper measure of Adams's interests. In his world there could be little place for the Potomac, as that region understood itself.

The Washington *Daily Advertiser* had reported on June 11, 1800, that the State, War, Navy, and Post Office departments—all but the Treasury—had been moved from Philadelphia. It sounded impressive, but the total number of employees was only 131. By departments, the numbers were Treasury 69, War 18, Navy 15, Post Office 9, State 7.* They had moved down during the previous month, and small as the federal establishment was, the cost of moving it was dear. Israel Whelan, the purveyor of public supplies, who had been in charge of boxing and packing the property of the departments, chartering the vessels and hiring the wagons to move the government from Philadelphia, reckoned the total cost of this operation at $15,293. An additional cost of $32,872 was accounted for by the moving expenses of the federal employees and their families. An intimate view of the removal is provided in the bill rendered by John Little, a Treasury clerk. With his family of nine, he had spent a leisurely six days on the road (the regular stage schedule between Philadelphia and Washington was but two days). He charged the government $100 for carriage hire and $72 for the expenses of his family en route. An additional $30 he calculated was due him for board and lodging in Philadelphia after his furniture had been shipped, and $30 more for similar expenses in Washington before he could set up house in the capital. Had he thought to include "dead rent" on his Philadelphia house, public stage charges for servants, the expense of coming to Washington in advance to procure a house, the cost of moving his furniture from the warehouse in Washington to his new home, or other items his more imaginative colleagues entered in their accounts, his bill might have exceeded its actual amount of $445. In the rather hazy bookkeeping of those days there was considerable difference between the amounts paid different employees. Nor was it surprising when a committee of the

* Thirteen employees moved were not attached to any department.

House of Representatives during the succeeding administration found that the entire sum had been paid without legal authority.

The departments were variously accommodated. The Post Office moved into three partly finished rooms on two floors of a rented building at 9th and E streets, and the postmaster general's family lived on the third floor. When the secretary of the treasury arrived in the city, he found the furniture and records of his department hopelessly crowded into the limited space of the one executive office building then completed; and in addition, he was not too happy to learn that seven employees of the State Department also were located in the Treasury Office. In a few months this situation was somewhat relieved when the State Department followed the example of the Navy Department and occupied one of the "Six Buildings" between 21st and 22nd streets on Pennsylvania Avenue, across from the War Department.

A great many of the federal employees thought well of their new location. While there was a great housing shortage and rents were high, provisions were plentiful and cheaper than in Philadelphia. They liked the situation and thought the place beautiful, and with good prospects for growth.

Small and unfinished the city certainly was. A recently taken inventory had listed only 109 brick and 263 wooden buildings. The population Census of 1800 found about 5,000 persons each in the established towns of Georgetown and Alexandria, and about 3,000 in the rest of the Federal District. Aside from Georgetown and Alexandria, the houses were scattered widely, and this produced an impression of even fewer buildings: the country thickened up rather than a city, as a later traveler observed. The lack of streets, fences, and gardens reinforced the desolate appearance. Of the largest single new settlement, on Greenleaf's Point, the dour Oliver Wolcott wrote his wife that he saw "fifty or sixty spacious houses, five or six of which are occupied by Negroes and vagrants, and a few more by decent people." There was no sign of business or industry in the Federal City. He generalized: "The people are poor, and as far as I can judge, they live like fishes by eating each other."

Nearly a million dollars had been spent in the nine years since the permanent seat of government had been chosen, but so huge was the task that there was remarkably little to show for it. Of the President's House, for which the young James Hoban had drawn the plans, the exterior only was largely finished. Over one year before the removal, Hoban had reported that the workmen were engaged "in cleaning down and painting the walls of the building, and striking the scaffolds." The Capitol, by the time Congress was ready to meet there, was quite unfinished, save for the north wing. The walls of the south wing were but a few feet above the ground, and only the foundations for the central section had been laid. The Congress met behind the pleasantly designed sandstone walls of the north wing, while the Supreme Court was finally allowed the temporary use of one Senate committee room. The only other public structure was the sole executive building, built for the Treasury Department east of the President's House, a modest Georgian structure economically built of brick, with fifteen rooms on each of its two stories. The designer had been George Hadfield, a man of quiet talents for whom might be claimed the title of first public architect.

The city itself was equally unfinished. The spacious avenue shown on L'Enfant's plan as a straight line from the Capitol to the President's House barely existed save as an allée slashed through the trees. In 1800, for nearly its entire distance, it was a "morass covered with alder bushes which were cut through the width of the intended avenue." Traffic made a wide detour to avoid the swamp. Four years earlier the commissioners had commenced building the avenue, but an irascible Irish landowner, David Burnes, had sown a field of grain there and the operation was postponed in deference to his crop. The entire plan for the capital city nearly foundered on the obstinacy of Mr. Burnes, who held the key sites and was not to be crossed; no less a person than Thomas Jefferson prepared an alternative sketch to the L'Enfant plan solely to avoid contention with the contentious Mr. Burnes. At the Capitol there was a sidewalk made of chips from the building itself that ran for a short distance only. But the main thoroughfare was just going under construction: a six-foot wide stone surface from the Capitol to Rock Creek, ditched on either side. In addition to this work, gravel was to be laid in the center of the avenue. Little of it had been done in 1800. Few other streets existed, save those carrying traffic out of the city, and all were dusty in dry weather and bogs in wet. The character of the city remained that of a forest.

After passing the summer in Massachusetts, President Adams returned for the opening of the fall session of Congress. On the 1st of November, 1800, the diary of Mrs. Thornton faithfully recorded, "The President, with his secretary, Mr. Shaw, passed by in his chariot and four, no retinue, only one servant on horseback." The president went immediately to the Executive Mansion, and on the 16th was joined by Mrs. Adams, who commenced the messy business of setting up house on the heels of the builders.

Five days after her arrival, Abigail Adams delivered her impression of the place to her daughter in a long letter which remains the principal contemporary account of Washington at that moment. Of all the amusing letters written by the gifted and experienced Mrs. Adams of her housekeeping difficulties in the various capitals of the world her husband's career took them to, it is the most penetrating.

After describing their arrival through the woods between Baltimore and Washington, and the unfinished aspect of the city, she continued: "The river, which runs up to Alexandria, is in full view of my window, and I see the vessels as they pass and repass. The house is upon a grand and superb scale, requiring about thirty servants to attend and keep the apartments in proper order, and perform the ordinary business of the house and stables; an establishment very well proportioned to the President's salary. The lighting the apartments, from the kitchen to parlors and chambers, is another tax indeed; and the fires we are obliged to keep to secure us from the daily agues is another very cheering comfort. To assist us in this great castle, and render less attendance necessary, bells are wholly wanting, not one single one being hung through the whole house, and promises are all you can obtain. This is so great an inconvenience, that I know not what to do, or how to do.

"The ladies from Georgetown and in the city have many of them visited me. Yesterday I returned fifteen visits,—but such a place as Georgetown appears,—why, our Milton is beautiful. But no comparisons;—if they will put me up some bells, and let me have wood enough to keep fires, I design to be pleased. I would content myself almost anywhere three months; but, surrounded by forests, can you believe that wood is not to be had, because people cannot be found to cut and cart it! Briesler entered into a contract with a man to supply him with wood. A small part, a few cords only, has he been able to get. Most of that was expended to dry the walls of the house before we came in, and yesterday the man told him it was impossible for him to procure it to be cut and carted. He has had recourse to coals; but we cannot get grates made and set. We have, indeed, come into *a new country*.

"You must keep all this to yourself, and, when asked how I like it, say that I write you the situation is beautiful, which is true. The house is made habitable, but there is not a single apartment finished, and all withinside, except the plastering, has been done since Briesler came. We have not the least fence, yard, or other convenience, without, and the great unfinished audience-room I make a drying-room of, to hang up the clothes in. The principal stairs are not up, and will not be this winter. Six chambers are made comfortable; two are occupied by the President and Mr. Shaw; two lower rooms, one for a common parlor, and one for a levee-room. Upstairs there is the oval room, which is designed for the drawing-room, and has the crimson furniture in it. It is a very handsome room now; but, when completed, it will be beautiful. If the twelve years, in which this place has been considered as the future seat of government, had been improved, as they would have been if in New England, very many of the present inconveniences would have been removed. It is a beautiful spot, capable of every improvement, and, the more I view it, the more I am delighted with it."

After twenty-five years of dangerous, nomadic living in all parts of the United States and Europe, the wife of a man with whom "the noble art of government was a lifelong passion," Abigail Adams might be excused a little waspishness. Still, the darkness of that first winter along the Potomac would have been relieved had she troubled to recollect the winter in Passy when, on the slender salary of an American minister, they were obliged to burn wood at two and a half guineas a cord to the amount of a hundred guineas a year. Or the winter in Bush Hill, when they burned forty cords in four months. But Mrs. Adams clearly was not in a mood for excuses or forgiveness. She was not to have much occasion to change that mood.

At the very entrance to the White House, the kilns used to bake the brick had not been removed; and nearby lay the pits improvised to hold the water used for mortar and plaster. Against the very sides of the building itself remained hastily constructed wooden shacks that still housed the carpenters and masons. Round about, the architect's wife, Mrs. Thornton, noted, was the litter of construction, odd bits of wood shavings and sawdust, rejected brick, old lime and mortar, and oceans of muddy clay. No trees or gardens had yet been planted to replace the virgin forest that had been destroyed to clear the site. Woods still fringed it in all directions, save to the south, where across the tidal marsh the white sails of riverboats could be seen. There were no footpaths or walks. The White House lot had no stables and none of those small outbuildings so indispensable for coal and wood, and for other purposes, although they had been requested nearly a year before.

Inside, a builders' chaos still prevailed. Five days after her arrival Mrs. Adams had been able to make no impression on it. The builders still occupied most of the spacious thirty-room house, and in it the president was able to claim but six rooms. In these the plaster was damp, the paint wet, and twelve fires were constantly necessary. Mrs. Adams complained that there were no mirrors, and not a twentieth part of the lamps that were required in the drawing room. To add to her exasperation, in the course of moving from Philadelphia many things had been stolen or broken, including half the tea china. And as late as November 27 Mrs. Adams's clothing and personal things that had been shipped by vessel had not yet arrived. In all these vexing and frustrating circumstances the Adams family and their official retinue lived and worked, comforted only by the thought that every day things would improve, and that in March they might leave.

On the 22nd of November, President Adams addressed the Congress, it may have been with a touch of irony. "I congratulate the people of the United States on the assembling of Congress at the permanent seat of their government; and I congratulate you, gentlemen, on the prospect of a residence not to be changed. . . . May this territory be the residence of virtue and happiness. In this city may that piety and virtue, that wisdom and magnanimity, that constancy and self-government which adorned the great character whose name it bears be forever held in veneration." It was fine talk to legislators who floundered through oceans of mud to reach the Capitol from lodgings jammed to the eaves, and who were struck off by Mrs. Adams in a hasty confidence to her daughter: "Congress poured in, but shiver, shiver."

Most of the difficulties at the new seat of government were not due to inefficiency or dilatory management, as Abigail Adams supposed in her letter, but to a simple lack of funds. The total national budget was $75,000,000 in 1800. Since the commencement of work on the public buildings of the capital there were seldom adequate plans and estimates, and when there were, the funds requested were not appropriated by Congress.

At session after session Congress, stimulated by rival sections who still coveted the seat of government, evidenced its utter lack of belief in the new Federal City and its disapproval of the building operations. It was frequently stated that the buildings would never be completed. The buildings were too large, it was said. The suggestion was reiterated that the President's House should be used to accommodate Congress, or the executive departments, and rented quarters be found for the chief executive. Those in charge of the work were constantly showered by criticism, and their decisions were shrouded in uncertainty.

The very building operations themselves were perplexing. The commissioners' files bulged with complaints from contractors, with letters explaining the delays and requests to use substitutes for the specified materials. The commissioners on their part were urging the contractors to greater efforts, smoothing out disputes between painters and carpenters, and wrestling with the financial problems of building. Their optimistic reports concealed their despair when advised in February, 1800, that the president's furniture would be moved to Washington in June, and they replied to Benjamin Stoddard that "We do not believe it will be possible to prepare the building for the reception of the President until October or November next. The plaster and paint must have time to dry."

As if these difficulties were not enough, when the commissioners tried to move the laborers from their temporary quarters in lean-tos against the President's House, they received the following letter:

"Gentlemen Commissioners: You cannot be ignorant of the utter Impossibility of Procuring houses for the Married or Lodgings for the Unmarried Carpenters employed at the President's House, Should your intention of removing the Buildings they at present Occupy be carried into effect. At this advance [*sic*] season such men as are able to build must use every Exertion to prepare for the day they are able to remove. Consequently the men will be all employed either for themselves or their friends and the President's House remain unfinished.

"If you persevere in taking the houses down we shall every man leave the employ on the return of this bearer by whom we expect your answer in writing. Signed by the Carpenters at the President's House."

From the perspective of Mrs. Adams, the worst of it was

the lack of bells in the President's House, and the president himself was obliged to intervene to secure them. The commissioners had not been unmindful of this necessary detail. On October 28 they had written Mr. William Herbert: "We are very desirous of having eight or ten Bells hung in the President's House during the present week and understanding that there are some articles in this line in Alexandria we beg the favor of you to use your Influence with the one you think best to come up immediately with eight or ten of his best bells and all the material necessary for hanging them."

Nothing resulting from this attempt, the Commissioners two days later wrote Thomas Fendall, a bell hanger of Baltimore, that they needed ten or twelve bells immediately "at the President's House and begg the favor of you to come down immediately with the necessary bells, &c., so that they may be hung the first of next week."

The results here were likewise disappointing, and the commissioners had to reassure the president's secretary on November 4. "We received your card of the 4th inst. relative to the necessity of hanging bells in the President's House. We were well aware of the want of them and endeavored to supply it by an application to a bell hanger in Baltimore who died suddenly and we were under the necessity of engaging another who is now employed in preparing Materials and will commence his operations on Monday next."

The following Monday must have passed without the desired bells, for on November 21 the commissioners wrote yet another Baltimore firm, Messrs. Harrison and Maynadier: "The bearer, Mr. Clark, goes to your City to procure Bells and Materials for hanging them for the President's House. We beg the favor of you to aid him in procuring these articles." On the same day Abigail Adams was writing her daughter: "Two articles we are much distressed for; the one is bells, but the more important one is wood. Yet you cannot see wood for trees."

The commissioners' representative doubtless could not find bells either, for after a long silence, the public building files contain a letter dated March 12, 1801, to William Blodgett saying that the commissioners have learned of a superior construction for a water closet in Philadelphia and they wish him to examine it since they require two for the President's House. The letter concludes: "We beg you also to inquire if a good skillful bell-hanger can be induced to come here to hang about a dozen bells in the President's House and at what price with and without the materials." And subsequently, on March 27, Mr. Blodgett was instructed: "We shall thank you to engage Mr. Hedderly to come down as soon as possible and bring with him materials of the best quality for 12 bells. We agree to pay the Price he fixes and pay the expenses expressed in his estimate though the whole taken together we conceive to be high."

By that time Mrs. Adams had left happily for Quincy, to busy herself with her raspberries and strawberries. The widower Thomas Jefferson, with his retinue of slaves, had occupied the White House.

The capital, like the nation, was new. It was also poor. Its legislature was parsimonious and still lacked a truly national spirit. The worm of bureaucracy was already at work. But even at the beginning the city of Washington, for all its apparent artificiality and its seemingly arbitrary location, had its roots firmly planted in the Potomac valley. If he cared to understand the new city, and could not look forward into the future, the president might at least have profited by a glance backward into what was already a very considerable past of nearly two hundred years.

Let us now amend his error.

TWO

THE CAPTAINS AND THE KINGS

As the seventeenth century opened, the Potomac was unexplored and unsettled. Perhaps the Spaniards had visited it, but their trace is indistinct. News of the Chesapeake Bay country had been gained in some way, however, and the decision of the Virginia Company to establish a settlement there reflected it. Their decision was also founded on decades of hard-earned colonizing experience. The kind of wisdom that had earlier attempted to keep Raleigh's settlement away from the angry waters of Cape Hatteras, and directed it instead to the hospitable entrance to the Chesapeake, would have provided Captain John Smith with summaries of the available information concerning that region. When he set off from Jamestown on his famous exploration of Chesapeake Bay in June, 1608, almost a year after his arrival in Virginia, it must have been with a very considerable idea of what he would find.

Most certainly he had seen the map drawn by John White about 1586, showing with tolerable accuracy the lower reaches of the James, York, Rappahannock, and Potomac rivers, and an elaboration of this by the Dutch cartographer De Bry. He could have mastered the few hundred words of the Algonkin vocabulary, and a number of the inflections that distinguished this universal language of the eastern seaboard Indians from tribe to tribe, in order to tap the knowledge of the Indians themselves. Further, before his exploration of the Potomac, Smith had seen the country during his captivity in 1607, when he had been taken as far as the Onawmanients, who lived below the Big Bend of the Potomac.

Among the later Elizabethan sea dogs, Captain Smith was definitely an amphibious specimen. If he was "admiral of New England," he was also quartermaster of Virginia. A solid, perky, energetic man, with a face full of whiskers and a radish of a nose, he was a born pusher. Coupled with his vanity, this energy made him difficult to get along with. He was forever claiming more than his due—and seldom getting it. Before Smith reached Cape Henry he was in irons for attempted mutiny. He was one of those hopeful men whose minds run in many tracks while his feet plod in one. "Brass without, but Golde within" is Smith's estimate of himself. His epitaph begins, "Here lies one conquer'd that hath conquer'd Kings, Subdu'd large Territories, and done Things."

Smith's capacity as a military leader is incontestable. But his importance to the young Virginia colony lay more in his superlative abilities as a quartermaster. His foraging and trading, and his strict rationing system, alone kept the colony alive from one ship to the next. He made the drones work. But it was his unscrupulous diplomacy and quick-witted stratagems in dealing with the Indians that brought the vital boatloads of corn to Jamestown to support the settlement in its darker days.

Less than a year after the settlement of Jamestown, on June 2, 1608, Smith set off on an exploration of the Chesapeake, "in an open barge neare three tuns burthen," with fourteen companions. Cruising up the eastern shore of the bay, they encountered squalls. The ship's sail was rent, and had to be patched with the shirts of the crew. Much time and effort were spent bailing. Their meager provisions, mostly corn and oats, were spoilt by salt water. Hostile Indians made it difficult to land, and their water supply ran low. The amateur crew toiled in the June sun over the heavy sweeps when the ship was becalmed. After two weeks of this, the ship's company began to grumble and complain, causing Smith to turn south. Years later he wrote that he spoke to the crew:

"What a shame would it be for you (that have been so suspicious of my tenderness) to force me to return, with so much provisions as we have, and scarce able to say where we have been, nor yet heard of that we were sent to seek? You cannot say but I have shared with you in the worst which is past; and for what is to come, of lodging, diet, or whatsoever, I am contented you allot the worst part to myself. As for your fears that I will lose myself in these unknown large waters, or be swallowed up in some stormy gust; abandon these childish fears, for worse than is past is not likely to happen: and there is as much danger to return as to proceed. Regain therefore your old spirits, for return I will not (if God please) till I have seen the Massawomeks, found Potomac, or the head of this water you conceive to be endless."

Turning south, and coasting down the western shore of the bay, on June 16 they came to the wide mouth of the Potomac—they reckoned it seven miles from bank to bank—and commenced its exploration.

For the first thirty miles no sign of life was noticed, but after passing many small bays and openings two Indians came out in a canoe to greet them. They were led up a little creek to a point where the woods along the shore were thick with what seemed thousands of Indians lying in ambush. A stop was put to this old game by firing muskets so that the shot grazed the water and the sound echoed from the hills, at which the Indians presently laid down their bows and made other signs of friendliness. Since these Indians spoke the language of Powhatan, it was quickly learned they had been instructed by that crafty king to ambush the party. Having no immediate cause for alarm they became very friendly, and a member of the crew, James Watkins, was sent six miles into the woods to the place where the king of the Onawmanients lived.

In other places along the Potomac, at Cecocawonee and Potowomek, a similar reception was encountered. But farther upriver at Moyaone, Nacotchtank, and Tauxenent the Indians greeted them in a friendly manner from the start. In this way they became aware of a capital political fact: the profound difference between the tribes on the lower river, subject to Powhatan, and those at the Piscataway, who were independent of his rule; the difference between the Indians who must be treated with care and those they could trust.

In his narrative of the expedition up the Potomac, Smith says the party went as high as they could with their boat, presumably to Little Falls above the Anacostia river. Certain it is that, as the stream narrowed and the banks became rocky, the Indians also became more numerous and more friendly. From them the party secured venison and bear's meat as well as corn. Fish, too, they caught in abundance, "lying so thicke with their heads aboue the water, as for want of nets (our barge driuing amongst them) we attempted to catch them with a frying pan: but we found it a bad instrument to catch fish with: neither better fish, more plenty, nor more variety for smal fish, had any of vs euer seene in any place so swimming in the water." And with the Indians they traded for beaver, otter, bear, marten, and other furs.

But of their search for gold nothing came: the one mine Matchequeon, king of the Potowemeks, could show them proved worthless. To be sure it was being worked by the Indians, with energy and efficiency. Seven or eight miles from the river it lay, a great rocky mountain into which the Indians had dug large holes with their shells and hatchets. The ore was then washed in a nearby stream, and the residue packed in small bags. This product, unfortunately, was of little use to the Virginia Company, since no market could be found in England for a dye that, when applied to the body, made the Indians look "like Blackamores dusted over with siluer." Nor could the Virginia Company expect much from the other Indian industry discovered, the manufacture of roanoke, white beads of lesser value than wampum.

Without finding gold or the secret of the Northwest Passage, Captain Smith's voyage up the Potomac may not have been considered of great immediate value to the profit-hungry officials of the Virginia Company. Yet discoveries of the first importance had been made, far more than would appear from his

True Relation, published in 1608, or his fuller and more considered account, *The Generall Historie of Virginia*, published in 1624.

To have made such an extensive reconnaissance and prepared a very complete map of this major river in one voyage was no

mean accomplishment. From this information major political conclusions could be drawn. They were immediately applied, we may be certain by examining the prescient instructions of the Virginia Company to Sir Thomas Gates when he set out for Virginia the following year, specifying that the colonists make friends with Powhatan's enemies, especially those in the vicinity of the Piscataway.

Probably the reasons for the unmistakable hospitality of the Piscataway Indians were not fully understood. Not merely were they disaffected from Powhatan's confederacy to the south; north and west they were constantly pressed by the Susquehannocs and the raids of the Five Nations. Already they had been forced up the river by Powhatan's tribes. Under these circumstances they looked upon the white colonists as prospective allies, and valued their offers of friendship and protection. The implications of this were not lost in Captain Smith's estimate of the situation, and they were promptly and correctly applied in London.

For decades the annals of the early Virginia and Maryland settlements are crowded with accounts of missions up the Potomac to the Piscataway to secure corn and other provisions when their own crops had failed, and the settlements were hemmed in by neighboring hostile tribes. With the distant Piscataway the Tidewater settlers traded when all other avenues were closed to them. That these Indians were so placed that they served as a buffer between the later Maryland settlements on the lower Potomac and the warlike tribes to the north and west also accounted for the relative ease with which the first Maryland settlements were established. The early governors of Maryland fully appreciated this and used every persuasion to keep the Piscataways located there, and gave them support in their struggle against the Susquehannocs.

What were they like, these close neighbors and friends of the first Potomac settlers? John White, the artist who accompanied one of Sir Walter Raleigh's expeditions to Virginia, drew a very intriguing sketch of the Indian village of Secoton he found in Carolina that must have been similar to many Indian villages along the Potomac. Secoton stood in a clearing on the north bank of the Pamlico, and the artist shows a dozen houses fashioned in the characteristic manner described by many other travelers: on a rectangular plan a framework of light saplings supports the bark cladding, capped by an arched roof also covered with bark or grass mats. Travelers frequently remarked that the houses resembled ovens.

To the right of the picture, in front of the village, are three cornfields, and in one of the lower corners a group of Indians are shown dancing around a circle of posts set into the ground. Opposite the dancers is a ceremonial fire, and close by is another building described as a tomb, quite possibly used as a place to prepare dead bodies for burial. In such a village we may easily imagine the friendly Indians lived.

As Smith described them, the Potomac Indians were tall and dark, and covered their bodies with a mixture of oil and paint, both as a measure of beautification and, more practically, to protect them from the attacks of mosquitoes. Different systems of body painting were developed, and tattooing seems rather to have been common among the women. They wore only aprons in the summer, and added capes of deerskin in the winter. With their jewelry they took much care, wearing many necklaces of shells and beads, bracelets and earrings.

On the assumption that the aborigines were just Englishmen with darker skins, most early descriptions err by attributing to the Indians habits and institutions they never had. A notable example is the persistent fiction of the Indian "kings." Each tribe had its chief, and each smaller group its chief man. But the universal use of the term "king" led to much confusion; there was nothing that could be used to designate one who stood higher than kings except "emperor." As W. B. Marye said, "Having begun with an exaggeration, they ended in an absurdity," in so describing Powhatan, the chief of the Piscataways, and other tribal leaders.

The Indians of the Potomac valley had a powerful influence on white settlement and subsequent white culture. Indian forms of agriculture, construction, and language, and even forms of government and oratory, were silently absorbed into white settlements all over the Americas. In the Potomac we can see the introduction of specific plants (corn, tobacco, potatoes), new words (hominy, opossum, succotash, wigwam), a new diet. From the first white settlement a new civilization began to be born.

Not least among these influences was the Indian trade. The very name Potomac in the Algonkin tongue is a verbal noun meaning "something brought," and as a designation for a place, "where something is brought," or, more freely, "trading place." Living, as they did, on one of the great natural trade routes east and west through the mountains, north and south along the fall line, with a highly developed and specialized culture, it was inevitable that the natives of this section should be great traders. Because of this they could have had the specialized industries the English found. It was equally inevitable that the English who ventured into this region should exploit these advantages, and it was natural they should take over much of the Indian trade with the interior tribes. It was, perhaps, their greatest gain, as the fur traders were to prove. The Indian trading methods were in many cases adopted intact, and the successful traders were those who could adjust themselves to operating under this system. The earliest white settlements were located at the established Indian trading centers. Henry Fleet located his trading post at Yowaccomoco, at the mouth of the St. Mary's River, a village in the course of abandonment by the retreating Piscataways. The settlement of Potowomek, lying on a peninsula between Potomac Creek and the river in Stafford County, became an early English settlement called New Marlborough.

The number of traders steadily increased in the seventeenth century, as did the scale of their operations. From the Indians they secured corn and other provisions, and furs. To the Indians they carried copper, hatchets, and trading jewelry. As trading goods, beads had become so important that among the French silk weavers and vintners and the German soapmakers sent over by the hopeful Virginia Company was also a group of Italians who set up a glass factory and made Venetian beads.

The Indians were eager and proficient traders. In 1619 the Potowomek Indians came to Jamestown and asked that two ships come and trade as they had a bumper crop of corn. Smith described one tribe as "the best Marchants of all other Salvages." All this is hardly strange because of the large volume of trade among the Indians themselves, and a considerable—if lost—history of dealing with casual white traders. Even in 1584 when Master Ralph Lane attempted to buy white pearls he was told the Indians saved them for other white traders, and would sell him only the black ones.

Numerous Indians had developed specialized resources and produced peake and roanoke (shells used for currency), arrowheads, body dyes, and other articles of Indian commerce. Trade operated over long distances. Almost from the very beginning, the English settlers on the Potomac were aware that the Indians upriver had been trading with the French in Canada. Smith noted hatchets, knives, pieces of iron and brass secured by such trade. Copper from Lake Superior has been found in Indian refuse heaps in the lower Potomac. Clever in trade, the Indians guarded their monopolies jealously. Powhatan, for example, cornered the copper market, and attempted to keep other Indians from trading with the Jamestown settlement.

To trade successfully with the Indians required much special training. Of the traders in the Potomac, all spent long periods in captivity, learning not only the language and customs of the Indians but mastering as well the essential political and economic facts of their lives. These were the favored gentlemen who almost alone made the colonial enterprise a profitable one. Consider Henry Spelman, inadvertently started by John Smith on his career as an Indian trader in the Potomac.

2

The sad tale of Henry Spelman reads like a ballad. In his *Relation of Virginia* he commences, "Beinge in displeasuer of my fryndes, and desirous to see other countries . . ."—what more ingenuous explanation can be desired as to why this third son of a distinguished and noble antiquary ventured into Virginia in 1609. How old he was is not known. And how Harry Spelman deserved his friends' displeasure is difficult to make out, for he seems to have been a quiet lad with a kind, almost magical touch with children.

Smith must have detected this immediately, for less than a

fortnight after Spelman had landed in Virginia he sailed with Smith to Powhatan's village and there was left with Powhatan's young son, who was captivated by his new English friend. Possibly they were about the same age. Spelman lived with the natives for six turbulent months in which he became the catspaw of intrigues, robberies, and massacres. With the unashamed opportunism of the very young, he had a genius for survival.

Providentially the chief of the Potowomeks came to visit Powhatan. According to Spelman's account, "he shewed such kindness to Sauage Samuell and myself as we determined to goe away with him." A further consideration was that Powhatan's "mind was much declined" from them. Powhatan must have had some further use for Spelman, however, for the party were apprehended in their departure and his Indian companion was killed. Henry Spelman escaped and spent the following year at the Potowomek village of Paspatanzie. The Spelman charm persisted, for when found by Captain Argall, who was trading in the Potomac, the Indians were so sorry to see him go they filled Argall's ship with corn. Spelman's observations of the Indians at Potowomek form the major part of his *Relation*, probably written on his return to England about 1613.

Spelman's *Relation* of his experiences among the Indians is written in a simple and orderly way. To describe their life he begins with religion, and goes on to geography, towns and buildings, marriage, christening, the sick and the dead, justice and government, criminal punishment, agriculture, etiquette, royalty, war, and recreation. Its charm lies in the innocence of the writer, and the intimacy of his knowledge of the Indian community in which he lived so long.

With such a beginning, it is hardly surprising that Spelman returned to Virginia after a visit at home. With such equipment, he rose to be a captain in the colony. Like Ensign Thomas Savage, who had likewise been sent to live with the Indians and learn their language, Spelman's abilities were keenly appreciated.

After his return to Virginia in 1613 he was mainly occupied with trading in the Potomac region, and is referred to in most of the accounts as the best interpreter in the colony. Presumably he was much occupied with his own interests, for he seems to have mingled little with the Jamestown settlers. The clearest view we have of him is in his last tragic moment.

Following the death of Powhatan in 1618, the relations of the Virginia colony with the Indians rapidly deteriorated. In 1622 the old chief's successor, Opachankano, plotted a sudden uprising that resulted in the death of nearly one-third of the colonists, and barely missed wiping out the entire colony of twelve hundred. In fearful retaliation the English were able to report of the Indians in the following year that "we have slain more of them this year than hath been slain before since the beginning of the colony." Meanwhile numerous wars had broken out among the Indian tribes themselves, confused further because the colonists were aiding those Indians who revolted against Opachankano. In this way Captain Madison and thirty men were dispatched to help the king of Potowomek, and Captain Croshaw and Captain Hamor became involved in the bloody burning of a large Indian settlement that stood at the junction of the Anacostia River with the Potomac.

Under these turbulent conditions, anything could have happened. Captain Henry Spelman must have known this better than most men when he was placed in command of the pinnace *Tiger* with twenty-six men and set out in April, 1623, on a trading expedition in the upper Potomac for the corn so badly needed in Jamestown. Perhaps the story is best told as it was evaluated by Captain Smith in England and related in his *Generall Historie*:

"Captain Henry Spelman, a gentleman that has lived in this country thirteen or fourteen years, one of the best interpreters in the land, being furnished with a bark and 26 men, was sent to trade in the Potomac river, where he had lived a long time among the savages.

"Whether he presumed too much upon his acquaintance among them, or they sought to be revenged for the slaughter made among them by the English so lately, or he sought to betray them, or they him, are all several accounts; but it seems but imaginary, for they who returned report they left him ashore

about Potomac, but the name of the place they knew not, with 21 men, five being left in the bark.

"Before anything had been suspected, the savages boarded them with their canoes, and entered so fast the English were amazed, until a sailor gave fire to a piece of ordnance only at random. At the report the savages leapt overboard, so distracted by fear they left their canoes and swam to shore. And presently after they heard a great commotion among the savages on shore, and saw a man's head thrown down the bank. Whereupon they weighed anchor and returned home, but how he was surprised or slain is uncertain."

Most people thought the head was Spelman's.

3

The Captain Argall who took the young Henry Spelman from the Potowomeks was himself a notable trader, although he is usually remembered as one of the long series of governors employed by the Virginia Company. By profession a sailor, he had the reputation of "a good Marriner," discovered the short transatlantic route via the Canary Islands, and later built the first ships at Newport News. He arrived in Virginia in July, 1609, captain of a ship, at a critical juncture in the affairs of the colony. Instructed by the owners of his ship "to truck with the Colony and fish for Sturgeon," there was little chance to carry out this mission. His ship was the first seen for many long months by the greatly diminished residue of eighty starving colonists—the third supply having been scattered and delayed by storms. It was virtually captured, and its large stocks of wine and other provisions enthusiastically consumed. "Though it was not sent us, our necessities was such as inforced us to take it," the colonists later explained. Argall's sympathies were clearly with the colonists, who provisioned his ship before sending it back to England.

Argall made other voyages to Virginia and soon established himself there, trading for corn in the Potomac. In the course of one of these expeditions in 1612, Argall was at Potowomek and learning that Pocahontas, Powhatan's celebrated daughter, was there, kidnapped her. Read Captain Hamor's account of this famous incident, complete with cues and prompt lines—although with modern spelling:

"Being at Potowomeks, as it seems, thinking herself unknown, was easily by her friend Iapazaws persuaded to go aboard with him and his wife to see the ship: for Captain Argall had promised him a copper kettle to bring her out to him, promising no way to hurt her, but keep her till they could conclude a treaty of peace with her father. The savage, for this copper kettle, would have done anything, it seemed by the relation.

"For though she had seen and been in many ships, yet he caused his wife to feign how desirous she was to see one, that he offered to beat her for her importunity, till she wept. But at last he told her, if Pocohontas would go with her, he was content: and thus they betrayed the poor, innocent Pocohontas aboard, where they were all kindly feasted in the cabin, Iapazaws treading oft on the Captain's foot to remember he had done his part. The Captain, when he saw his time, persuaded Pocohontas to the Gun-room, feigning to have some conference with Iapazaws, which was only that she should not perceive he was in any way guilty of her captivity, so sending for her again, he told her before her friends she must go with him, and compound peace between her country and us before she should ever see Powhatan; whereat the old Jew and his wife began to howl and cry as fast as Pocohontas that upon the Captain's fair persuasions, by degrees pacifying herself, and Iapazaws and his wife with the kettle and other toys, went merrily on shore; and she to Jamestown.

"A messenger forthwith was sent to her father, that his daughter Pocohontas he loved so dearly, he must ransom with our men, swords, pieces, tools &c he treacherously had stolen."

Argall's tougher policy with the Indians brought some results, but the consensus is that the subsequent marriage of Pocahontas to the tobacco planter, John Rolfe, in 1614 brought more lasting peace.

In 1614 Argall was responsible for making peace with the

Chickahominy, and also, before returning to England in June of that year, dispersing some French settlements. He returned in 1617, this time as governor of Virginia, and ruled through growing Indian outrages, until thievishness caused his recall in April, 1619, and he returned to England for what seems to have been the last time.

Among the party of twenty-one men Henry Spelman had taken ashore with him when he lost his head at Anacostia was a smart lad named Henry Fleet. Not all of the party shared Spelman's fate: at least Fleet was held in captivity for several years, principally at Yowaccomoco. On Fleet's escape he returned to London, and immediately began to exploit his adventures in what will be recognized as a characteristic manner. A reporter in the London press wrote:

"Here is one, whose name is Fleet, newly come from Virginia, who having been lately ransomed from the Indians, with whom he hath long lived, till he hath left his own language, reporteth that he hath oftentimes been within sight of the South Seas, that he hath seen Indians besprinkle their paintings with powder of gold, that he has likewise seen rare precious stones among them, and plenty of black fox, which of all others is the richest fur."

One is hardly surprised to find eager British merchants falling over themselves to finance Fleet's return to America, and in the fall of 1627 he sailed for Virginia in the *Paramour*, on a trading expedition. Four years later he is heard of sailing in the ship *Warwick*, and a little later had added two smaller ships to his trading enterprise, a pinnace of 20 tons picked up in New England, and a bark of 16 tons built on the James. He was prospering, but there is considerable doubt his London backers ever saw their money again.

Unlike Smith, Spelman and Argall, Fleet's trading activities were not primarily in relation to the Virginia settlements. He traded with the English colony in New England, with the West Indies, and also with London. Nor was his activity limited to the lower Potomac; he was the first to tap directly the rich fur resources of the upper Potomac and the interior. Finally, in his well-developed acquisitive personality and double-dealing he reflected the fully matured trader type: a proper representative of that body of Potomac traders who followed the fur-trading animals as they retreated west.

The great trade in furs that was to distinguish early English aspirations in the New World had already commenced. For nearly a century the fur trade was decisive in forming the pattern of early exploration, trade, and settlement. The Potomac beaver trade was known even to the enterprising French, of whose activities in the Chesapeake Bay region Ensign Savage was complaining as early as 1621, the year Fleet was captured. By the time of his return six years later, trade had mushroomed and William Claiborne was firmly established in the business. In 1633 Father White was told that on the Potomac "a certain merchant in the last year exported beaver skins to the value of

FISHERMAN'S HOUSE,

TIDEWATER

40,000 gold crowns, and the profit of the traffic is estimated at thirty-fold." The fur trade was carried on under strict license from the Virginia Company and, later, the proprietors of Maryland, and lent itself well to monopoly as well as an inevitable amount of illicit trading and conflict.

A very clear picture of the details of this trade is given in Henry Fleet's Journal, in which he describes one of his trips from England to the fur region of the Potomac in the fall of 1631. When he arrived at his Yowaccomoco trading station, he states, the Indians were "burning [curing?] beaver skins, as was their custom." He arranged with the Indians to preserve the furs during the next winter, and told them he would return to trade in the spring.

Loading his ship with eight hundred bushels of Indian corn, Fleet was on his way out of the capes guarding Chesapeake Bay when a storm was encountered. He was obliged to anchor in the James River until it blew out (being careful to avoid the Virginia authorities, because his papers were not in order) and early in January set his course for New England, after having crushed a near-mutiny on board. After trading in New Hampshire and in the Massachusetts Bay during the winter, he returned to the Chesapeake, arriving in May, where through poor luck he was becalmed off the English settlements.

Because of adverse winds and calms Fleet arrived in the Potomac much later than he had planned, and at his first stop found that the Indians, having been told he was lost at sea, had already disposed of their beaver skins to other traders. Alarmed at the consequences of this rumor, Fleet immediately dispatched his brother upriver to reassure the Indians, and sent two Indians by land to visit the various settlements and tell them of his arrival. These arrangements were made too late, since the rival trader, Charles Harman, had "gotten 1500 weight of beaver and cleared fourteen towns."

Sailing up the Potomac to the head of navigation, Fleet was able to pick up 114 skins at Potowomek and 80 more at Nacotchtank. Here he developed the possibility of overland beaver trade with the Susquehannoc Indians, and the tribes in the upper valley of the Potomac, sending out messengers with presents, and urging these Indians to come down below the Great Falls of the Potomac where he would be in his ship.

Anchored at a point between the palisaded banks, "two leagues short of the Falls," where the city of Washington now stands, Fleet described the river at this point: "This place without all question is the most pleasant and healthful place in all this country, and most convenient for habitation, the air temperate in summer and not violent in winter. It aboundeth in all manner of fish. The Indians in one night commonly will catch thirty sturgeons in a place where the river is not above twelve fathoms broad. And as for deer, buffaloes, bears, turkeys, the woods do swarm with them, and the soil is exceedingly fertile." In a shallop he rowed up the river until he could hear the sound of the falls.

Within a week Fleet heard that a hundred and ten Indians, loaded with 4,000 weight of beaver skins, were on the way. However, great efforts by the Anacostia Indians, who naturally wished to monopolize the fur trade at the head of navigation, were largely successful in stopping them. Fleet conferred with the Anacostia Indians but refused their offer to deliver the furs from the upper river—a decision he later regretted.

Shortly after his return to the ship, an Indian was seen on the riverbank, holding up a beaver skin on a pole and trying to attract their attention. Fleet did not understand the language he spoke, but he made signs that he wished to trade, at which the Indian disappeared into the woods, and presently appeared with five other Indians including an interpreter. They traded for skins, and Fleet learned to his great chagrin that these Indians drove hard bargains. As he said, "These people delight not in toys, but in useful commodities." They demanded "hatchets and knives of large size, broadcloth, and coats, shirts and Scottish stockings. The women desire bells, and some kind of beads."

They demanded to see Captain Fleet's trading goods, admittedly not worth over one hundred pounds, and then scorned them. Fleet was piqued, for he says, "They had two axes, such as

Captain Kirk traded in Canada, which he bought at Whits of Wapping, and there I bought mine, and I think I had as good as he." Up against hard-boiled Indians who had been trading with the French in Canada, fifteen days' journey distant by way of the Conococheague, the Conodoguinet and the Susquehanna to the Great Lakes, Fleet could make little progress.

Drifting down the Potomac, Fleet apologized to the Piscataways for trading with their enemies, and employed sixteen of them as his agents to locate skins in the upper river. This shortly produced eighty skins and promises of more later.

While waiting in the river, a ship arrived, carrying Captain John Utey of the Virginia Council and Fleet's rival, Harman. After hospitalities were exchanged, Utey said, "Captain Fleet, I am sorry to bring ill news, and to trouble you in these courses, being so good; but as I am an instrument, so I pray you to excuse me, for, in the King's name I arrest you, your ship, and goods, and likewise your company, to answer such things as the Governor and Council shall object."

It would have been a simple matter to overpower Utey, whose small crew were almost entirely indentured servants.

Fleet considered this a little, but finally agreed to go to Jamestown, as he explains, "wanting truck, and having much tobacco due to me in Virginia, I was unwilling to take any irregular course, especially in that I conceived all my hopes and future fortunes depended upon the trade and traffic that was to be had out of this river." He sailed for Jamestown and arrived there September 7, 1632.

The official temper at Jamestown during the administration of the avaricious Governor Harvey and some useful observations on traders are given by the Dutch sea captain De Vries, who wrote in his book of *Voyages:* "The English there are very hospitable, but they are not proper persons to trade with. You must look out when you trade with them, Peter is always by Paul, or you will be stuck in the tail. If they can deceive any one, they account it among themselves a Roman action. They say in their language, 'He played him an English trick.' "

They said English but they meant American, for the Indian trading ways of the Potomac had already left their mark.

Fleet must have told Utey about the seven thousand Indians with the forty beaver skins apiece, for he found the governor and council exceedingly interested. Further "every man seemed to be desirous to be a partner with me in these employments." A quick meeting of the court, a gubernatorial wink, and Fleet was again free. Back he went, up the Potomac.

The unscrupulous Fleet does not appear to have been at ease with the Virginians. Perhaps there were too many other fur traders involved or he was not sure of his arrangements, for he eagerly greeted the 24-year-old Leonard Calvert on his arrival with the first Maryland settlers early in 1634 and escorted him on a personally conducted tour of the Potomac. Calvert was profoundly impressed with Fleet's intimate knowledge of the river; and there were, after all, few others upon whom the Catholic governor could have depended for such guidance. Still, one is hardly surprised that Fleet persuaded Calvert to place his first settlement and the capital of Maryland at his old trading post at Yowaccomoco, where only three years before Fleet had a warehouse bulging with furs.

What all the considerations were is not known, but Fleet's protection in the fur trade by the young Maryland colony must have been one of them. He was also granted 2,000 acres of land opposite St. Mary's City where he later built a house, and he sat in the Maryland Assembly of 1638. Under the protection of his new friends, Fleet immediately commenced a complex but transparent agitation to eliminate rival fur traders who, he claimed, were inciting the Indians. When the charge was investigated, the chief of the Piscataway indignantly denied it, called Fleet a liar, and regretted that Fleet was not there so that he could tell him so to his face. He expressed amazement that anything Fleet said would be taken at its face value by the whites. At this observation, the Virginians present explained that the Marylanders "did not know Captain Fleete so well as we of Virginia, because they were so lately come." After the investigation was concluded, Fleet admitted that the charges had been false, and weakly excused himself by saying that they had not been made under oath.

By 1645, however, the tides of fortune must have changed, for Fleet was back in Virginia trading in the Rappahannock and signing up Indian allies for the Virginia government, and in 1652 he sat in the Virginia legislature from Lancaster County. Almost the last thing we hear of him is a license he was issued that year "to discover and enjoy such benefits and trades, for fourteen years, as they shall find out in places where no English have ever been and discovered, nor have had particular trade, and to take up such lands, by patents, proving their rights, as they shall think good." That much rope might have satisfied him.

THREE

THE TIDEWATER FRONTIER

AFTER HIS CRUISE TO THE PISCATAWAY WITH CAPTAIN FLEET Leonard Calvert, the first governor of Maryland, wrote that he found not a single open field on the north bank of the river. All he could see was forests and water. In this he agreed with the accounts of other seventeenth-century travelers. To the eye of the trader and shipbuilder, Captain Argall, these forests contained "the goodliest trees for masts that may be found elsewhere in the world." At home Great Britain no longer contained a single tree large enough to make a keel or a mast for a man-of-war, as John Evelyn was about to write in *Sylva.* In this part of the New World it seemed that every tree was a mast. For many generations the Potomac Indians had been in the habit of burning the virgin forest to improve the browse for deer. The absence of underbrush gave the landscape a parklike appearance, and a man on horseback could ride almost anywhere through the thickest woods, the great trees rising like columns around him.

Through these forests, along the footpaths of the Indians, any considerable amount of wheeled transportation was out of the question. Broad tidal inlets and marshes made roadbuilding difficult. The waterways themselves were better and more natural thoroughfares, and the settlers promptly took to them, at first in light barges and pinnaces, prefabricated in England and assembled in Virginia, and only a little later in good ships of their own construction. Seafarers most of them had been, and it was easier to make a boat than a cart. The Indian dugout was quickly adopted; so were the lighter canoes. Clever and imaginative shipwrights soon commenced to build a distinguished line of light-draft, serviceable and handy sloops, well adapted to these waters, whose busy sails flecked the Potomac almost to the end of the nineteenth century. The early Potomac world was a floating world, and ships were prominent in wills and inventories of Tidewater planters.

The ease of transportation along the broad waterways made for a rapidly spreading pattern of settlement throughout the Chesapeake Tidewater country. By 1634 when Governor Calvert arrived to plant the new colony at St. Mary's, the south bank of the Potomac and, indeed, the presumably uninhabited territory granted the Calverts north of the Potomac already contained a scattered number of troublemaking settlers, most of them Virginians.

As English settlement in the Chesapeake Bay country expanded, it followed the well-established water routes of the earlier traders. The settlements in the lower Potomac were first located on the sites of Indian towns. On the spit of land between Potomac Creek and the Big Bend of the river, opposite Maryland Point, the Indian settlement of Potowomek—which had seen Smith imprisoned, Spelman living, Pocohontas captured—gradually became a white community. On the Maryland shore, the initial settlement of St. Mary's was located on the site of the ancient Piscataway town of Yowaccomoco.

Settlement spread rapidly along the waterways. Below the Big Bend, the first Virginians appear to have settled at the mouth of Coan Creek. Northumberland County was created in 1648, but only five years later it was divided and Westmoreland County was established. In this section of the Northern Neck were later born that galaxy of Revolutionary figures—lawgivers, heroes, philosophers, presidents—that cannot be equaled in any corresponding time and place in our country, the home of the Washingtons, Carters, Lees, Fitzhughs, Monroes, and scores of less renowned clans.

The middle years of the seventeenth century were the great years for settlement on both banks of the lower Potomac.

Ordinarily the settlers were recruited from the second and third generation of Virginians along the James, the York, and the Rappahannock, forced by the workings of the law of primogeniture to seek new homes for themselves. The Potomac lands were not highly prized in the beginning. The region came late into the Virginia colony's world, and lay at its edge. Good plantations could be found without difficulty along the earlier settled rivers. The Potomac lands were termed "backs," much as the later residents of the Potomac in turn referred to the "back lands" they owned in the upper valley and farther west. There was a touch of snobbery, too, that reflected marked differences in culture. Even a century later it had not been overcome. The antagonism that still remained was reflected in the will of a spartan Potomac planter who forbade his sons to mingle in James River society "lest they should imbibe more exalted notions of their own importance than I should wish any child of mine to possess."

The Potomac settlements of those early years might be described as a genuine frontier. First came the pioneers, the traders, and the trappers, rapidly passing west in pursuit of quick riches. Following them were the individual families, generally with meager resources, to break the virgin soil and build their homes. Afterwards came the large planters, eager for new tobacco lands, with their retinues of indentured servants and their slaves. Up the valley they advanced, like figures in a frieze.

Not all were Virginians. Some came directly from England, the bulk of them from the seafaring and commerical classes, or from the old upcountry yeomanry. John Washington and Andrew Monroe, men who founded distinguished lines, were representative.

Marylanders came, too, at first to the original Northern settlement of Chicacoan, and later to establish the towns of Machadoc and Nomini. They had been disappointed at the wavering outlook for religious freedom in Calvert's colony. Other immigrants came even from New England. To the Maryland Protestants in Virginia were soon added a few Quakers, some early Presbyterians (whose first congregation was called "Potomake"), French Huguenots, and possibly a sprinkling of the Swedes and Germans who early found their way into Maryland—and out of it.

Finally, there came from England those exiled Cavaliers of whom so much has been made in Virginia history and romance, the royalists whose numerical importance has been so overestimated. These followers of the executed Charles I arrived in driblets in the 1650's, until the end of that decade when the reason for their leaving England had largely disappeared with the restoration of that king of landlords and merchants, Charles II. Fresh from England, and settled in generally congenial surroundings, they must have had a considerable social value to the pioneer region, but the extent of their contribution awaits disinterested evaluation.

2

From the broad entrance to St. Mary's River on the Maryland side of the Potomac, the immigrant settlers on the *Ark* and the *Dove* looked in on waters whose charm persists to this day. To the north spread a broad, winding waterway, punctuated with low headlands and estuaries, and boldly framed by high wooded banks rising slowly to elevations of a hundred feet above the water. Six miles above the entrance to these waters on the right they found their landing at Yowaccomoco, the Indian village nearly abandoned by the retreating Piscataways, where Henry Fleet's trading post stood. Arrangements had previously been made by Governor Calvert that the remaining Indians would occupy one part of the town and the new settlers the other. Thus, from the beginning, there was not only good water, wood, and cleared fields with some crops already planted, but even houses left by the Indians that could be used for shelter and for divine services.

On March 27 Governor Calvert landed at St. Mary's with his party, amid the solemn thundering of cannon. The friendly relations already established with the natives immediately began to produce results: Indians showed the settlers how to hunt for deer and turkey, and how to catch the strange fish and oysters; the trade for corn produced such an abundance of grain that in the first year a thousand bushels were sent to New England to be exchanged for salt cod. Indian women showed the settlers how to make corn pone. From Virginia they were able to secure hogs, poultry, cows, and beef cattle. They found the fields fertile and the pasture good. The cattle and hogs fattened easily on the browse and mast in the woods. From the Indian corn the Maryland housewives commenced to make beer. From home they had brought meal and other provisions as well as fruit trees and seed. When autumn came, their spiritual leader, Father White, could well claim that the settlement had made as much progress in six months as the settlers at Jamestown had in as many years.

The construction of houses also began, and having selected the spot with an eye to its defense, it was further strengthened with fortifications. Bricks were made in the first year of the settlement, and a corn mill built.

Little time was spent in the expeditions of earlier years or the quixotic agricultural ventures of the Virginia Company. After nearly twelve months, Father White could observe, "The Mineralls have not yet beene much searched after." On the contrary, he considered that "Butter and Cheese, Porke and Bacon, to transport to other countrys will be no small commodity."

Although in London the planners of St. Mary's City had entertained some definite notions of urbanism, and specified a contiguous settlement of row houses, in the New World way the community was actually laid out regularly but in a free and open manner. The research of Henry Chandlee Forman shows the early colonial capital to have been "a country neighborhood," two and a half miles from one end to the other, and within this area were located by 1678 only some thirty houses.

At St. Mary's can be seen the fully matured English colonization technique. Thorough study had been given the neighboring experiment at Jamestown as well as earlier and later settlements; and at least all of the important figures concerned had conned John Smith's lengthy accounts of the region. The Calverts were scarcely inexperienced in colonization, and their carefully selected settlers came in family groups, bringing all the supplies and tools needed for their establishment. A careful preliminary survey had preceded the general choice of a site for the settlement, and a detailed survey with expert native assistance was made before the final location was selected. The title to the land was definitely settled, Indian claims were extinguished, and the alliance of neighboring tribes secured. Small ships for local travel and trade were brought from England, and ordnance for a fort. The colony arrived early in the season, in time to put in crops and to build against the winter. Trade was commenced immediately with Virginia and New England, as well as with the Indians. If anyone doubts the degree of organization that prevailed, he has only to consult the detailed instructions to settlers or examine the printed forms of indenture, bills of lading, and other instruments brought from England. Finally, in London the proprietary exercised constant control of the settlement, and championed it during the uncertain years of establishment, receiving regular and detailed reports and sending copious instructions.

In their efforts to carry out Lord Baltimore's instructions to preserve unity and peace in religious matters, the colony was the first to venture into the stony fields of religious equalitarianism. Toleration was a political necessity for the colony, but it was not enough to proclaim it; in Maryland toleration had actually to be practiced. When schools were established the children of Catholics and Protestants had each their own schoolmaster. The Catholic chapel was opened to Protestants for worship until separate churches could be built, and the rights of each were strictly enforced, often against powerful influence. Thomas Gerrard was censured and fined 50 pounds of tobacco for locking the chapel against the Protestants; and Thomas Cornwallis's overseer was fined and committed to the sheriff for speech offensive to the Protestants. The Governor's Council was composed of Catholics and Protestants, who managed to work smoothly together.

Meanwhile, on the Virginia shore, without benefit of impressive and formal declarations, an even more genuine advance in tolerance and self-government was being made. The Northern Neck was ideally fitted to play a unique role in the settlement of the Potomac region. Far from the established settlements of Virginia, what happened on this frontier was of little concern to the rest of the colony. Much as the lands on the Severn were little regarded at first by Maryland's proprietor, so we find the Northern Neck early becoming a refuge for the disaffected and the persecuted. It was to the Northern Neck that the first dissenting Presbyterians came; later came the Protestants who had been disappointed with the promises of religious freedom in the Calverts' colony in Maryland; and later still came the Huguenots from England, France, and other countries. The lower Potomac became in fact, if not yet in law, all that Plymouth, Providence, Philadelphia, and the other celebrated sanctuaries of religious freedom and tolerance later were in principle: a place which tolerated all forms of Christian worship.

3

Because the waterways made accessible so much more land than could possibly be used, the price of land all during the seventeenth century was absurdly low. This was one of the most important factors leading to a generally thin pattern of settlement and stupendously large landholdings. It also encouraged large-scale land speculation and much absentee ownership by residents of England, who never saw the lands to which they held title, as well as by residents of the older and more populous sections of Virginia. These early trends were steadily reinforced as the colony became irrevocably committed to the culture of tobacco, and the plantation system was buttressed by the cheap labor provided by indentured servants and the growing number of Negro slaves. The ability of small oceangoing vessels to reach deep into the Tidewater country aided the system of large plantations. Land was granted by the thousand-acre unit. Holdings of tens of thousands of acres were not uncommon, much of this land being held in reserve to assure a supply of virgin soil when the older fields had become exhausted. Among plantation owners land speculation became universal. Land was as cheap as it was abundant and accessible. When the freehold system became established on the Northern Neck, as contrasted with the "head right" system prevailing elsewhere in Virginia, a thousand acres could be bought outright for five guineas.

The career of the Maryland colony's leading military figure, Captain Thomas Cornwallis, shows one of the common ways early fortunes in land were established. Aged thirty-one in 1634 when Maryland was first settled, Cornwallis had already a fund of experience and some positive ideas. He was a member of a family that included distinguished literary and political figures, but himself had little property. In Maryland one of his first undertakings was to lead an expedition against Virginia traders operating in Maryland territory. A little later Cornwallis himself was licensed as a trader. He took up large landholdings, some of which he secured as reward for his services to the colony, the rest for bringing in additional settlers. In 1640 he built the largest and best brick house in the colony on St. Peter's Freehold at St. Mary's City, a good portion of which was later destroyed, during his prolonged absence from the colony at the time of Ingle's rebellion.

Cornwallis steadily expanded his activities and his landholdings until he became the wealthiest man in the Maryland colony. To his holdings on the lower Potomac he added 4,000 acres along the river above Port Tobacco. In 1642, eight years after he had come to Maryland, he surveyed his progress, and wrote that "By God's blessing upon his endeavors, he had acquired a settled and comfortable subsistence having a comfortable dwelling house furnished with plate, linen hangings, bedding, brass, pewter, and all manner of household stuff, worth at least a thousand pounds, about twenty servants, at least a hundred breed cattle, a great stock of swine and goats, some sheep and horses, a new pinnace about twenty tons, well-rigged and fitted, besides a new shallop and other small boats."

On his return to Maryland after ten years' absence in England, Cornwallis was building again, for records show him contracting for 60,000 bricks. While in England he had married, and this possibly had some bearing on his decision to return to England in 1658. He continued to live there, "a merchant of London," until his death in 1676.

Forceful, sometimes testy, occasionally mistaken—even in matters of church and state—Cornwallis was one of the strongest and most successful figures in the early settlement of the Potomac. Loyal to the Calverts, and a stout defender of the colony, from the beginning Cornwallis also showed himself a forceful champion of the civil rights of the colonists. He early wrote Lord Baltimore that he considered security of conscience the cornerstone of civil policy in Maryland. He took a strong stand against the absolute powers granted the monarchical Calverts, and participated in the 1646 uprising against the proprietary. He led the movement of the colonists to legislate for themselves.

Lord Baltimore, in the manner of a sovereign (which, in law, he was), had sent over in 1637 a code of laws he desired the legislature to accept. Governor Calvert convened the legislature and suggested the laws be accepted after a single reading. Cornwallis immediately objected, rallied the legislature behind him, and when the vote was taken a large majority refused to accept the laws. Seeing how his proposal was received, Calvert promptly adjourned the legislature, and reconvened it some days later. The delegates then resolved that all laws should be read three times on three different days, "and they also expressed a wish that all bills might emanate from their own committee." When Calvert attempted to adjourn the legislature again to prevent further action, Cornwallis pointedly objected, saying "that they could not spend their time in any business better than this for the country's good." On the next meeting a month later Calvert found them still obdurate. With the good sense he frequently showed, Calvert then advised his brother that he would assent to all laws enacted by the provincial legislature, subject to final approval by the proprietary. The process that was to lead the Potomac settlers to newer and bolder attitudes and customs had already commenced.

4

The Indian trade and the fur trade, the traffic in lands and the closely allied indenture system, the rise of tobacco planting, political favors and place-holding—these were the pattern of the early years of the Potomac. They were very much a part of the English world in the seventeenth century, and everywhere they were leading to new patterns of political expression, new views of man and the state. Of these the Potomac saw its share, and if the episodes that marked the breakup of feudalism and medieval institutions were more dramatic here, so were the abundance of cheap and fertile land, the remoteness from the centers of established authority, and other contributing factors, not the least of which was flux.

After thirteen years of conservative but steady rule, Governor Leonard Calvert died intestate. The affairs of the colony had never been in greater difficulty. An armed force under Richard Ingle, a fiery Roundhead, had landed and seized St. Mary's. The saintly Father White had been kidnaped and returned to England in chains. Thomas Cornwallis's estate had been largely destroyed. Kent Island, well inside the Maryland line, had again been captured by the Virginia trader, Claiborne. All that the Lords Baltimore had endeavored to build up, including the fortunes of their Catholic settlers, was in jeopardy.

Instead of a will, Governor Calvert, in the colony's uncertain hour, made a last statement to witnesses present at his deathbed, giving instructions for the management of his estate and the conduct of his numerous official duties. As heard by Thomas Green, the man he appointed as his deputy, Calvert turned to Margaret Brent and said, "I make you my sole executrix; take all and pay all." Later the room was cleared and Mistress Brent and Governor Calvert conferred privately.

Had matters been otherwise, Maryland then and there might have been absorbed by the Virginia colony. The remarkable Mistress Margaret Brent had come to Maryland in 1638 with her sister Mary, and two brothers, Giles and Fulke. She was then thirty-eight years old. The Brents brought over a large party of servants, nine of whom (five male and four female) were attached to the sisters. This entitled them, under laws then current, to 800 acres of land. But going to Governor Calvert, Margaret Brent presented letters from Lord Baltimore, and claimed for her sister and herself land equal to that given the first arrivals. The larger areas were granted, and subsequently unusually large favors in land grants and high offices were given the Brent family. Among other lands, they held the "Sister's Freehold" in St. Mary's City. The Brents continued to bring in new settlers, in this way constantly adding to their landholdings. This was a common practice. Among others, Thomas Cornwallis brought in an average of seven or eight yearly during the first decade of the colony.

Margaret Brent was the first woman in Maryland to hold land in her own right, and the first to exercise fully the powers of an attorney. A strong-willed and capable seventeenth-century spinster, it appears likely that Calvert was attracted more by her loyalty and capability than by her femininity. Catholic, royalist, loyal to the proprietor and sufficiently disinterested to be entrusted with power—the combination was unique. Her ability was unquestioned, her actions decisive. She had a reasonably sound knowledge of the law. She not only cared for her own affairs wisely and competently, but she also represented her sister in legal matters, and served as agent for her brother Giles. In her defense of Calvert's property she proved herself vigorous and aggressive. Every episode in her career shows undoubted force of character.

She rallied behind the new deputy governor, and guaranteed the pay due Calvert's little army from his estates. Cleverly she maneuvered through these bad times, at first paying the troops with corn from Calvert's land and his rents and, when these were exhausted, with his cattle. During these desperate moments Captain Edward Hill, the very man Calvert had deposed, was watching eagerly from Chicacoan, across the river, demanding the payment of debts due him and hoping for some sign of disturbance that would justify his reappearance in

Maryland. That he found no opportunity was due, historians unanimously record, to Mistress Brent.

After establishing herself as Calvert's executor, Mistress Brent petitioned the Provincial Assembly to appoint her attorney for Lord Baltimore, to collect his rents and manage his estates. This granted, she proceeded to become involved in more lawsuits than anyone else in the colony, and from this litigious career to gain more as well. In her capacity as sole executor, she then petitioned the Assembly for a seat, pointing out that she deserved one vote as Calvert's executor and another as the lady of a manor. The Assembly of 1649 seemed not unwilling to grant this request, and to the proprietor wrote a strong endorsement: "As for Mistress Brent's understanding and meddling with your estate, we do verily believe, and in conscience report that it was better for the colony's safety at that time in her hands than any man's else in the whole Province after your brother's death, for the soldiers would never have treated any other with that civility and respect, and although they were ready at several times to run into mutiny, yet she pacified them, till at last things were brought to that strait that she must be admitted and declared your Lordship's attorney by order of court." But Lord Baltimore in England had other ideas.

Rebuffed in her plea, Mistress Brent gradually became disaffected with Maryland and soon followed her brothers to the Virginia side of the river. From there she explained to Governor Stone that "I would not intangle my Self in Maryland because of Ld Baltemore's disaffection to me and the Instruccons he Sends aft us." She settled above Aquia Creek on family lands that ran almost all the way up to Alexandria. Her plantation she named "Peace." The date of her death is uncertain, probably in 1670, but before the end it is possible that romance did not wholly pass her by. At the age of fifty-seven, Leonard Calvert's attorney made her last appearance in court to state that "Thomas White lately deceased out of the tender love and affection he bore unto the petitioner, intended if he had lived to have married her, and did by his last will give unto the said petitioner the whole estate of which he was possessed."

Early settlers like Cornwallis and Margaret Brent were exceptional only in the magnitude of their landholdings and the rapidity of their rise to wealth and influence in the colony. The same pattern can be traced at the opposite end of the social scale in a modest couple, Daniel and Mary Clocker. Daniel Clocker came to Maryland in 1636 as an indentured servant of Thomas Cornwallis, and the girl he married, Mary Courtney, had been brought to St. Mary's as a servant by Margaret Brent. In England the Clockers would have been tied to a feudal system for life, but facing the abundant lands of the New World they took root and flourished.

Once his freedom had been gained, Daniel Clocker's position in the colony grew steadily. He even sued his old master, Cornwallis, for not giving him the corn, clothes, and other allowances due him in accordance with the customs of the country. In 1641 he was employed by the colony for a three weeks' expedition, receiving 75 pounds of tobacco for his services. In 1642 he served in the Assembly, and is later mentioned in colonial records as sitting on juries and serving as a sheriff's assistant. His regular employment seems to have been that of a builder, for there are several references to his being engaged in building and repairing tobacco barns. He appears to have been a frugal man, and to have invested in land, for in addition to the 50 acres for himself and an equivalent amount of land for his wife received upon the expiration of their indentures, he also owned 100 acres on the Patuxent River and a piece of property near St. Mary's called Clocker's Marsh, which he later sold.

When he came to build a house, the land he chose was that given his wife by Mistress Brent, a pretty site at the mouth of St. Andrew's Creek in the old St. Mary's Forest overlooking the water. The architectural historian, Henry Chandlee Forman, who has made a careful reconstruction of Clocker's Fancy, the typical seventeenth-century dwelling Clocker built that is one of the few surviving examples of a modest house of the time, suggests the location was chosen to make it easy for Mary Clocker to continue in Margaret Brent's employ. Clocker's Fancy stands immediately behind the Sister's Freehold, where the Brent sisters lived, and is one of the two surviving buildings in St. Mary's.

The house Daniel Clocker built is worth noting, for it was the common type of small Maryland dwelling for many years. Few like it have survived. Here we can easily reconstruct everyday life in the seventeenth century along the Potomac.

Rectangular in plan, under a gable roof, Clocker's Fancy had two floors. The first floor was divided into a living room and a dining room, each with a fireplace. The second floor, hardly more than an attic with small square windows for ventilation at either end, had two bedrooms under sloping ceilings. From one corner of the living room a steep, winding staircase led to the upper floor. At either side of the living-room fireplace were cupboards. Perhaps the living-room fireplace was originally used for cooking, but Dr. Forman presumes that a kitchen was later provided in a separate outbuilding between the house and the smokehouse, which is still standing. This is the kind of house in which most of the early Potomac settlers lived. It is probably better than most, for it had a pegged wood floor. The ax and the adz built houses like this. Within them furnishings were generally primitive, beds hardly more than narrow shelves fastened to the walls, benches and stools, chests and tables. The fireplace was fitted for cooking in kettles and roasting on spits. Sometimes ovens were built into the fireplace. Meals were served on wooden plates or in earthen bowls, and eaten with wooden or pewter spoons. The house was heated entirely by the fireplaces, a small one having been provided for the east bedroom. While the gable end walls were built of brick, the front and back of the house were of framed timber, filled with loose brick for insulation and covered with clapboards. Small farmhouses no larger than Clocker's Fancy were the ordinary dwelling of the Potomac Tidewater.

Almost from the beginning, however, the plantation house appeared, set on a prominent headland overlooking the river or one of its estuaries. It belonged, typically, to one of the great landholding families. On the Northern Neck of Virginia, one of the first plantation houses was built by an early Cavalier settler, the first William Fitzhugh, who came out from England, sailed up the river to Chotank Creek, and settled there in 1670. In addition to the management of his famous plantation, Bedford, Fitzhugh practiced law, owned ships, held public office, and may have imported slaves. He married 11-year-old Sarah Tucker of Westmoreland and sent her to England to be educated. He was perhaps the earliest of the Potomac grandees.

We have a fair view of this Potomac settler sixteen years after his arrival, when he described his frontier plantation in a letter to an English friend. It then comprised a thousand acres, on which he had built a dwelling and some other buildings, all surrounded by a strong locust-post palisade. He owned twenty-nine slaves, most of them born in Virginia. Aside from the inevitable and omnipresent tobacco, the plantation supported livestock, an orchard of 2,500 apple trees, a gristmill and a cider mill. Colonel Fitzhugh also manufactured apple brandy. For efficient operation the plantation was divided into three units. At this time Fitzhugh also owned 2,500 acres of land up the river, which he was holding for speculation. From his landing a ferry ran to the Maryland shore. His sense of luxury was marked for that early date. In letters to his agent in London, orders for silver plate predominate. His living room was unique among the river houses, and was hung with tapestries.

Christmas of the same year found a Huguenot refugee from the Dauphiné named Durand, in company with twenty other pleasure-seeking Virginians, at Bedford. The stranger wrote: "The Colonel's accommodations were . . . so ample that this company gave him no trouble at all; we were all supplied with beds, though we had indeed to double up. Col. Fitzhugh showed us the largest hospitality. He had store of good wine and other things to drink, and a frolic ensued. He called in three fiddlers, a clown, a tight rope dancer and an acrobatic tumbler, and gave us all the divertisement one would wish. It was very cold but no one thought of going near the fire because they never put less than the trunk of a tree upon it and so the entire room was kept warm."

The French were in evidence even earlier, for when the

planter's son, Henry Fitzhugh, at the age of hardly twelve, was sent to England to school, his father went to some length to explain that, although the lad might seem dull in English, it was because nearly all of his instruction had been in the French language, and according to the French rules of Latin grammar. Therefore, he was placed under a French schoolmaster near Bristol and provided with "little money to buy apples, plums, etc." Fitzhugh was heard to say his children "had better never been born than ill-bred."

This first Fitzhugh is a proper representative of the Cavalier influx. He kept a fine house, was celebrated for a gay and lavish hospitality that became traditional in the Chotank region, and prospered as the country became settled and his investments brought large profits. One of the earliest Virginia capitalists, he differed from the Cavalier settlers in seeking income from sources other than the land. He had received a thorough training in law, probably by reading in chambers in London, and showed his talents to excellent advantage in the famous Beverley case. Like many others who clipped the corners of the law in those early days, he was singed a bit when accused by the Virginia Burgesses of misrepresenting claims for emolument; but like others with substantial influence, he was never tried. A man who scorned the masses—he called them rabble—he died rich but not greatly loved.

Another early Potomac settler in Virginia was the immigrant John Carter, founder of the long Carter dynasty, who came to Virginia about 1650. He rapidly established himself in the economic and political life of the colony and rose to membership in the Crown Council. Under his second son, however, the fabulously great family fortunes were created until, at his death in 1732, he was able to will 300,000 acres of land, a thousand slaves, and £10,000 worth of other assets.

The first Carter, who settled at Corotoman, married five times and had numerous descendants; nor were his children less active in propagating the race of Carters: his celebrated grandson, who engaged the diarist Philip Fithian to tutor his children at Nomini Hall, fathered seventeen. Through intermarriage the clan spread, like a dye, over much of the Virginia aristocracy and even into Maryland.

Born in 1663, the bewigged figure of the second Carter, Robert, may stand as an emblem of the first full generation of American-born planters, yet the English influence was still dominant, and the great wealth of the Carters was founded on the massive land grant that Thomas Fairfax had acquired. As Fairfax's agent Carter bought backwoods tracts of land, opened them up, and ordered the Indians liquidated. At his death some of his lands lay even beyond the Blue Ridge Mountains. His management of the Fairfax lands in the two periods 1703-1712 and 1722-1724 culminated in 1726 when he leased from Fairfax the entire Northern Neck for £450 a year. Carter soon managed to become the largest landholder on the Northern Neck and, upon his death, "the inventory of his estate made a prodigious catalog." Over the eulogistic inscription on his tomb was once irreverently chalked:

> Here lies Robin, but not Robin Hood,
> Here lies Robin, that never was good,
> Here lies Robin that God hath forsaken,
> Here lies Robin the Devil has taken.

5

These assorted characters, Cornwallis, Brent, Clocker, Fitzhugh, Carter, and hundreds not unlike them, with their indentured workers and increasing numbers of Negro slaves, were pouring into the Potomac all during the seventeenth century. Their concern was with land, prosperity, and security. Nor were they markedly interested in change. The institutions and ways of life in the England they had left had not been found very objectionable—if there had only been enough land and wealth to go around. On the fringe of the New World, remote from established seats of government and religious control, the old habits generally persisted. But if there was no visible desire for an original way of life, originality was thrust upon them, and the new experiences most certainly brought a marked degree of freedom.

From the beginning the population had a heterogeneous character. Rich and poor, gentle and lowly, Cavalier and Roundhead, Catholic and Protestant, Englishman and Virginians all rubbed against one another in their new home. As they became acclimatized to the new environment, the old skin of habits peeled off, and soon they began to see differently. The feudal institutions of the Old World were quietly discarded, and new ones took their place.

Persistence of the religious forms of the established church in Virginia or adherence to the religious forms believed in by the Catholics and Protestants in Maryland was a logical and early characteristic of the Chesapeake Bay settlements. Yet in notable distinction from England and the Massachusetts Bay settlements, there was an unmistakable air of religious toleration. No one was ever put to death for matters of a spiritual nature. Even the witchcraft scare, which spread over the entire Protestant world in the seventeenth century, like a spiritual measles, met calm resistance and found no victims in the Potomac valley.

In Maryland, despite clashes between Catholics and Protestants, the record justifies Cecil Calvert's expedient of admitting Protestant and Catholic settlers on equal terms. Nor does it diminish the luster of Maryland's reputation as the home of religious toleration to know that toleration was founded on political and economic necessity: Calvert could scarcely have excluded Protestants nor could he have successfully populated his domain wholly from the slender ranks of English Catholics.

In Virginia, always more observant of the British proprieties and forms, the Anglican Church appears to have been regarded essentially as an instrument of state, but as such it was handicapped by the absence of any bishop to confirm or ordain loyal ministers. Under Virginia law ministers were paid in tobacco, and in practice were hired by the year, resulting in independent vestries and practical Congregationalism. It has been shrewdly observed that in parishes where the tobacco grown was of poor quality it became impossible for the established church to secure ministers, and in these localities the dissenting faiths established themselves. The loose organization of the Anglican Church in Virginia made it easy to carry within the official structure of the church a number of dissenters—Puritans and Huguenots—who could thus comply with the formal conditions of establishment yet retain substantial religious independence, and to this circumstance has been traced the low-church characteristics of Virginia Episcopalianism: until the memory of men and women now living there were Episcopal churches in which communion was taken in a seated position, and where a surplice was never seen.

In the first Potomac settlements in the seventeenth century we can also see the immediate decline and eventual disappearance of another feudal institution it had been assumed could be transplanted in the New World: the manorial courts, or courts baron and leet as they were called. In the records of the old Maryland manors, it has been well said, we may see "the manorial courts handling trivial business in amateurish fashion, demonstrating the futility of the scheme." Power might be asserted over slaves and indentured servants, but under the release offered by vast tracts of empty lands it was impossible to retain the traditional powers over manorial tenants. As a manorial figure, the lord proprietor of Maryland himself soon lost most of his powers. Similarly, quitrents, tenancies, rent rolls, freeholders and copyholders, all were tried before withering away under the adverse surroundings in the Potomac.

From the beginning the Virginians were unalterably opposed to the introduction of the manorial system, but shortly after the granting of Maryland, the fleeing Charles II made a grant of the entire Northern Neck to seven of his lords, with powers to create manors similar to those granted Lord Baltimore. Only a few years later the area was again granted to the Earl of Arlington and Lord Culpeper, and subsequently there fell by marriage into the possession of Lord Fairfax and his heirs title to all that immense domain between the Potomac and Rappahannock rivers—to the still-unknown headwaters! A manor, indeed, yet the colonial representatives of the Fairfaxes were too wise to press their manorial rights.

Under the extremely decentralized pattern of the Chesapeake colonies in the Potomac valley, the hand of provincial government was weaker, taxes were almost forgotten for long periods, law enforcement was almost entirely a local matter, and the courts grew casual in procedure and lenient in the penalities handed down. A difference began to mark the Potomac settlements from those in other parts of Virginia, and this individuality translated itself into various independent attitudes toward all institutions, molding them to the conditions of life in this time and place. A sectionalism was born.

6

Land tenure best showed the change. Cheap and abundant, land was also the prime form of wealth. Its value changed rapidly under the impact of immigration and new crops. It became the principal vehicle of investment and speculation. These New World conditions forced a modification of the more rigid feudal systems of tenure, reflecting conditions of a stable society and designed to hold the population to the land. From the Norman Conquest on, free tenants, cottars, and villeins had composed the manorial system. The free tenants, or yoemen, were obligated only for military service and the payment of quitrents. The cottars obtained the use of small garden-sized pieces of land and performed services for the manor lord. The villeins were the principal farmers of the manors, and in exchange for rents or services obtained the use of long strips of land amounting to from as little as thirty to more than a hundred acres. Satisfactory as this system had been for centuries in England, it already was disintegrating there and did not fit conditions in the New World. But where in England efforts were being made to modify it by developing new forms of leases, in the Potomac the pressure from the first was in the direction of freehold. That was the only way the undercapitalized, land-poor owners of large proprietary grants could liquidate their holdings. That was what the smaller settler wanted. It fitted the needs of the colonial companies and the proprietary.

In Virginia a system of "head rights" prevailed as the common method of granting land. This assured to each immigrant 50 acres of his own selection, to be held on payment of a quit-rent of a shilling a year or 12 pounds of tobacco. Larger tracts were assembled by additional grants of 50 acres for each dependent brought into the colony, and a similar area for each servant. For special services and benefits to the colony, grants of not more than 2,000 acres were frequently made. Nominal amounts of land were also granted to indentured servants when they became free. This system with its freight of quitrents early began to break down in Virginia; in the remote Potomac valley it does not seem ever to have been in effect.

In Maryland terms were similar but more liberal. To each immigrant the proprietor gave 100 acres, 100 acres for his wife, and 50 acres for each child, and 100 acres additional for each male servant and 60 acres for each female servant. Servants were granted 50 acres at the conclusion of their indenture. These provisions, designed to attract large numbers of settlers, led to the system in which each immigrant holding of more than a thousand continuous acres—later increased to two thousand—was made a manor, and its owner constituted a lord with feudal rights and privileges. Any immigrant bringing more than five settlers between sixteen and sixty years of age received 2,000 acres.

That was how it began. It began to change immediately, and on both banks of the Potomac the systems of land tenure came to be virtually the same, and markedly different from those in other parts of Virginia and Maryland. Aside from the remoteness of the Potomac lands from established seats of authority and other circumstances that have been mentioned, a further differentiating factor existed on the Northern Neck. Here the confused and fluctuating ownership clouded land titles to such an extent that there was literally no legal way in which the first settlers could have gained title to their lands. By the time a Fairfax had arrived to take possession of the vast domain he had come to own, settlement had been firmly established and the residents had strong, independent ideas about the possession of land—and the payment of quitrents.

Opposition to the manorial system was not coincident with opposition to large landholdings, for the lands on both sides of the Potomac were rapidly divided into huge plantation units, their sizes ranging from a thousand to tens of thousands of acres, with frontages of miles along the river and its tributaries. Great plantations rubbed gently against the fences of the many smaller plantations of the freed indentured servants that were reckoned only by the hundred acres. The largest were really spectacular.

In seventeenth-century Virginia, Richard Lee of Northumberland County accumulated 20,000 acres; William Fitzhugh owned over 50,000 acres; and only a little later the magnificent Robert Carter owned the staggering total of 330,000 acres. The Brents, who came from Maryland to the Virginia side, held nearly 10,000 acres. Later, George Mason's holdings came to some 15,000, and George Washington's lands in the valley may be accounted modest at 8,000 acres—although both held far more along the upper Potomac and in the Ohio country. Across the river in Maryland, that "wisest and best" settler Thomas Cornwallis accumulated over 8,000 acres; Leonard Calvert's holdings ran well above 6,000 acres; and a host of others held from 4,000 down to 1,000 acres.

As the valley filled up, these large original grants soon became smaller. The large estates were often divided among children, or were broken up and sold for other reasons. When land values increased, a further incentive to its sale appeared. In Maryland, by 1683, land was no longer granted by the much-abused head right but was sold by the proprietor at a price of 100 pounds of tobacco per 50 acres. By then the average size of plantations had fallen to around 250 acres, although some of the original large grants remained intact.

The early Potomac settlement was from the beginning essentially English in its population and institutions, but from the beginning as well change was evident. While homogeneous, the population was not without significant dilution; and just as important, among the English themselves were major grounds for division. Social stratification and the formation of social classes proceeded along lines different from those in England. The rewards were greater for enterprise than for investment and holding, and whether it was the fur trade, the land trade, the tobacco trade, or other activities, it produced a different quality of leadership than was found in the landed aristocracy in England. Tied to markets in England, it also looked to the west for its future and its expansion.

In the prosperity and relative freedom of the lower Potomac was shaped a civilisation at first different and later profoundly antagonistic to that of the mother country—and even to the Tidewater capitals and ports of Virginia and Maryland. There developed a characteristic blend of old and new institutions and ideas. The Potomac was becoming a seedbed for future Revolutionary leaders. As long as its western momentum was unchecked and as long as the great design of empire offered security and gains commensurate with its obligations, its restrictions, and its losses, the inevitable conflict was postponed. But, with the frustration of the tobacco planters under the shackles of British mercantilism, and the restraint imposed by the crown on western settlement; with a corrupt, confused, and unsympathetic government that refused to undertake necessary improvements or the defense of western settlements—to say nothing of the taxes which became odious when they did not represent value received—the Potomac planters began to stand shoulder to shoulder in defense of their way of life. Earlier than the third quarter of the eighteenth century they were not ready. A common cause had not been formulated, and the colonists could not act together. This can be seen in their unsuccessful efforts to control the tobacco trade. It can be seen, too, in their failures in handling one of the oldest colonial problems—the Indians.

7

As the settlement of the lower Potomac increased, so did friction. When the Indians were pressed back, intertribal wars flared up among the Piscataways and the remnants of the Susquehannocs, and the Iroquois. Senecas and Oneidas filtering through the mountains were captured and tortured by the Potomac Indians, and their threats of reprisal became real. Between the Maryland and Virginia settlements there were intermittent boundary disputes, quarrels over trading rights, religious friction, and conflicts growing out of the differing policies of the two colonial governments. Between the white settlers and the dwindling Indian population individual clashes grew more frequent. But the great Indian insurrections, like those of 1622 and 1644 in Virginia, were never repeated, and on the north bank of the Potomac the settlers were never troubled by large-scale Indian disturbances. In the years since the English had landed, nearly two-thirds of the Potomac Indians had disappeared. Disease, warfare, and migration finished off most of them. Toward the end of the century, however, the two Potomac colonies collaborated in one of the ugliest slaughters in the annals of the river, a disorderly lynching that had far-reaching consequences.

Settlement had brought hogs and cattle, and a frequent cause of trouble was the accusation that the Indians were killing the settlers' stock. The Indian viewpoint was crisply and officially stated: "Your hogs and cattle injure us; you come too near us to live, and drive us from place to place. We can fly no further."

In the winter of 1675 in Stafford County, near the Big Bend, a herdsman, Robert Hen, was killed by some hostile Indians. A party of Virginians under Colonel George Mason and Captain Giles Brent pursued the Indians. They crossed the Potomac and in the ultimate confusion of all wars, big and little, they lost their quarry and instead attacked two parties of friendly Susquehannocs. Before explanations could be made, one group had been wiped out and fourteen Indians in the other killed.

Further murders of whites by Indians led almost immediately to a fresh pursuit, this time a joint operation of Virginia and Maryland militia. The Virginia troops were under the command of Colonel John Washington, then a resident of twenty years in Westmoreland County, and an old hand with the militia. He had settled on Pope's Creek and commenced to grow tobacco. Married three times, early records also show he sat in the House of Burgesses.

To command the Maryland troops the Governor's Council had sent one of its members, Major Thomas Truman. He had come to Maryland to live upon lands granted him by Lord Baltimore, and took his seat in the Council. Aside from his objection there to the political Presbyterianism of Francis Doughty, little else is known of his activities or views. Since the operations were conducted in Maryland, Truman was placed in command of the entire force.

The two bodies of troops, five hundred from each colony, converged on the unfinished Susquehannoc fort at the mouth of the Piscataway. Its sturdy wall of earth and logs was protected by a ditch. Without artillery to breach the walls, there was little possibility of the militia taking the place by storm. Between the impatient and hotheaded Virginians and the excessively cautious approach of Major Truman there was little possibility of agreement; seldom, indeed, has divided authority resulted in a campaign worse managed.

Truman had providently brought the colony's best interpreter, and had with him representatives of five different tribes of friendly Indians. Under a flag of truce five chiefs of the Susquehannocs came out of the fort to parley. They protested their loyalty and innocence; they showed papers and a medal given them by Calvert, and accused the Senecas of having committed the murders. By this time the Virginia forces also had arrived, and it was agreed that the Senecas should be pursued.

Before the combined party could set off the next morning, one Captain John Allen arrived, bringing the bodies of some recently slain settlers. Inflamed by this sight, the soldiers, who were not interested in differences among the Indians, came to the end of their patience. They demanded that the Indians "be knocked on the head." The Virginia officers, under Colonel Washington, grew so violent in their demands that Major Truman finally consented to the execution of the five Susquehannocs who had stayed with his party after the agreement to pursue the Senecas, and to whom he had promised no harm should come.

The entire party then turned on the Susquehannocs in their fort, who numbered only one hundred fighting men—perhaps five hundred Indians in all—and besieged it for nearly seven weeks. During this time the starving Indians seem to have lived mainly on horses captured from the besieging force. They killed an astonishing total of more than fifty whites. In the end all the Indians escaped by the simple expedient of crossing the Potomac, when it froze, into Virginia. In their bloody retreat through Virginia they killed the overseer on Nathaniel Bacon's plantation. The indignant Bacon then aroused his neighbors to such strong protests at the Virginia capital that armed rebellion finally resulted from the bitter dissatisfaction of the frontier settlements with the lack of protection given them by the colony.

In Maryland the reverberations of the affair at Piscataway produced an extended investigation. A pamphlet entitled *A Hue and Cry Out of Virginia and Maryland* voiced the patience-exhausted attitude of the frontiersmen and attacked the proprietary. Major Truman was impeached by the Council on three charges. But although he was found guilty and deprived of his seat in the Council, and despite violent criticism of the course he had pursued, he escaped further punishment. Assembly and Council could not agree. The objection was made that Truman's conduct, if not condemned, would set an example of a breach of public faith that could only be disastrous in future relations with the Indians—a point of view never shared in Virginia. In the confused condition of the testimony and the statements contained in the records of the Assembly and Council, it appears that, while Truman was considered guilty, "the general impetuosity of the whole field, as well Marylanders as Virginians, upon sight of the Christians murdered," and the need "to prevent mutiny of the whole army, as well Virginians as Marylanders," were considered extenuating circumstances.

By the end of the seventeenth century, nearly all of the surviving Potomac Indians had moved west to the mountains or into Pennsylvania. Soon afterward their old reservations along the river were divided and sold by the proprietary. From Anacostia to Point Lookout, hardly an Indian remained to be seen.

FOUR

THE TOBACCO CIVILIZATION

ONCE THE FUR TRADE HAD MOVED WEST, SUBSTANTIALLY THE only export of the Tidewater plantations below the falls of the Potomac for nearly a century was tobacco—*Tob°*, as the much-used word was commonly abbreviated in innumerable contemporary diaries, letters, and ledgers. Upon this plant the fortunes of the lower Potomac in the late seventeenth and eighteenth centuries were built. Like water for irrigation in a dry country, tobacco alone made land valuable. Tobacco became the coin of the region. From its profits came all the planter's luxuries, his fine carriages and Madeira wines, his London suitings and silver plate, the wages of his tutors and ministers, his books and paintings. With it he paid his taxes. On the narrow base of this single crop the plantation system of the lower river established itself, came to a brilliant climax, and declined.

The crop determined the fate of the land and the character of its people as well as their fortunes.

To meet the special requirements of tobacco production the land was carved into plantations. The quality and flavor of the leaf is the essential object of all tobacco culture, and the crop's sensitivity to minor differences in soil, drainage, climate, and other local conditions meant that even within tobacco-growing regions not many acres were ideally suited to its production. Thus, on a plantation of a thousand acres as few as thirty might actually be under cultivation. Like other monocultures, tobacco grown without manures or fertilizers or even crop rotation rapidly exhausted the soil. Early realization of this dictated those large reserves of virgin soil on every plantation that could be tapped when the yield of original fields commenced to decline. It was also necessary to have a large acreage for the plantation forests, which supplied the necessary hogsheads for shipping the leaf, and to have access to the river and the rolling roads down which the heavy casks were delivered to warehouses and ships. To house and feed the plantation establishment with its army of indentured workers and slaves, some other agriculture was necessary, however subordinate. So it was that, although a single worker could tend only about three acres of tobacco, the other factors in its cultivation made necessary the large total acreage of the tobacco plantation, much of it by design left in forest.

At each step the cultivation of tobacco needed large amounts of hand labor, a fact hardly less true today, three hundred years later, than at the moment of its first extensive planting. The minute tobacco seeds were sown in carefully prepared beds, usually planted in the early spring and kept under a covering of cloth or brush to protect the young plants from frost. When the late spring rains came, the plants were set out in previously prepared hills, and the long process of cultivation began. Apart from clean cultivation during its growing season of about four months, tobacco demanded constant pruning to prevent the plant from branching and blooming. This promoted the growth of large leaves. Constant hand labor was also necessary to protect the plants from the tobacco hornworm.

In harvesting tobacco, the plant was first cut and allowed to wilt in the field. Then the individual plants were hung on rods and carried to the drying barns where the curing process was hastened and assured by controlled ventilation or by carefully watched fires. When quite dry, after a spell of moist weather, the still-intact plants were taken down and the leaves were stripped. After sorting and grading they were finally prized into huge hogsheads weighing from five hundred to a thousand pounds, ready to be rolled to the ship.

The growing of tobacco was an exacting job, and one that

in the many phases of its cultivation had to be pursued the year around. It thus differed from rice, cotton, or other staples. In the nursery, the fields, and the tobacco barns, it required careful, concentrated attention. Large numbers of workers, and their close supervision were needed. The tobacco itself was steadily improved under the spur of higher prices for premium crops. It encouraged enterprise, industry, and ingenuity, and some other, less desirable, characteristics.

Of the culture of the tobacco plant itself little remains to be known, but of what this culture did and continues to do to the lives of the men, women, and children who raise it little is yet understood. A wiser and later Calvert wrote Lord Baltimore in 1729: "In Virginia and Maryland Tobacco is our staple, is our All, and indeed leaves no room for anything Else; It requires the Attendance of all our hands, and Exacts their utmost labour, the whole year around; it requires us to abhor Communities or townships, since a Planter cannot Carry on his Affairs, without Considerable Elbow room, within his plantation. When All is done, and our Tobacco sent home, it is perchance the most uncertain Commodity that Comes to Markett; and the management of it there is of such a nature and method, that it seems to be of all others, the most lyable and Subject to frauds, in prejudice to the poor Planters."

Concerning the tobacco farms of the lower Potomac today, a wayfaring sociologist could write a textbook with each phrase of this quotation as a chapter heading, so little have things changed. A whole folklore has been created in this culture. Of the long season and the long hours one says that "it takes thirteen months a year to grow." Because the careful hand labor of women and children in suckering, working, and curing is taken as a matter of course, planters complacently say, "Only large families of poor whites can raise tobacco." It is still the crop that "wears out men and land." To Jefferson it was "a culture productive of infinite wretchedness." Indeed, few contemporaries could be found—certainly after the middle of the eighteenth century—to say a good word for it. But it continued to be grown, and time stood still. The southern Maryland and Northern Neck counties are unique in our country as places where an identical agriculture has been practiced for three centuries without substantial interruption. If anyone wishes to see the result, he can find it there.

If the human as well as the natural needs of the tobacco culture are understood, it is evident that by far the larger part of the people who settled the Potomac could not have been plantation owners; rather they were indentured workers, and Negro slaves, or they belonged to the steadily increasing yeomanry, largely recruited from the indenture system, who owned small, independent plantations. With wealth, a class of professional men and merchants also appeared.

By the middle of the eighteenth century few of the large original land grants remained intact; and although land was still the preferred form of wealth, it was likely to be held in the upper valley. The outstanding plantation seats were interspersed with smaller farms, equally enthralled with tobacco; the growing number of shipping centers and the expansion of the tobacco factor system reflected their needs. Despite the colonial capitals of Williamsburg and Annapolis—neither of which was on the Potomac—the region continued to be conspicuously lacking in towns.

Diluted as the plantations were, the social accent and characteristic ideals of the Tidewater region, nevertheless, were largely determined by a leadership firmly lodged in the planter class. The social ascendancy of the Tidewater patricians, although challenged by merchants, western settlers, and other insurgent elements, long went unchecked. Indeed, the imprint of the planters was to be the lasting one on the character of the region.

In the single century in which it first flourished, the tobacco civilization stamped the landscape, originated social habits, gave birth to ideas, and created a type of personality that gave the people of the lower Potomac region a distinctive character. In its composite form, and not without contradictions, the character formed in the mid-eighteenth century persists today. The traditional openhanded hospitality and genial sociability of the region, the love of outdoor life and its activities, loyalty, the high sense of honor and chivalry, the sporting blood, and the taste for speculation—all are survivals of the tobacco era. So, too, are an arrogant and empty pride, a musty conservatism, and the various forms of ancestor worship peculiar to the region. Good manners and sharp trading, hard riding and soft living, have gone hand in hand. By progressive dilutions, the habits of a class have become the characteristics of a region, no longer pure but recognizable beyond doubt.

2

The characteristic mode of plantation life first appeared in the lower river. The Christmas party at William Fitzhugh's Bedford, one of the most typical early plantations, was representative. Situated in the center of a region that later acquired its generic name from Chotank Creek, Bedford was gradually surrounded by plantations equally large and, soon, equally luxurious. First among them were the ones Fitzhugh built for his five sons. The second William Fitzhugh lived at Eagle's Nest; another son lived at nearby Boscobel on Potomac Creek; while the three others were farther up the Potomac or nearby on the Rappahannock. At Cedar Grove, just west of Chotank Creek, lived Richard Stuart. His neighbors to the west were the Alexanders at Cedar Grove and the Ashtons at Chatterton. So much of the community was within easy carriage drive of St. Paul's Church.

On the shore opposite Mathias Point a wide tidal inlet leads to the plantations of Port Tobacco Creek, and only a short distance away is the broad Wicomico. On these waters lived the Smallwoods at West Hatton, the Harris and Jenifer families, the Hansons of Mulberry Grove, and the Causines. At the head of Port Tobacco Creek were those two noted colonial physicians Dr. Craik and Dr. Brown, and the Stone plantation, Habre de Venture.

On the Virginia shore, opposite the Wicomico, were further celebrated plantations in the neighborhood of Pecatone, where the Washingtons, Fauntleroys, Turbervilles, Corbins, Monroes, and Ashtons had their seats. Here is the celebrated and still miraculously preserved Stratford, home of the Lee family, and Chantilly. And finally the Nomini Bay area, home of many Carters, and beyond the many other plantations of the lower river.

A more intellectual tone was claimed for this district, but the social life was not materially different from Chotank.

From Chotank north, up the river, one encountered the numerous Brent family, the Mercer strongholds around Potomac Creek, more descendants of Carters and Lees and Washingtons, until one enters that Fairfax-land which a distinguished local historian once mourned as "lost in Washington-land." And across on the Maryland shore were more Smallwoods, many branches of the Hanson family, Stones, Semmes, and other names that have become famous in Maryland history.

Chotank was a generic name for this region, an area with indistinct and ever-expanding boundaries, perhaps more a state of mind than a physically limited territory. Originally designating the centrally situated area occupied by the Fitzhughs, the name spread by common consent to include the neighboring Virginia and Maryland plantations. As the years passed, social intercourse and marriage strengthened the unity of the district and it acquired its distinct personality and flavor.

In this region and during these years the Potomac was not a separating boundary but a unifying influence, the principal avenue of travel and trade, and the common bond. Life on one side of the river was much the same as it was on the other: the same crops, the same plantation houses, the clear-paned brick churches, and eventually the same kind of cousins. Inhabitants of this homogeneous region moved freely from one side to the other. Just as the Brents migrated to Virginia, we find Richard Lee settling at the mouth of Port Tobacco in Maryland. While almost every rich planter who lived on the river had his own sloops or barges rowed by trained and often uniformed Negro crews, the region was also unified by a celebrated ferry that ran from Pope's Creek on the Maryland shore two or three miles to Mathias Point in Virginia, at about the spot where the Morgantown bridge now crosses the Potomac.

Operated by the Hooe [pronounced *hoe*] family, this ferry was the essential link in the shortest route between the capials of Maryland and Virginia, and also between the upper and lower Potomac settlements. It was crossed by the first post route, and frequently used by Washington, Mason, and others on their way to the plantations in Westmoreland County or the colonial capital of Williamsburg. As travel by land became easier, this ferry became a principal north-south link of the colonies. Travelers by the hundred, including many who kept diaries or wrote letters, crossed there and invariably commented on it, and on the Hooe family, and the open sailboat in which they and their bound horses were carried across the Potomac.

The ferry made Chotank accessible, and fertilized it with new ideas and personalities. It had been pioneered by William Fitzhugh, to whom permission to run a ferry at about this spot was given by the Virginia legislature in 1705. The Hooes had come up the river shortly after 1700 and settled at Barnsfield, at the southern terminal of the ferry. They seem to have begun as tobacco farmers, and married into the Chotank neighborhood, but operating the ferry soon became their main interest.

Gradually the Hooe hospitality grew until it acquired a legendary character. Their home became virtually a tavern. They not only carried travelers across the river, but fed them, provided beds, served up liquid collations, fed and stabled their horses, gave out directions and neighborhood gossip, and supplied copious information on the condition of the roads, and the comings and goings of travelers. The Hooes acquired a singular fame for confabulation, perhaps cemented by the account of an astonished English traveler who recorded:

"Here we were not a little diverted at a reply made by the owner of this ferry to a person enquiring after the health of one of his nearest relatives . . . 'Sir (said he) the intense fragility of the circumambient atmosphere had so congealed the pellucid aqueous fluid of the enormous river Potomac, that with the most eminent and superlative reluctance, I was constrained to procrastinate my premeditated egression to the Palatinate Province of Maryland for the medical, chemical and Galenical coadjuvancy and co-operation of a distinguished sanative son of Esculapius, until the peccant deleterious which it has ascended and penetrated, from the inferior matter of the Athritis had pervaded the cranium into pedestrial major digit of my paternal relative in consanguinity, whereby his morbidity was magnified so exhorbitantly as to exhibit an absolute extinguishment of vivifications.' "

Over Hooe's ferry and the roads and trails of Virginia and Maryland went not only legitimate travelers, but a horde of plantation cousins, forever visiting one another, attending the various social functions of the region that seem, in retrospect, endless in their number and variety.

Social habits acquired architectural expression. Some plantation houses were designed with removable paneling to permit several rooms to be united into a single large ballroom, and eventually special rooms or separate buildings were provided for social occasions. At Pecatone, as early as 1670, four neighboring planters contracted to build jointly a "banqueting hall" on a site where their respective estates joined. Special large bedchambers were permanently set aside for guests, and children and servants were banished to outbuildings.

The social activities commenced with the young. For the children of the Northern Neck plantations there were the lessons in dancing given by the stern Mr. Christian and in music by Mr. Stadley. For these occasions, the young of Chotank were collected in one plantation after the other as the teachers made their way from place to place.

For the grownups a gay round of balls, horse races, boat races, and other frivolity was provided. The men amused themselves at fox hunts, cockfights, and drinking parties—euphemistically called "barbecues"—and an occasional boxing match, as well as gunning, fishing, and other sporting pursuits. The captains of tobacco ships and naval officers commanding convoy ships were invariably entertained, and found it obligatory to return the compliment. At one ship's dinner on the *Beaufort* forty-five ladies and sixty gentlemen guests were present. But the typical plantation entertainment was dining and visiting, and for these the slightest pretext was sufficient.

Each Sunday after the service there was a sociable hour on the church steps and in the churchyard, in the course of which dinner parties were made up. Housewarmings, complete with a ball and supper, were mandatory not only when new houses were occupied but even when old houses were repaired. A coachful of guests was an ordinary occasion at Nomini Hall, and Philip Fithian, the diarist who tutored the Carters, casually reports a dozen or so guests in addition to the numerous Carter family as not exceptional.

While the Potomac tradition of hospitality and entertainment originated in the Chotank neighborhood, its social manner was not greatly different from the similar neighborhoods of Pecatone, between the Yeocomico and Machodac rivers; St. Mary's, Port Tobacco, or Marlboro; or the Pohick district between Occoquon and Great Hunting creeks where the Fairfaxes, Masons, and Washingtons lived. The plantations of Hayfield, Woodlawn, Belvoir, Gunston Hall, Hollins Hall, Mount Eagle, Rose Hill, Bell Vale Manor, and Round Hill in the Pohick neighborhood were as closely knit and contained as gay a life as Chotank ever did. And there were the same comings and goings on both sides of the Potomac. Especially close relations existed between the Diggs and Washington families. The Hansons and Smallwoods, outstanding Maryland families in this region, moved much in Pohick society.

Not only were there other Chotanks along the Potomac, but Chotank re-created itself in fact as well as in the minds of the Potomac people. On land near Manassas that the London notary, Nicholas Hayward, had once purchased for a Huguenot settlement, the sons of the Fitzhughs, Alexanders, Stuarts, and others re-established themselves on either side of Cedar Creek. Land titles here had been obscured by the terms of Lady Fairfax's will and subsequent litigation until 1741 when Richard Foote subdivided the district into 200- and 300-acre plantations, and leased them for three lives at an annual rent of 350 pounds of tobacco to the sons of Chotank, who transplanted their families to virgin soil and created a new community with all the characteristics of the old Chotank.

The recollection of a pleasant life lingered. A generation later, in the Alexandria *Gazette* under the nostalgic title "Recollections of Chotank," the lament could be read: "Can I ever forget the happy days and nights there spent; the ardent fox hunt with whoop and halloo and winding horn: And . . . the old family bowl of mint julep, with its tuft of green peering above the inspiring liquor—an emerald isle in a sea of amber—the dewy drops, cool and sparkling, standing out upon its sides—all, all balmy and inviting? And then, the morning over and the noon passed, the business of the day accomplished, the social board is spread, loaded with flesh and fowl and the products of the garden and the orchard . . . But Chotank, like many other parts of the Old Dominion, is not now in its 'high and palmy state.' Some fifteen or twenty years ago it obtained that celebrity which makes it famous now. The ancient seats of generous hospitality are still there, but their former possessors, so free of heart, so liberal and blessed withal with the means of being free and liberal, where are they? 'And echo alone answers, where are they?'" Up the Potomac, and west to Kentucky, and south to Tennessee and Mississippi, the vision was borne, and there the Chotank ideal still may live.

3

It is easy to carry away the impression of a more leisured and luxurious society in Tidewater Potomac than in fact existed. The nostalgic recollections and the favorable descriptions of many travelers are deceptive, and they are often punctured by small facts. One overlooks, for example, the spartan nature of the accommodations for guests. Six or eight persons to a bedroom was not uncommon. Chastellux reported: "They make no ceremony of putting three or four persons in the same room, nor do these make any objection to their being heaped together." The plantation houses themselves seldom contained more than three sleeping rooms, for the master and his lady, and for guests. Men and women guests were commonly segregated to bedrooms hardly more than barracks. Children and servants were expected to sleep in separate buildings apart from the main plantation house.

The chintzy, highly correct version of the colonial household that modern interior decorators and museums have presented scarcely ever existed in fact. Even the very rich Carters had to wait some months to get a broken window pane in a bedroom replaced. Their table was waited upon by a 12-year-old boy. An unprejudiced English wayfarer considered the interior of a plantation house and discovered "all kinds of incongruous accidents are visible in the service of the table, in the furniture, in the house, in its decorations, menials, and surrounding scenery." Chastellux commented that "the chief magnificence of the Virginians consists in furniture, linen, and plate," and said, "they want nothing in the whole house but a bed, a dining room, and drawing room for company." Visitors found that the china and silver were often assembled from unmatched pieces. When china was smashed the nearest replacements were months distant in London.

The diet of the great plantation houses seems to have been only a little above that of the common countryside. Ham was "as inevitable as the table." "Hospitality," expostulated a visiting Continental, "What do you mean by hospitality? Beef, mutton, ducks, geese and turkeys; a bed and a dish of tea." Except for the planter's imported Madeira and an occasional bottle of brandy, liquor was ordinarily home distilled and often presented such unpredictable and proudly experimental surprises as persimmon brandy.

The plantation houses themselves were not extraordinary. Of all the Potomac plantation houses, a recent authority found only Pecatone, Stratford, Nomini Hall, Gunston Hall, Mount Vernon, and the Carlyle House in Alexandria worth inclusion in a comprehensive survey of eighteenth-century Virginia architecture. On the Maryland side the notable mansions would be fewer. At that Latrobe had looked at Mount Vernon with sympathetic and acclimatized eyes, and could only report it "extremely good and neat, but by no means above what would be expected in a plain English country gentleman's house of £500 or £600 a year. He then thoughtfully added: "It is, however, a little above what I have hitherto seen in Virginia." Most of the larger houses had only four rooms and a hall on the ground floor. Gunston Hall is well described as "really a large cottage in architectural effect." Architectural interest, in most cases, was represented by carved-wood or plaster-ornamented interiors, usually copied from standard books of designs, and executed by indentured craftsmen, or by mantels, wallpaper, paneling and moulding and other decorative elements often brought like the furniture from England.

Not architectural quality but, as all travelers noted, the commanding site with its view over the river, the agreeable and indigenous materials, and the emphatic expression of a pleasant way of living—these were the telling characteristics of the eighteenth-century plantation houses of the Potomac. Their owners liked them to be respectable, but they did not insist that they were in the latest mode. It is worth remembering, too, that most of the large houses were in a constant state of evolution. In thirty years Mount Vernon, not unrepresentative in this respect, underwent half a dozen major alterations and additions, carrying it from a one-story cottage to a semblance of its present appearance.

The young grew up on the plantation, formed by its peculiar environment. Families typically were large—ten or a dozen —and Lord Adams Gordon counted the Virginia women "great breeders." "Our Land free, our Men honest, and our Women fruitful," a common toast ran. Even among the prosperous, "incessant child bearing, hasty remarriage, and large infant mortality was the rule." Family life in the collective sense, as the next century idealized it, hardly existed. The use of wet nurses was common among those who could afford it, despite the gibes of moralists who accused mothers of attempting to keep their figures. Children were suckled by slave women, "and then cared for by Negro nurses until they were old enough to be turned over to tutors and governesses." The influence of Negroes on the very young was a topic of constant speculation. "I cannot be reconcil'd to hav'g my Bairns nurs'd by a Negro Wench," Boucher wrote. "Seriously that is a monstrous Fault I find with ye people here, & surely it is the course of many Disadvantages to their Children." It was contended that speech, manners, and morals were corrupted by the intimate association with Negroes. Jefferson lashed out against another consequence of slavery: "The parent storms, the child looks on, catches the lineaments of wrath, puts on the same airs in the circle of smaller slaves, gives a loose to the worst of passions, and thus nursed, educated, and daily exercised in tyranny, can not but be stamped by it with odious peculiarities." A double standard of morality developed certainly, and Jefferson hardly exaggerated the system's notable ravages to individual personality.

Escape from the plantation world came most commonly when children were sent away to school, for the Potomac had few schools of its own before the Revolution. In Maryland the Jesuits had attempted some systematic instruction and in Westmoreland there was Mr. Campbell's school, which, it has been dubiously claimed, James Monroe, James Madison, Thomas Marshall, and George Washington attended. There was also the College of William and Mary, and St. John's at Annapolis. A Catholic lad of wealth and promise like Charles Carroll might be sent to Saint-Omer in Belgium, and the best educated lawyers in Virginia were usually those who had read at the Inner Temple.

For those who stayed at home the larger part submitted to the discipline of a plantation tutor, although some were taught by their fathers or gained their knowledge from long acquaintance with the classics that lined the library shelves of the more prosperous plantation houses. Even modestly circumstanced families attempted some education of their children at home. The English traveler, Richard Parkinson, found one such planter near the Potomac instructing his children. "He told me he had been so troubled to get his children educated, that at last he had found more satisfaction in doing it himself than pursuing any other method."

Plantation life itself was an education—of a sort. It produced, certainly, superb horsemen and splendid shots, men who knew dogs and woodcraft. An uncertain knowledge of farming and an ability to direct slaves also came in this way. More significantly, and a paradox because of the isolation of the plantation, was the educational value of the conversation in the plantation house, and the diversified experiences of its operation and management.

The genius of the place was for people. The development of

great personalities, manners, social codes, leadership, ideas were the typical achievements of Tidewater Potomac society, rather than great agriculture and industry, the building of cities, monuments, inventions, or other works. Transitory and ephemeral as these might seem, the valley's accomplishment was seen by the nation in Washington, Madison, and Monroe. More intimately it could be detected in the quiet talents of Dr. Gustavus Brown of Rose Hill, remembered for the garden he created, with its hothouse and irrigation system, where "he assembled a rare and most valuable collection of roses and other flowering plants from all parts of the world." At the worst it produced a museum of harmless eccentrics, like the Hooes, or the Turbervilles, but at its best it has seldom been equaled.

The river was a bond, a unifying rather than a dividing influence. It led to social and ideological unity, and to a unity of blood through such intermarriages as Robert Carter and Ann Tasker, George Mason and Anne Eilbeck, or Colonel Fitzhugh and Mrs. Rousby; and for many lasting friendships. The awareness of a common culture and common economic interests on the Maryland and Virginia shores laid a firm foundation for the unity that was later to arise in the stress of Revolutionary times. The entire valley depended upon the common crop, and looked to England rather than to other parts of America to supply the necessities as well as the luxuries of life. When a shipment of tea arrived from London it was a signal for the neighborhood to arrive and sample it. But the time was to come when tea would be a despised beverage, and a New Jersey tutor in Virginia would write "they are now too patriotic to use tea."

Potomac civilization in the tobacco age was rich and diversified. Although aristocratic in structure, it was numerically founded on a great yeomanry of farmers, seafarers, and perhaps above all, merchants who had immigrated over many decades. As it grew, these elements became amalgamated. In its golden day its plantations produced as admirable and fascinating a group of men as have ever been brought together. The dignified intellectual capabilities of a John Taylor were offset by the more masculine military exploits of the Fitzhughs and Lees. The

Tory wealth of the lower river produced a long line of thrifty Carters, but also the pleasure-loving barons of Chotank. The Scotch acquisitiveness of William Carlyle was counterbalanced by the farseeing statecraft of George Washington and the disinterested patriotism of George Mason.

As the soil thinned and the income from plantations grew less, the traditionally lavish hospitality became a nightmare and a curse. At Nomini Hall, Mrs. Carter reckoned the household and guests annually consumed 27,000 pounds of pork, and 20 beef cattle, 550 bushels of wheat and limitless corn, 4 hogsheads of rum, and 150 gallons of brandy. To escape guests, one of the Fitzhughs, related to the entire tobacco-planting aristocracy, deserted his too-prominently situated house on a main road and moved to an inaccessible spot in the wooded hills of Fairfax County. On his retirement to Mount Vernon George Washington found there was no surcease from visitors, and described his home as "a well resorted tavern."

Poor relations and distant cousins increased in number, and developed the art of interminable visiting at one house after another up and down the river. One of the lesser Lees—a crusty bachelor who traveled with his body servant—emerged from his Piedmont retreat annually and managed to spend ten months of every year enjoying the Potomac hospitality of his relatives and friends.

By the end of the century a thoughtful architect, Benjamin Henry Latrobe, in an effort to explain what had happened, committed to his journal some reflections in the form of an imaginary conversation with a Dr. Scandella. This "Venetian gentleman of the most amiable, fascinating manners and of the best information upon almost every scientific subject, who speaks English perfectly" had complained of the absence of Virginia hospitality. In replying Latrobe thought Virginia had passed beyond that stage of society, explained by Dr. Adam Smith in his *Wealth of Nations*, where isolation, plenty, and the desire for some distinction led the planters to exchange hospitality for the gratification of their curiosity, and pointed out that the increasing accessibility of towns had contributed to the decline of hospitality. "Strangers are still welcome," he wrote, "although they are now no longer collected 'from the highways and hedges and pressed to come in.' " He concluded that the traditionally expensive forms of hospitality, designed to impress visitors, interfered with the development of more sensible and agreeable forms of entertaining, and hopefully looked forward to the time when "the literary hospitality of Europe" would prevail.

4

That trouble was ahead for the plantation system could be detected as early as the first quarter of the eighteenth century, and before 1750 planters were actively taking steps to meet present and anticipated difficulties. Their first step was to improve the cultivation of tobacco, and the conditions of the tobacco trade. This led to experiments of agricultural reform, the conservation of soil fertility, crop rotation, and eventually diversification. Planters attempted to develop other sources of income to relieve the perils of an exclusive dependence upon tobacco, and these included other exports, the trade in western lands, the professions, raising Negro slaves, new crops and cattle, and industry.

At the plantation other activities necessary to the self-sufficient life could be developed commercially, and the planters were not slow to seize upon them as business opportunities as well. The cooperage of hogsheads for the tobacco led logically to the manufacture of cask staves, which figured largely in the West Indies rum and molasses trade; and soon "turnery and joynery" and other forms of wood became important in the Maryland exports. Salt grew commercially important. The construction of small craft was expanded, and shipbuilding became a more active pursuit.

The reports of colonial governors refer to the steadily growing number of ships built in Maryland and Virginia, and the increasing amount of trade carried in American bottoms. In his celebrated complaints against the tobacco commerce, the poet

Ebenezer Cooke had recommended that the Potomac planters raise hemp, grain, sheep, cattle, and urged shipbuilding in these optimistic words:

> Materials here, of every kind
> May soon be found, were Youth inclin'd,
> To practice the Ingenious Art
> Of sailing by Mercator's Chart . . .
>
> Nothing is wanting to compleat,
> Fit for the Sea a trading Fleet
> But Industry and Resolution.

Not only ships, but ship's fittings, rope, and naval stores were produced in small quantities.

Like George Washington, other planters manufactured ship's biscuit. Councilor Carter sold it by the schoonerload, charging from 15 shillings a hundredweight for the brown biscuit to 33s. 4d. for the best white biscuit.

Flour- and gristmills were built, and on some plantations the demand for the product was so great that the production of the plantation itself was not sufficient and additional grain was purchased. In 1774 Carter was attempting to buy a third of his mill's requirements from James River plantations. These activities commenced at a moderate scale all during the eighteenth century, were greatly stimulated by the Revolution, and came to their climax with the increasing European demand for American grains and the repeal of the corn laws in England.

Finally came a whole series of miscellaneous commercial activities the planters found profitable, depending on their resources. Nearly all of the many ferries on the Potomac were owned by plantations, and place names still often remind us of them. Hallowing Point, near Mason's Neck, originated from the "halloo-ing" for the ferry on the opposite bank. Most of the plantations on the water soon had warehouses in which they stored tobacco for the smaller planters inland, and some of the planters gradually began to accumulate a store of the necessary articles demanded by the smaller planters, and even to manufacture some of these items. In this way the larger planters came to exercise some of the functions of the despised tobacco factors, and some even lent money against the security of tobacco in the warehouse.

Manufacturing was commenced in part from the universal desire for economic diversification, but it was an outgrowth of these plantation industries. Since the plantation owners held the principal reserves of capital, they naturally played the major role in the financing of such enterprises as the Baltimore Iron Works, just as they later were the prime financiers of the Ohio Company and the Potomac Canal. Iron was manufactured for export in 1732. The plantation smithies first engaging themselves in the production of frying pans, nails, hinges, and other ironmongery, later turned to cast-iron plows and similar necessities of farm life.

The transition of the tobacco planter to diversified farming and industry can be seen in the lives of many men who were contemporary with the Revolution. Most clearly, perhaps, it is seen in the career of Councilor Robert Carter, who, on the eve of the Revolution, engaged a young divinity student from Princeton College, Philip Fithian, to tutor his children. From Fithian's journal of his experiences and observations at Nomini Hall, his letters, and from the correspondence and voluminous account books of Robert Carter is created a more vivid picture of life on the home manor at Nomini Hall than of any other Potomac plantation. This pleasant portrait reconstructs a mode of life that was common throughout the Potomac Tidewater on all the larger landholdings in their golden age; and from Fithian's sympathetic description of Robert Carter we can see at once the shadow of decay in the traditional plantation system and the beginnings of a new order of affairs.

His wealthy family sent young Carter off to the College of William and Mary at the age of nine. Later he had the pleasure of reading law for two years at the Inner Temple in London. When he returned to Virginia in 1751 he could find no one as charming as Miss Ann Tasker of Maryland, whom he married; they had seventeen children.

The vast Carter properties totaled 70,000 acres and 500 slaves, and were divided into so many plantations they were named for the signs of the zodiac. Each was supervised by its own overseer. On them Carter tried every known device to escape the yoke of tobacco, including most of the novel crops of his day and the newer agricultural methods. Carter's ideas tended to be those of a businessman rather than a farmer, and ran toward impressive systems.

His ideas of efficiency on the plantation were erratic and often bizarre: the oxen he used as draft animals were identified by large white numerals painted on their flanks. He took great satisfaction in inventing an elaborate scale for computing the difference between Maryland and Virginia currencies, explaining that it was a necessity in his business. Aside from his flour, textile, and salt mills, numerous other ventures in trade engaged his attention. His wife owned a one-fifth interest in the Baltimore Iron Works, and Carter was actively interested in the development of iron on Occoquon Creek.

The household and plantation ran like clockwork. A bell weighing over sixty pounds that could be heard for miles punctuated the day. It rang at eight for breakfast; at nine for the children to go to school and the workers to come from the fields for breakfast, and at ten for them to return to work; at noon it rang for the school play hours, and at two for everyone's dinner; at three it announced the return to school and labor.

With the severity of a barracks inspector, at breakfast Carter often quizzed the children on their progress in their studies. He threatened to leave all his property to the child who showed the greatest abilities in business. He acquired strange opinions for a planter, and began to take on the hue of an early industrialist. He even forsook the sacred Ohio Company, that hope of so many Potomac planters and statesmen; the company's treasurer, George Mason, had to dun the rich but indifferent Carter to pay his assessment.

A true child of the Enlightenment, although less gifted than Franklin or Jefferson, he was skillful in mechanics and engineering and his library was liberally stocked with treatises on these subjects. He manufactured gunpowder, and experimented with the production of distilled spirits. He could run accurate boundary lines with a theodolite, and was an able builder. The library also contained books on geography, astronomy, law, music, history, the ancient and modern classics, travel, natural history, and science; but Fithian also found many practical books on mathematics, assaying, local government, manners, agriculture, architecture, engineering, medicine, as well as French and Latin grammars and dictionaries. Such resources in Potomac plantation homes were not unusual in the eighteenth century.

Music was Carter's passion. His library was well stocked with scores, and he had a great variety of good instruments: harpsichord, pianoforte, organ, musical glasses, guitar, violin, and German flutes. He practiced regularly and considered dedicating one room at Nomini Hall wholly to this purpose. Nor was this an individual performance only: he often played little concerts with other members of his family. The theory as well as the practice of music interested him, and he took care that his children were instructed in the art. Nancy played the guitar, Ben the flute (he was bribed to practice). Priscilla also played. Even the tutor Fithian was swept into it, and learned to play the flute well enough to get through a sonata with Colonel Carter. Under the leadership of the intrepid colonel, they experimented with part singing with and without instruments. His waiting man was sufficiently accomplished to play for the ladies of the household to dance, and many Negroes had some accomplishments on the fiddle. But here, too, this innocent activity was touched with the Carter inventive urge. The tutor recorded in his Journal: "The Colonel shewed me after Dinner a new invention which is to be sure his own, for tuning his *Harpsichord* & *Forte Piano*: it is a number of whistles, of various Sizes so as to sound all the Notes in one Octave."

As life advanced, Carter questioned the institutions of planter civilization, and began to free his slaves pursuant to a characteristically elaborate schedule. Each year, beginning in 1791, the fifteen slaves nearest forty-five years of age were to be freed, as well as any boys reaching twenty-one or girls reaching eighteen. He calculated all female slaves would be freed by 1809, and all male slaves by 1812. In his deed of manumission he explained that this regular schedule would result in the "least possible inconvenience" to his neighbors. But within five years from the time he commenced, Carter was receiving anonymous letters protesting that the freed slaves were "indolent, thievish and demoralizing to the neighboring slaves" and demanding that they be removed to another state "where they can do no harm." One is hardly surprised that he left the Potomac valley and chose to live in the brisk commercial atmosphere of Baltimore. To enter manufacturing or trade or to move west was the principal alternative of the Potomac planters at the end of the century. Those who chose to remain faced financial ruin.

Fithian's sympathetic portrait of Carter was drawn in the years 1774 and 1775, and shows him at the pinnacle of his transition from planter to industrialist. His farming activities were then at the high noon of their movement toward diversification, but the industrial pursuits that were later to engage him were already visible clouds on his economic horizon.

Another possible response to the declining fortunes of the tobacco trade was seen in the career of George Mason, the Potomac planter and intellectual who best exemplified the eighteenth-century virtues of enlightenment and humanism. As a planter Mason revealed the best qualities of his class and generation. His abilities as a farmer were unquestionable. He stood in the shadow of no man in his zeal for the predominant interests of his age in the west. To George Rogers Clark he was almost a father. As treasurer of the Ohio Company he brought to that unwelcome task not the money-minded interest displayed by Pennsylvania's western venturers, but the searching vision of statecraft. And it is all the more remarkable that, unlike Washington and many others, he did not own vast holdings in the upper Potomac and the Ohio from which he might expect to gain directly by western development.

The fourth Mason of his name in the Potomac, his family had worked their way up the river from Westmoreland on the Northern Neck, where they had first settled in 1655. Perhaps he was related to the Captain Mason who came to Jamestown with John Smith. From Westmoreland the Masons moved to Paspatanzie at the mouth of the Accokeek, and then to Gunston Hall, where they presently found the Fairfaxes to the south of them along the river and the Washingtons to the north. They owned lands in Maryland, too, and ran a ferry.

George Mason's father was drowned in a sailboat while crossing the river, and like many in those days when fathers worked hard and died young, the young Mason was reared by his mother and a capable uncle, John Mercer, who became his guardian. A lawyer who managed to become involved in most of the interesting enterprises of his day, including the select society of the Ohio Company, Mercer had a library of fifteen hundred books, fully a third of them on law, in which the young Mason read to good effect. Through his guardian, Mason's early membership in the Ohio Company allied him with the Washingtons, Fairfaxes, Carters, Lees, and other men of the Potomac who saw in western expansion an inevitable destiny for the region as well as a political obligation.

At twenty-five George Mason was married, and soon afterwards a brother returning from England brought the means to build Gunston Hall: an indentured architect, William Buckland, later to leave his unmistakable Palladian signature on Maryland's colonial capital, Annapolis, and a host of Tidewater plantation houses.

Mason commenced the already traditional career of a young planter, but he applied himself to it with remarkable diligence. He personally managed the plantations, and this may have been the underlying reason why, after years of public service, he still regarded himself as a private gentleman rather than a public figure. Nevertheless, one might say he was born a public figure, and he performed all those local offices that the ambitious and rising often look upon with contempt. For twenty-five years he was a trustee of the town of Alexandria; for longer than that he was a justice of the peace; and he served faithfully in all the usual lay functions of the Anglican Church.

From this apprenticeship he was sent to the House of Burgesses in 1759, and after being frustrated to the point of indigestion he learned the art of snoozing through the dull, interminable legislative sessions. He resigned before the end of his term.

Never a great parliamentary figure, Mason seems to have been at his absolute best in a small company of men where his talents for compromise and formulation came to full flower. He did not return to Williamsburg until the momentous session of 1775, but his influence out of office was perhaps greater than it was in office. Again he resigned from the office to which he had been elected before his term ran out.

Mason's third session at the Virginia capital was in 1786, but his old enemy, the gout—in his case "gout of the stomach"—incapacitated him.

Remembered fondly in Virginia for his authorship of the Virginia Bill of Rights, Mason was far more than a brilliant provincial figure. His mind ran to large designs and broad principles. He never compromised and, although decades often passed, his judgements generally proved correct. Even his objections to the Constitution of the United States were subsequently recognized, as the Bill of Rights and various early amendments demonstrated. Perhaps his greatest achievement was moral: he condemned the constitutional quibble that later led to the Civil War, and opposed the spread of slavery to the western states, freed his own slaves, stood for emancipation and the education of the Negroes.

The evolution of the plantation system, the failure of tobacco, of escapes from tobacco, of western settlement and land speculation, drove Mason from the relatively quiet life of Alexandria to the larger theaters of state and national life. Where men like Carter found solutions that preserved their individual fortunes, Mason attempted to find a way to preserve the tobacco civilization itself. Naturally he did not succeed, but in making the effort he accomplished something quite as important.

5

All along the Potomac in the eighteenth century tiny villages sprang up, each strategically located to tap the tobacco trade of a small hinterland. On the Maryland side they were located at Leonardtown, Upper Marlborough, Port Tobacco, Piscataway, Nottingham, and Georgetown; in Virginia, at Alexandria, Dumfries, Colchester, and other spots. In each, typically, would be found a merchant with his office, store, and warehouse, ready to buy tobacco and sell goods from London. Usually Scotch, these merchants were the representatives—called factors—of large firms in London or Glasgow. For the small planter who marketed a hogshead or two a year, or whose plantation was not handy to a Tidewater wharf, these merchants at the end of the rolling roads were indispensable; and even some of the larger planters dealt through them. The factor's store contained almost everything the planter might need, whether hardware, dry goods, clothing, or notions. Here the planter could buy a quire of writing paper, an anchor, or a bridle.

Around these trading centers some development took place, although it would be hard to describe any of them as towns. The descriptions of travelers are seldom impressive, even at the height of their prosperity, and little is known of the kind of life the people led. Fragments of information are enticing: a play by a troupe of visiting actors in the ferry town of Piscataway; the tutor Fithian stopping at the same place amusing himself with "two young Ladies Daughters of the Landlady, rather gay & noisy than discreet, very forward in discourse, both in Love with Scotch Merchants & both willing to be talked to"; theatricals in an old barn at the edge of Alexandria seen by the architect Latrobe; a glimpse of odd Major Fouchee Tebbs at Dumfries, a character fresh from the pages of Smollett.

But the towns came to nothing. Not the largest had ever more than eighty buildings, and except for the two towns at the fall line, all the rest disappeared with the collapse of tobacco. Dumfries faded away even before the river silted up the harbor, and Alexandria and Georgetown were saved only by tapping the hinterland and becoming wheat ports. In all of St. Mary's City only two houses remain today; and in Port Tobacco only one.

Of the trade itself, it was usually on the unsatisfactory basis of credit accumulated against the planter's yearly crop, and in consequence the small planter was nearly always in debt, and the rich planter frequently so. When prices were depressed or the crop poor, planters and factors alike found themselves in difficulties. To avert this, the independent American merchants—many of them planters themselves—fostered diversification and developed new mercantile outlets and gradually took possession of the tobacco trade. Closer to the sentiments and interests of the tobacco growers, they capitalized on decades of dissatisfaction with the factor system and drove it into the ground. Many of the factors themselves became merchants.

Alsop had advised in 1660, "Let the Factor whom you employ be a man of Brain, otherwise the Planter will go near to make a Skimming-dish of his Skull," and described the planters as "a more acute people in general, in matters of Trade and Commerce, than in any other place of the World." He said that any man who became a merchant in Maryland "must have more of a Knave in him than Fool." This observation was repeated with minor variations for a hundred years. Benedict Calvert in 1729 complained that the tobacco merchants "labor only with the pen; the planter can scarce get a living, runs all the risks." Dr. Charles Carroll threatened the London merchants with the fable of the hen [*sic*] that laid the golden eggs.

The bad blood existing between tobacco merchants and planters continued, and the animosity to London tobacco factors turned out to be the animosity of all debtors to all creditors. Interminable disputes took place over the correct weight and quality of the tobacco sold. Factors called the tobacco trash, and planters complained because the factors did not pay premium prices for high-quality leaf. Lawsuits were endless; and suits for debt ranked high among them. Even substantial planters fell out with merchants who overcharged them for warehouse rent, or insurance, and did not protect the tobacco against plundering when stored in warehouses. And in the view of one representative factor "without the wiles of a serpent, the innocence of the dove would be betrayed and crushed" in dealing with planters; he concluded "a man of business must have a good share of the devil in him."

Fithian observed that "all the Merchants & Shopkeepers in the Sphere of my acquaintance, and I am told it is the Case through the Province, are young Scotch-men," but was soon to hear raving and ranting against the Scotch. Vehemently one of the Lees swore "that if his Sister should marry a Scotchman, he would never speak with her again; & that if he ever Shall have a Daughter, if she marries a Scotchman he shoots her dead at once!" When a Scotch merchant, Callister, went into business for himself he found that his long service as representative of a Liverpool firm so prejudiced planters against him that he failed.

The word "foreigner" became an epithet. The Scotch accent was regarded as an abomination, and Scotch tutors deplored. This silly talk continued, despite the fact that the Potomac was thickly settled with Mercers, Carlyles, and dozens of other eminently respectable planters and merchants of Scotch origin.

From the first big collapse of the tobacco market early in the eighteenth century, the Potomac planters struggled, not without success considering the primitive economic tools available, to control the tobacco market. They made strenuous efforts to increase productivity, and to keep down the rising costs of growing tobacco. They experimented with restricting production, hotly debated the economics of regulation, and burned their fingers in politically immature schemes for price control. In lower Maryland, Henry Darnall organized eighty other planters in an effort to bring order into the marketing of tobacco—only to find that the tobacco merchants in England were unable to achieve the necessary control there to make the scheme work. A government inspection system was inaugurated in Virginia and later adopted in Maryland; and soon the tobacco ports of Dumfries, Port Tobacco, Piscataway, and the fall-line towns of Georgetown and Alexandria had their Scotch tobacco factors and merchants, wharves, warehouses, and "naval offices" that regulated the trade, and graded and inspected the tobacco. Despite these promising developments, the colonial tobacco trade steadily declined, although the exports of the Potomac River plantations continued to be higher than those of adjacent areas. More than a quarter of the entire Maryland production of tobacco moved through the Potomac inspection station in the middle years of the century, amounting to about eight thousand hogsheads annually. Intensive discussion of the problems of the tobacco business ranged from minor recommendations that the hogsheads not be rolled to the wharf—a practice that damaged the leaf—to comprehensive proposals for a wholly regulated market.

Recognition of the soil exhaustion that accompanied tobacco culture also came early. We have seen how the planters sought to secure large areas of land to cushion their operations against the threat of worn-out fields. But now the discussion and practice of soil conservation itself expanded. New tools were imported or invented. Jefferson landed at Norfolk from France with the first drawings of the moldboard plow in his portmanteau. Serious experiments in crop rotation were begun, and in a few years produced a frenzy of debate in the popular press. The early conservationist, Philip Mazzei, told Jefferson there should be a law restraining landlords from cutting trees indiscriminately and damming rivers; but received the prompt reply that the idea was repugnant and a curb on the liberty of the landowners. Barnyard manure was spread on tired fields, and marl came into common use, and much was claimed for these practices.

As more efficient methods of cultivation were adopted, horses and oxen were used to draw the primitive plows. Washington was overjoyed when Lafayette presented him with the first Spanish asses to reach the United States, and began to breed mules. But the cultivation of tobacco remained then, as it is today, essentially a hand operation.

From the newer ideas of crop rotation, however, notions of a diversified agriculture began to emerge. To the early cultivaion of corn, the planters soon added wheat. Originally planted only to supply flour for pastry, wheat soon became a major crop. On the thin topsoil that overlay the clay slopes at Mount Vernon, George Washington grew wheat that, when sold in Alexandria, made ship's biscuit that was famous the world over —and rye that supplied his less celebrated distillery. The increasing number of cattle accounted for the introduction of mangel-wurzels, turnips, and other root crops in the rotation. The soil-building virtue of peas was discovered. Beef cattle grew in increasing numbers, and began to appear prominently in inventories and wills. Orchards and vineyards were planted more widely. With these developments, simultaneously with the decline of the tobacco trade, a lively business also sprang up in shipping corn, wheat, and livestock to the West Indies, a trade mostly carried in American bottoms.

Crop diversification was seen not only as a means of perpetuating the tobacco culture, but in getting rid of the exclusive dependence upon one crop and the lean years that resulted when tobacco was poor in quality or flooded the market. It also promised some freedom from the tobacco factor. One of the reasons why western settlement was greeted eagerly by thoughtful people was the extent to which the Piedmont lands would be suitable for crops other than tobacco.

In analyzing the efficiency of tobacco production and trying to get costs down, the labor element was seriously studied. From the beginning it had been recognized that the indentured white servant was more productive than the Negro slave, was more readily acclimatized and involved a relatively lower capital investment. To this was due the large yeoman class in the colonies on which later frontier settlement drew. Long before the Civil War most of the Potomac planters who were not engaged merely in slave breeding—for Virginia had then become celebrated for its slave farms—had turned against the institution. They would willingly be rid of it—if only they knew how to liquidate such huge investments in an orderly way. Individually many of them, like Washington, Mason, and Carter, solved the problem by manumission. Opposition to slavery was the backbone of Mason's celebrated objection to the Federal Constitution.

Negro slavery was condemned not only on economic grounds: moral factors were seriously considered, too. The rationalism of the day left its mark on the views on slavery held by the Potomac planters, as well as their views on government. And more subjective influences were present. Councilor Robert Carter, of Nomini Hall, when he began to waver amongst the Baptist, Roman Catholic, Swedenborgian, and other faiths, also began to develop serious misgivings about slavery.

Revealing entries appear in private papers, such as that Colonel Landon Carter made in his diary in 1776: "A strange dream this day about these runaway people. One of them I dreamt awakened me; and appeared most wretchedly meager and wan. He told me of their great sorrow, that all of them had

THE ROCKS,

GREAT FALLS

been wounded by minutemen, had hid themselves in a cave they had dug & had lived ever since on what roots they could grabble and he had come to ask if I would endeavor to get them pardoned should they come in, for they knew they shd be hanged for what they had done. I replied a good deal. He acknowledged Moses persuaded them off and Johnny his wife's father had helped them to the milk they had to wit 4 bottles. He was to have gone with but somehow was not in the way; declared I had not a greater villain belonging to me. I cant conceive how this dream came into my brain sleeping, and I don't remember to have collected so much of a dream as I have done in this these many years. . . ." Apprehension became a general mood.

6

All during the eighteenth century the Potomac continued to be true to its name, a river of traders. In the wake of the corn trade, the fur trade, and the tobacco trade came the trade in diversified crops that became typical of the next century. By the time of the Revolution this transition was well launched. In the 1780's wheat, oats, barley, and corn were becoming predominant crops in the lower reaches of the river. The expansion of Alexandria and Georgetown was directly related to their new function as wheat ports. Tobacco continued to be produced, but the prosperity of the entire region no longer was wholly dependent upon it.

These activities in the middle years of the eighteenth century continued to engage the planter's closest attention. His operations were extremely complex, and to succeed the planter's life could hardly be the gay, leisured round of fox hunting, cockfights, horse races, and loo that has so often been presumed. The planter thought deeply about agriculture and economic problems. He sent off to England for the latest edition of Muir's *Book-keeping Modernized*, and for works on agriculture and engineering; subscribed to the *Maryland Gazette* or the *Vir-*

ginia Gazette, and received the farmers' almanacs. This practical literature took an honored place on his shelves beside the classical authors and scientific books that came with the Enlightenment. From captains of the tobacco ships he received not only the luxury goods of London, but letters and indispensable business news. An active curiosity prevailed and visitors who could gratify it were much appreciated.

While most large planters employed overseers and foremen to direct their vast operations, a few like George Mason personally undertook the heavy work of direct supervision. But only a little reading in the letters and account books of Washington, Richard Henry Lee, Robert Carter, Charles Carroll, and other great planters, or a glance into their libraries, offers convincing proof that these men, too, gave careful attention to the management of their great estates. Soundly educated in the classics and accustomed from youth to moving in a well-organized and (for all its plantation isolation) a remarkably cosmopolitan world of affairs, they thought deeply about the problems facing the Potomac planters. When they left off worrying about the high cost of hauling and spreading barnyard manure or the price of bar iron and turned to the clear speculative logic of Montesquieu and John Locke, it was with their own problems, political as well as economic, in their minds.

Agriculture, place-holding, landownership, and the professions were insufficient bases for the tobacco civilization of the lower Potomac, and no one recognized it more keenly than the planters themselves.

For all its ultimate tragedy, this civilization must be characterized as progressive. It made an impressive effort to understand its problems and work out adjustments to them. Steady progress was made in improving tobacco growing, in dealing with the evils of tobacco culture and its commerce; great efforts were expended to diversify agriculture; complementary trade and industry were zealously pursued, and manufacturing commenced. But mark well the difficulties in escaping from the universal tobacco culture, for they are a measure of the boldness and enterprise of the men who led the way out of that morass.

Before the Revolution, Councilor Carter and his wife studied their accounts and discovered that "the plantations were earning at a rate so low that the value of the slaves alone if liquidated and lent out would have brought a greater income as interest." This was the first clear fissure that later widened to a chasm that engulfed all of the great plantations. At Washington's death that careful farmer's estate at Mount Vernon was in ruins. "No Virginia estate," he wrote, "can stand simple interest." George Mason would have been financially unable to attend the Constitutional Convention in Philadelphia had not his expenses been paid by the state of Virginia. Madison tried vainly to get a loan from the United States Bank and was forced to sell his land and stocks and capital to meet his debts. Even Jefferson had financial embarrassments, and sold his property by lottery. Monroe found himself financially ruined when he left the White House. That half-crazy prophet of secession, John Randolph of Roanoke, predicted that masters would soon run away from their slaves and be advertised by them in the press. Decay was everywhere in Tidewater Potomac as the eighteenth century ended.

FIVE

POTOMAC ABOVE THE FALLS

To the falls of the Potomac above Georgetown, the river formed a broad highway; but there it was equally a barrier to travelers who wished to cross. Above the falls the river became shallower, islands appeared, and here and there were places where a man on horseback, or a pack train, could ford the stream. At one such place, above the juncture of the Shenandoah and the Potomac, the immigrant Robert Harper established a ferry in 1734. Here the natural lines of travel through the upper river basin lay north and south in the great folds of the land, wide limestone valleys running from the rich agricultural counties of Pennsylvania deep into the Carolinas, and leading to the virgin headwaters of those southern streams which penetrated the mountain barrier and flowed to the Mississippi. Long before the tide of settlement from the lower Potomac had reached much above the falls, other settlers—from the north—were penetrating the upper river valley.

First among the newcomers along the upper Potomac were the Germans and the Scotch-Irish pioneers, who had early been welcome colonists in Pennsylvania. They had been settled on the extreme western frontier of that province, where they formed a convenient buffer between the restless Indians and the placid Quaker settlements. But the new immigrants almost at once began striking out for themselves, generally caring little for colonial boundaries or the alleged owners of wilderness land titles. Long before this southward thrust of settlement met and merged with the descendants of the Tidewater tobacco farmers, slowly making their way upriver, its separate identity was clear.

Among these new people were novel religious sects, tiny clouds torn from the larger storm mass of the Reformation. They included the Ulster Presbyterians, followers of John Knox, who had already been ground between the Established Church in England and the Catholic landowners in Ireland. They included German Lutherans, and that Reformed Church which has been well described as "a German edition of Presbyterianism." But they also included many small sects of Pietists, who struggled to preserve freedom of conscience and individual communication with God through prayer: the Mennonites, the Moravians, the Labadists, and others. They struggled to realize in the New World a dream conceived in the Old. Most of them had experienced the persecutions of civil and religious authority in many European nations, and even in some American colonies, before finding sanctuary in the quiet mountain valleys of Maryland and Virginia. They had acquired strong and original ideas of the relation between church and state. They cultivated sturdy views of the church itself, believing in representative forms of church government, the election of ministers, and strong congregationalism. Convinced of the necessity of reforming life to permit a better relation of the individual to God, their fundamental interest lay in securing favorable conditions for such reforms. In search of suitable places to settle, their scouts and missionaries ranged widely through the mountains. It was the Presbyterian, Francis Makemie, who first reported, "The best, richest, and most healthy part of your country is yet to be inhabited, above the falls of every River, to the Mountains."

The lands above the falls of the Potomac had been exploited long before they were ever settled. Since the time of Captain Fleet, the fur traders had brought out pelts. Following the buffalo trails and Indian traces, they sketched maps and kept journals, noting the good springs of mountain water, the rich soils, the tall stands of timber. French Jesuit missionaries, if we can believe the evidence in Champlain's map of New France, had somewhat explored the upper Potomac and its southern tributary, the Shenandoah, even before 1632. An earnest and capable geographer, John Lederer, crossed the mountains in 1669 from the valleys of the Rappahannock and the James, and came to the Shenandoah before turning south. Occasional other wanderers passed through the land on moccasined feet. From their sketchy reports of the geography some of the land had already been described and royal titles granted, as vague in their indications of boundaries as the celebrated land Charles II had granted on the Northern Neck.

Shortly after 1700, the Swiss prospector, Louis Michel, traversed the fertile territory between the falls of the Potomac and the mouth of the Shenandoah. He came down the Monocacy valley from the north, and crossed the Potomac at Point of Rocks, where the Piscataway Indians, halting for a few years in their westward retreat, had built fish traps among the Conoy Islands, and then resumed his journey up the valley of the Shenandoah beyond the forks of that tributary. By 1707 he had made a temporary settlement at Harpers Ferry and was engaged in surveying the mineral resources of that area, and the next year had moved his camp to the mouth of the Monocacy. George Ritter, another Swiss from Bern, was soon persuaded to come. Presently the well-traveled Martin Chartier, who came out from France as a house carpenter, and had spent over forty years in the Mississippi valley with La Salle and other French explorers and traders, joined the little settlement.

By the spring of 1711 they had found few minerals, but were sufficiently well established to make a profound impression on a distinguished traveler, the Swiss baron, Christoph de Graffenried, who came to survey the region with a view to colonization. De Graffenried, who had been advised in London by Louis Michel, had taken out from England a party of Palatine refugees and settled them in North Carolina, where they had founded the city of New Bern. Disease and other misfortunes had overtaken the colony, and since de Graffenried's letters of patent from Queen Anne also permitted him "to take lands in Virginia above the falls of the Potomac River," he had journeyed there to survey that territory.

Ascending the Potomac on the Maryland side from the head of navigation at Rock Creek, following a chord of the river's bend, de Graffenried came to the settlement of Martin Chartier, who lived with his native wife in a cabin at the Monocacy's mouth. "After that," the enthusiastic Swiss baron noted in his journal, "we visited those beautiful spots in the country, those enchanted islands in the Potomac river above the falls. And from there, on our return, we ascended a high mountain standing alone in the midst of a vast flat stretch of country, called because of its form Sugar Loaf." From its summit to the west three mountain ranges could be seen. He noted the great profusion of swans and ducks that made their home along the river, and sketched a map showing the principal features of the land and the habitations of Chartier, and the Piscataway Indians, and indicated the details of an especially fine tract of land along the Virginia shore between the Monocacy River and Point of Rocks.

But de Graffenried's "Project to Establish a Colony Along the Potomac River in Virginia and Maryland," came to nothing, and like the other maps that had been drawn, his became shadowy above the junction with the Shenandoah, and wandered off into general perspectives of mountain ranges, or rude sketches of turkeys, ducks, deer, or other four-footed animals less easily identified. A sure and accurate knowledge of the region awaited the coming of more permanent settlers. They were not long in arriving.

Into the region of the upper Potomac, presumably uninhabited but even then sprinkled with an anonymous advance guard of settlers, Governor Spotswood of Virginia led his well-publicized expedition, the "Knights of the Golden Horseshoe" —a most luxurious pilgrimage of gaily dressed travelers from Tidewater to the interior, with fine wines to drink, servants, and tents to sleep in each night. He forcefully drew attention to the resources of the area, and in his company were names that became great in the later story of the upper valley:

Beverley, Mason, Brooke, and others, whose descendants carefully handed down the little souvenir horseshoe with the motto "Sic juvat transcendere montes," which had been presented by the governor in 1716. But skilled as he was in such byplay, Spotswood had a larger and more practical vision. He wanted to call attention to the immense western land claims of Virginia under its charter, an area that reached to the still unknown western ocean and included "the island of California." More immediately, he was concerned with land speculation and the development of an ironworks at Germanna, where the helpful de Graffenried had arranged to import a dozen families of German ironworkers to commence operations.

Working their way down the valleys from Pennsylvania, following the fertile limestone soils that skirted the mountains, came the Germans, the Scotch, and the other early settlers. The decade of the 1730's saw the Dutch fur trader and frontiersman, John Van Meter, cross the Potomac and settle down with his family in the South Branch valley. The Welshman, Morgan ap Morgan, received patents to Virginia valley land and moved in. The German, Joist Hite, came with sixteen families and settled a hamlet they called Fredericktown (later Winchester), in the Great Valley. These pioneer settlements were gradually augmented by frugal Germans, discouraged in fighting for land titles with the heirs of William Penn, and others, disappointed in New York. Near one of the principal pack fords of the Potomac, their first settlement was called Mecklenburg (today Shepherdstown). Its origins have been traced to 1727.

As the Pennsylvania Germans filtered from the northern colonies into western Maryland they joined other Palatine colonists, who came direct from Germany and were landed at Annapolis or Alexandria. Soon they were found along the Monocacy, the Antietam, the Shenandoah and its tributaries, but the first settlements and churches were on the well-traveled road that skirted the first range of mountains and paralleled the Monocacy River. Lutherans and Reformed congregations were represented in the settlements near Creagerstown and Frederick, and Moravians founded the celebrated religious center of

Graceham at the foot of the Catoctins. In the Cumberland valley, easily reached through Crampton's Gap, Jonathan Hager accumulated patents to nearly 2,500 acres of land and founded the valley metropolis of Hagerstown.

The lovely valley of the South Branch of the Potomac early attracted settlers, as did other southern tributaries, the Opequon, Back Creek, Tuscarora Creek, Cacapon, and the western streams running into the South Branch. Easy access along the Potomac or over North Mountain wove the early settlements of the Shenandoah and the South Branch into a single community.

The ability to create towns and trading centers was marked, but the German immigration was essentially a movement of farmers. Their proverbial industry and thrift was mirrored in what these people did with the fertile soil they found: while they still lived in neat but primitive two-room log houses, their big stock barns were filled with sleek cattle. Their churches were solid and numerous, and noted for the purity of their sectarianism. Their ministers and missionaries traveled hard and wide and left descriptions of an attractive and romantic society, free in religious practice, culturally distinct in its isolation. Such seasoned colonial observers as William Eddis marveled at this wilderness civilization.

In its virgin state this upland Piedmont country was not so heavily forested as the Tidewater country or the tangled peaks to the west. It was punctuated with natural meadows and open tracts that had been burned by the Indians. Cattle raising in these meadows, their native grasses waist-high, became an early form of agriculture; wheat was the principal crop.

In this region the land had been patented just in advance of settlement, and the prosperity of the growing population built the fortunes of such favored Marylanders as the Dulanys and the Carrolls, who had secured title to the land in the Monocacy and Cumberland valleys. Daniel Dulany brought over additional German settlers, often paying their passage, and founded towns like Frederick which early became important agricultural centers. Laid out in regular gridiron patterns to facilitate the sale of lots,

THE HEADWATERS

these valley towns were soon built up with solid timber and brick houses of middle-class artisans. The famous manor of Carrollton, where Charles Carroll never lived (although he added it to his name when he signed the Declaration of Independence), occupied the entire lower valley of the Monocacy.

But the lower half of Carrollton Manor, the portion nearest the Potomac, was settled chiefly by the yeoman from Tidewater, descendants of the indentured servants. Mingled with this population were Quakers and other dissenting groups. Steadily they moved up the river in search of richer, cheaper lands where they could begin afresh. By the middle of the century such settlers accounted for nearly half the population in the valley of the Monocacy, the balance being Germans with some increment of Scotch.

To the growth of this inland region may be traced the rise of commercial cities, the towns two centuries of legislation and exhortation had failed to create by fiat in the Tidewater region. At the falls of the river quiet tobacco ports, Alexandria and Georgetown, took on new life. The new commercial center of Baltimore, where only twenty-five houses stood in 1750, owed much of its growth to the granary of the Monocacy, which formed an agricultural hinterland, connected by wagon roads to the headwaters of the Patapsco River, and so to the port city of the Chesapeake.

Of the many links between the lower and upper rivers, the most significant was the ownership of the land. In both Maryland and Virginia the provincial officials were well aware of the importance of the western territory, indistinct as their impressions may have been of its precise character and resources. From the seventeenth century on, their fixed policy was to bring the frontier forward as rapidly as possible, modifying this in detail as the decades passed in the interest of securing a more compact and more easily defended array of settlements. Disputes among the three colonies of Virginia, Maryland, and Pennsylvania in their western reaches were continuous, caused partly by the lack of adequate surveys of the land and contradictions in the royal grants and partly by the avarice of the provinces and their proprietors, and this added fuel to the desire to settle the contested districts.

In Maryland and Virginia the original intent of the proprietary and royal patents of land in the interior appears to have been to grant land directly to the colonist as a means of encouraging settlement. After the early abortive attempts to establish a fixed Indian frontier at the falls of the river, efforts were made to reduce the demands for military protection on the border—the issue that had led to Bacon's Rebellion and many later disputes—by granting land in large tracts to those who would undertake to settle a certain number of colonists. As such tracts were granted, the patentee's interests were allied with those of the granting authority, and formidable communities of interest were thus built up over the years. Land grants also tended to become political rewards to well-situated followers, and much of the land thus came to be first held by influential Tidewater families. From these owners the land was usually let out on long leases, or sold in fee simple, subject to nominal quitrents. As the income from frontier lands increased, many owners deserted their tobacco plantations and migrated to the interior, where they endeavored to re-establish the aristocratic pattern of life they had left along the lower reaches of the river. Under these systems of land tenure, conditions much like those of the lower valley should have been perpetuated. But this aspiration slipped from their grasp in the great spaces of the interior.

Not until 1755 was there such a thing as free land in the west, land a man could legally acquire by the simple process of settling upon it—and even then it was only to be found far to the west, over the mountains. To impoverished settlers, who had sold themselves for a period of years as indentured servants in exchange for their passage money, the cost of land was a major factor. They kept moving until they found land cheap enough to take up, or land on which they could squat without interference and perhaps eventually secure "tomahawk rights" —by thus marking the trees that surrounded their claim. The long drawn-out contests over land ownership, such as that between Lord Thomas Fairfax and Joist Hite, or the boundary dispute between the Calverts and the Penns, clouded titles to such an extent that the territory affected became ideal for settlement by the debt-ridden frontiersman who was willing to take a chance. The confused overlapping grants that existed in much of the inland region, grants that when plotted on a map looked like a shingled roof, contributed further to land difficulties of owner and tenant alike. As settlement moved forward, the price of previously settled land increased, and a further incentive appeared to sell and move on.

The flow of settlement thus moved not only within the great natural land forms—along the rich limestone soils of the valleys, through the water gaps, avoiding the pine barrens of the south and the heavy forests of the west—but within an economic framework of land costs and titles as well. The restless spirit of the frontier, everywhere noted, was at first due not so much to the innate psychological temper of the frontiersman as to his necessity. In time an elaborate rationale was created, and the impulse to move on was reinforced. To move to virgin land where the hunting was better, where settlement did not press so closely on one's heels, where land was cheaper, and where it might be more fertile—all these optimistic and expansive hopes of the frontier culminated in the great drive to the west.

That drive found one great natural objective in the headwaters of the Potomac, where the Indian trace to the Youghiogheny and thence to the Ohio had early been found. It found another natural objective in the headwaters of the Cumberland and Tennessee rivers, flanking the mountain barrier and leading north to the Ohio. These two great geographically predestined routes crossed in the upper Potomac. Here fords, ferries, and small towns were soon found in the region between Harpers Ferry and the pack-horse ford near Shepherdstown.

By the middle of the eighteenth century reporters described hundreds of wagons, traveling generally in small groups, along these main routes. In the big canvas-covered Conestogas, as they tilted over the rolling landscape of the high foothills, were the entire possessions of families. They had moved once, and they could always move again. A restless spirit grew naturally from individualism, independence, skepticism of authority, from the impoverished and debtor character of the frontier. Governor Dunmore wrote: "They acquire no attachment to Place; but wandering about seems engrafted in their Nature; and it is a weakness incident to it that they should forever imagine the Lands further off, are Still better than those upon which they are already Settled." From the depths of experience, he lamented that laws and proclamations could not control the frontiersmen.

In the perspective of the Tidewater capitals of Williamsburg and Annapolis, the frontier early acquired a strange complexion. The "back settlements" were recognized as a homogeneous region, and for all the political and geographical separateness that was found in the west, the entire area stood as one. Peculiarities of dress and of speech as well as of ideas marked their few representatives who appeared in the established seats of government.

A leather-clad people who slept on bearskin rugs, whose diet was corn pone and deer meat, whose religions were dissenting faiths, were slowly welded together to form one body, diverse but united. The insecurities of the frontier they faced with individual fortitude. Yet separated in the canoe-shaped valleys of the interior, divided by barriers of language and faith, moving and restless, generally poor and deeply in debt, imperiled by Indians and wild animals, they made common cause. Seeking military protection, free lands, good roads to bring their wheat to market, freedom from religious and civil authority, they found their principal Tidewater friends were the great landowners. Save for such eccentric loyalists as Fairfax, the great propertied men of the inland region stood to a man with the western settlers. When the great revolt came, the Masons, Washingtons, Johnsons, and Carrolls stood with the west against the crown.

As the procession of westward-bound settlers swirled and eddied among the mountain valleys, there were many who took up land and remained. The fortified stone houses, and log houses and big barns of the Pennsylvania Germans were to be found up and down the Frederick and Cumberland valleys, along the Shenandoah and the South Branch. The industry and frugality of these people became celebrated. Their red cattle and fields of corn and wheat fitted easily into the Piedmont landscape. Their industries, their arts, their songs still give the region its unmistakable flavor and individuality.

Even with the primitive material conditions of life, an impressive cultural shape loomed among the German settlers. Their churches and ministers were outstanding, and had wide influence through the entire frontier. The travels and journals of the missionaries and ministers, Schlatter, Stoever, and Schnell, through the upper Potomac in the early eighteenth century are absorbing accounts of devotion and even heroism. Likewise they helped sustain the native language and served as a medium of maintaining the close connection with Germany and Holland. Good elementary church schools were early found in this district. From early German printing presses at Hagerstown and New Market came a remarkable flow of primers, text books, hymnals, Bibles, and almanacs. In music their contribution was great, encouraging and perpetuating a body of primitive choral works—some dating from the thirteenth century—whose significance to our folk art has not yet been fully appreciated. The Mennonites and Moravians of the Valley of Virginia, especially, developed a tradition of shape-note religious song that forms a deeply beautiful and influential member of our body of folk music, and was collected as early as 1816 by the Mennonite farmer, Joseph Funk. In their decorative arts and crafts, they established the famous Strassburg pottery and helped shape the austere forms of Amelung glass. Their ornamented birth certificates, painted furniture and chests are still to be found in Maryland and Virginia.

Usually to be found a little farther inland, forming a western fringe of the long, cultural peninsula that dropped down from Pennsylvania, were the Scotch-Irish. Even more than the Germans they developed the Indian trade and set up outposts in the mountain glades where whites and Indians might come. Tough and independent, when asked for land titles they replied that the provinces "had solicited for colonists and they had come accordingly." They stuck like Scotch burs. Frequently they were forced out, and moved to the next convenient unclaimed land. A questing people, a border people, at home in the highlands, they moved into the Cumberland valley and beyond, or passed into the narrow defiles to the south and broke land where even the Indians had not lived. They penetrated the mountain wall at Pittsburgh and uniquely made that city the first capital of the transmontane west.

2

In the chain that early bound the Tidewater region and the upper Potomac valley, the most important link was the Ohio Company. The dissatisfaction, uncertainty, and restlessness that prevailed in the lower Potomac found many expressions besides the desire for a more stable tobacco economy, a diversified agriculture, and the new commercial and industrial pursuits. The region awakened to its virgin resources to the west, and to the significance of the traffic in lands that rapidly increased in value. A representative Tidewater planter, George Washington, told the secret of his success toward the end of such a long and profitable career, explaining that new countries are the best in which to lay the foundation of wealth, since "lands which, comparatively speaking, are to be had there cheap, rise in four-fold ratio."

Among the many Potomac planters who found in the cheap lands of the west a solution to their problems, Thomas Lee was the first and perhaps the greatest. He has been too hastily discarded by historians and biographers.

Born near the Potomac's mouth in Northumberland County in 1690, Lee was the fourth son of Richard Lee. By the time the family estate had been divided, his share amounted to only 150 acres in Northumberland County and a thousand acres more in Maryland, a rather meager inheritance for those days and hardly a beginning for a man with the ambitions of Thomas Lee.

We may assume that he sought the appointment as Lady Catherine Fairfax's resident manager for those vast acres that began with the Northern Neck and ran back to unknown limits. He was recommended for the post by Thomas Corbin, and succeeded "King" Robert Carter, who had profited largely from the office but failed to produce the income from quitrents expected by the owner.

The young Lee threw himself into his new work, traveled widely through the Fairfax properties in the upper Potomac, and inaugurated large schemes to stimulate land settlement by colonizing Pennsylvania Germans in the valleys of the upper Potomac and the Shenandoah. As he traveled he also acquired lands in his own name, including 4,200 acres in Fauquier and several thousand acres in Loudoun County, where later the town of Leesburg was built. He became interested in the commercial possibilities of the west, talked long and profitably with the men who were trading with the Indians, and came to understand better than any other man the strange political situation of western Virginia. By 1732 when he sat in the Virginia council he concluded, "the French are intruders into this America." His earnest talk of rich virgin lands, rapid settlement, prosperous back-country traders, hostile French and Indians—the Indians were now only a memory in Tidewater—fascinated and impressed some of the tobacco planters. Even the secretary of the Pennsylvania colony, who had good reason to know, thought Lee had a "plotting head."

But Virginia's western frontier was long, and Thomas Lee ran ahead of his time. His friends on the Potomac remained a minority in the Virginia Council, and others with notions about the west could more readily secure the governor's ear. Lee persevered. He acquired more information and experience in the west; he conferred with the shrewd Pennsylvanians; he listened to the frontier talk.

As one of the Virginia commissioners at the Lancaster

treaty meeting of 1744, Lee had participated in the ceremonials and symbolism of the Indian treaty meetings, the taproot of American oratory and drama. To treat successfully with the Indians, the white representatives found they had to adhere strictly to their conventions. Before the large audiences of Indian and white families who gathered at the treaty negotiations, the standards of formal oratorical expression were established on a high level, protocol was rigorously obeyed, the Indian style of address was generally accepted, and with it the Indian's conciseness, his gravity, his rhythm, and often his sharp and witty realism. At Lancaster, Lee had been steeped in one of the significant formative experiences of the frontier. From this irrevocable conversion, he also emerged with a new vision that hardly stopped at the Mississippi.

Among his neighbors on the Potomac, Lee spread the new gospel of the west. He became smooth and confident in the art of Virginia politics.

When Lord Thomas Fairfax and his cousin William came to Virginia in 1747, Lee was ready. After twenty years his "plotting head" was amply furnished with plans. The next year King George's war drove off the Pennsylvania traders, and the Virginians moved in to capture their monopoly. In 1748 the Ohio Company was organized, a royal charter and land grants secured, and the grand machinery of British imperialism, well oiled with self-interest, began to move. One more year and Thomas Lee became acting governor of Virginia, to be succeeded by the friendly Governor Dinwiddie who held shares in the company. The Potomac barons had finally attained power.

When Lee organized the Ohio Company he brought to it all he had learned from his highly educational experiences. The company's twenty shares were closely held by those Potomac planters Lee had interested in the west, and by a shrewdly selected group of other men who were in strong positions to assist the company. The list begins with Thomas Lee, his father, two of his sons and a son-in-law. It included John Tayloe, whose daughter had married another of Lee's six sons. George Fairfax and his cousin John Carlyle owned shares. Then there were the

Washingtons, Augustine and his sons Lawrence and George. Next came the influential lawyer John Mercer, with his three sons, his nephew George Mason, and a cousin James Scott. Inevitably that relative of the Lees, Corbins, Tayloes and Fairfaxes, Councilor Robert Carter, was among the shareholders.

The company's position in Maryland was defended by Nathaniel Chapman, Jacob Giles, and James Wardrop—all residents of Maryland, but with substantial holdings of Virginia lands and important western interests. In the less strategic direction of North Carolina a share was held by the lieutenant governor, Arthur Dobbs. Another useful member was Robert Dinwiddie, soon to become governor of Virginia.

The company also included as shareholders that gamecock of the Maryland mountains, Thomas Cresap, who acted as the principal company official in the west, and in London the important merchant John Hanbury and three other well-situated persons.

To secure their royal charter this powerful and skillful group nimbly dodged the obstructions placed by the colonial governor, rival land companies, and conflicting sectional interests in Virginia. They asserted their superior patriotic interests in defending the crown's western claims against the French, and proposed to survey, settle, and fortify this valuable increment to the empire. They reminded people of the Virginia colony's royal grant with its distant western boundary. Thus the company emerged dedicated half to the patriotic support of imperial interests and half to the pursuit of Indian furs, trade monopolies, and the well-understood profits from real estate—500,000 acres of it. The Ohio Company flourished in the warm sunlight of British expansion, but withered when the shade of cautious imperial retrenchment later fell over the land.

Gone were the days when a lone trader could set off with a packful of beads, bells, and baubles, and expect to come back with his canoe loaded with beaver. He could not compete with the prices the large companies offered the Indians; indeed, he was lucky if he could keep his scalp. The Indians were all wearing matchcoats and Scotch stockings, eating the white man's rations and using his tools and weapons. The capture of fur-bearing animals grew highly efficient. The trading companies refused to sell rum—because it kept the Indians from properly attending to their traps. Trading posts had become expensive forts, with artillery and swivel guns. It cost the Ohio Company an initial £20,000 just to get started in the business; and even then, when important trading operations were to be undertaken, they had to hire men from the rival Pennsylvania companies.

To represent this company at the scene of its principal operations in the upper Potomac, men of high courage and few scruples were required. Not only must the company's representative face the rigorous competition of the fur trade. He must deal in diplomatic situations of first-rate importance with the French and Indians, to say nothing of the three jealous provinces whose interests converged and clashed at the headwaters. For its purposes the Ohio Company could count itself fortunate in finding Thomas Cresap.

A dark Yorkshireman, who had tried tobacco planting and gone broke, Cresap had wound up on the frontier. Virtually an agent for the Calverts and for Daniel Dulany, in tricky and obscure deals in land and settlement, he was properly seasoned to the company's taste. Thomas Lee had met Cresap. The Washingtons had rented him a farm, and George Washington as a young surveyor had visited him. He had been engaged by the Fairfax interests to determine the source of the Potomac, and drew the first clear map of its headwaters. Such men were impressed with Cresap's abilities. Not merely did he know the country and its inhabitants; he could be relied upon. So indispensable was Cresap considered that he became one of the original shareholders of the Ohio Company, the only American who did not come from the select tobacco-growing aristocracy of the lower Potomac. After the death of Thomas Lee in 1751, Cresap's judgment on western matters went almost unchallenged by the company.

Hardly less important than Cresap was Christopher Gist. Like Cresap, he had lived in Tidewater Maryland, engaged in tobacco planting, and failed. The facts of their remarkably parallel careers are not distinct, but it appears that, like many small tobacco planters, both fell too heavily in debt to extricate themselves. Gist moved southwest to a farm on the Yadkin River, and acquired that universal passport of the frontier, the ability to make quick, accurate surveys. No one excelled him in his ability to strike out into virgin territory and return with fact-crammed notebooks and accurate maps. He knew land values, mineral values, transportation factors, military strategy, and was faultless in choosing valuable tracts of wilderness lands to be patented and settled, locations for trading posts and forts, and the routes to be developed as roads. The English fought the French and Indian War in the Potomac largely on Gist's maps and his information on the terrain. In the years before the war with the French that Virginians believed inevitable, Gist was invariably ordered to note military and political factors as well as the details of land and trade. His missions for the Ohio Company from 1750 to 1752 laid the practical foundation for their land and trading operations, and the information thus gained for the modest investment of £200 alone would have justified the entire existence of the company.

With Gist other famous frontiersmen were associated with the Ohio Company's effort, and all of them were frequent visitors to its western base at Cresap's. Andrew Montour and William Trent were employed by the company. That prince of Pennsylvania traders, George Croghan, his partner Barney Curran, and the Maryland trader Hugh Parker came often to the settlement at Shawnee Old Town. The aging and distinguished Conrad Weiser, who had lived as a boy with the Mohawks along the Schoharie, entered the Ephrata monastery, and ultimately wrote a frontier autobiography of outstanding value, served as interpreter for the company and knew Cresap well. With these frontier figures mingled the Indians, among whom were found such sharply defined individual types as the tragic Logan, the celebrated Queen Aliquippa, the so-called Half-King, the French-Indian mother of Andrew Montour, and many others. On a stage populated with such outstanding characters, their host had no difficulty in holding his own.

3

Cresap's preparation for becoming the western agent for the Ohio Company had begun when he settled in the controversial area on the western bank of the Susquehanna that was hotly contested by the Calverts and the Penns and where the Lords Baltimore had first colonized the peppery Ulster folk. It was hardly by accident that Cresap had been given large tracts of land here in 1730, nor that these grants made him the Maryland settler farthest north, a spearhead at the 40th parallel, the boundary Maryland claimed. One is not surprised to find that this alleged farmer and ferryman was a Maryland magistrate, a captain of Maryland militia, and a Maryland land agent.

Well aware of his role, he built a small fortress of a house and commenced to terrorize his Pennsylvania neighbors. They called him "the Maryland Monster." He was a professional troublemaker, a feuder, a murderer. Legends were built around him in the gossipy frontier taverns. When taken prisoner he was manacled with iron handcuffs, but no sooner had the smith completed his work than Cresap raised both hands together and "gave the smith such a tremendous blow upon his black pate that it brought him to the ground." After innumerable clashes with Pennsylvania authorities, when he was arrested and taken to Philadelphia, he spoke an authentic line of frontier humor, and pronounced that provincial capital "the prettiest town in Maryland."

In prison, distinguished Maryland lawyers, the elder Daniel Dulany and Edmond Jenings, defended him, and after nearly a year in jail he was released.

Dulany was heavily engaged in land operations in the Monocacy and Cumberland valleys, and he soon gave Thomas Cresap large grants of land among his settlers on Antietam Creek, and a generous loan of £500. Over a spring on his new land Cresap commenced building the only kind of house in which he ever appears to have been comfortable—a fort. Land speculation and the fur trade with the Indians engaged his attention. A business misfortune, when his first cargo of furs was lost at sea, broke him in this venture, and again he moved farther up the Potomac to Old Town. Here, at last, he had found the conditions in which his special genius could flourish.

Always clever in managing the Indians, Cresap soon found that he had located in a matchless spot. The territory was marked on the maps as "abandoned Shawnee lands," but other Indians had moved in, and the fact was discreetly overlooked by surveyors—since grants of lands occupied by the Indians were invalid. At Old Town the Indian paths from north to south and from east to west converged and intersected. Here the furs were plentiful and of the highest quality. Cresap cultivated the passing Indian war parties. His lavish hospitality earned him the name "Big Spoon."

White travelers who came that way were entertained. George Washington was a frequent visitor. English soldiers and Moravian missionairies noted Cresap's hospitality. Numerous settlers passed through to the west, and soon more cabins were built in the surrounding country. Gradually Cresap's house became another fort; warehouses were built. "Cresap's" began to appear as a name on maps and adventurers began to mention him in their diaries. Members of the Ohio Company noticed him as well.

When the company was organized in 1747, Cresap was well established at Old Town and was doing well in land speculation and the Indian trade. After he became the company's principal agent in thc west, his position gave him a central place in the effort to break open a commercial route from the head of navigation on the upper Potomac to the virgin lands of the Ohio valley.

For many decades the French and the English had disputed the title to the Ohio lands, but in the absence of a resolution of this dispute both sides traded with the Indians. The Pennsylvania traders had long monopolized the eastern side of this lucrative trade. The firm conviction upon which British official support of the Ohio Company rested was that successful trade and settlement in the Ohio would put an end to the French claims, and the company was expected to accomplish this result.

Cresap commenced to lure the Indians away from his old enemies, the Pennsylvanians, in the most obvious way—by offering them trading goods at lower prices. He also blackguarded the Pennsylvania traders and tried to convince the Indians that they were being cheated. He poured out implausible tales of his pity for the aborigines, and how he proposed to sell them goods at less than cost. He bragged loudly about the Ohio Company and its intentions, its roadbuilding plans and schemes to settle hundreds of families in the Ohio country. These injudicious remarks were promptly turned against him by the Pennsylvanians, who pointed out that the proposed white settlements would kill the Indians' game, and slyly argued that the projected road would afford a route by which the southern Indians could attack the Ohio tribes.

In consequence of such bickerings, the Indians became suspicious of both groups of English traders, and turned to the French.

To recover this lost ground and establish a clear basis for future operations in the Ohio valley, a treaty was considered indispensable, and at Logstown in 1752 the Ohio Company gained the right to bring in settlers and build forts east of the forks of the Ohio. On the basis of this agreement, the company commenced building an advanced base on the site Cresap had chosen at Will's Creek (today Cumberland), and laying out the road over the Laurel Mountain following the route chosen by Christopher Gist and the Indian guide Nemacolin. This road led to the Monongahela and thence to the forks of the Ohio, and became the strategic pivot on which military contests and future development of the region turned. At the forks of the Ohio they planned the construction of a strong fort, and Thomas Cresap, Christopher Gist, and William Trent were placed in charge of this work.

While the French were ceremoniously marking out their claims by burying lead plates along the Ohio River and constructing a chain of forts, the English elected the more practical course of establishing possession, and were busily filling up the region with as many settlers as they could entice into the "Big Woods." The long-delayed and inevitable clash began to draw close in 1753 with young George Washington's fruitless mission to the Ohio, thence to the French fort near Lake Erie.

Military operations commenced when the French easily took possession of the Ohio Company's unfinished fort, and renamed it Fort Duquesne, a fact duly reported by Cresap (who was conveniently absent from the fort) in a letter to Governor Dinwiddie. On Washington's return to Williamsburg, he was ordered to take out a hastily improvised force and recapture the company fort at the Ohio, but after a small initial success against a French scouting party he was bottled up at Fort Necessity, an improvised palisade defense in the Great Meadows.

The indignity of this military defeat has been stressed in general histories as a cause of the war, but the heavy loss to the powerful Ohio Company of their two expensive forts and the connecting road between them has not been sufficiently noted. Nor has the role of the company and its shareholders in directing

General Braddock's plan of campaign the following year been well appreciated. When that choleric and incompetent British officer arrived at Williamsburg he met with Governor Dinwiddie, journeying thence to Alexandria, where he set up headquarters in the home of John Carlyle and conferred diligently with other Potomac barons. All these personages were shareholders in the company. So was the young provincial officer who was added to the general's staff, George Washington. A satisfactory plan of campaign was settled upon.

In truth the Ohio Company had become almost a part of the government of Virginia. It owned the principal military installations and roads, and possessed the most comprehensive knowledge of the prospective theater of war as the result of Gist's explorations and the work of Cresap and his agents. At the meeting of provincial representatives, the diplomatic wiles of even Benjamin Franklin, seeking to advance the cause of an approach to Fort Duquesne through Pennsylvania, were useless against the Ohio Company's overwhelming desire to secure the advantage of Braddock's approach through its territory and the improvement of routes it had pioneered, to prime the pump of colonization.

From their point of debarkation at Alexandria, and reinforced by two regiments of colonial troops, Braddock's force of two thousand men set off for the upper river. The main body of troops marched up the Virginia side, and the trains took the route through Maryland. The two parties met at Cumberland, where the ubiquitous Thomas Cresap had been instructed by Governor Sharpe of Maryland to have sufficient wagons and rations to supply the army. But Maryland's interest in a war "to protect Virginia real estate" was lukewarm, and her peace-loving farmers of the interior preferred to keep their wagons. Like his previous instructions to construct a road from Will's Creek to the forks of the Ohio, which had resulted in only a meager clearing through the woods, Cresap again failed to provide the needed wagons, and the casks of beef he sold the army proved to be unsalted and so spoiled they had to be buried.

From the numerous accounts of the army's prolonged stay

at Will's Creek, Cresap also seems to have been a greedy sutler who charged high prices for wine and spirits sold to the thirsty soldiers and their numerous inevitable female companions. His unpopularity reached its zenith among the troops. A reasonably fair reference from among the many journals of the Braddock expedition is this:

"May 8th [1755] Ferried over the River into *Maryland;* and March'd to Mr. Jackson's, 8 miles from Mr. Cox's where we found a Maryland Company encamped in a fine situation on the Banks of the *Potomac;* with clear'd ground about it; there lives Colonel Cressop, a Rattle Snake Colonel, and a D——d Rascal; calls himself a Frontiersman, being nearest the Ohio; he had a summons sometime since from the French to retire from his Settlement, which they claimed as their property, but he refused it like a man of spirit . . . he hath resolved to keep his ground."

There is hardly need here to review the tragedy of the Braddock expedition. It failed for the lack of everything: plans, supply, preparation, training, strategy, and tactics. General Edward Braddock failed to take the most ordinary precautions against ambush required by contemporary European military practice, and failed to heed repeated warnings from American advisers. He "threw away the book." Not only did he and his staff despise the colonial troops, but in the face of the enemy he displayed an arrogance and stupidity that brought him to disaster and a wagon-tracked anonymous grave. And with his failure the Ohio Company commenced its decline. It had not fulfilled the conditions of the charter by settling the lands originally granted, and no number of petitions and memorials ever succeeded in recapturing its once-powerful position.

Yet significant as General Braddock's defeat was in the mind of a nation seeking to contrast the British regular and the colonial soldier, mature reflection has since convinced many of Braddock's success. If his mission was to force out the French and prepare the way for English settlement of the Ohio, his accomplishment in completing the 12-foot road that Gist and Nemacolin had pioneered to the forks of the Ohio ensured that result. His decision "to level every mole hill and bridge every brook" may have given the French time to prepare, but it ensured eventual victory. The solid column of settlers that eventually followed Braddock's road raised the cabins in the clearings that staked out the western American claim forever. When George Washington, who had dreamed of this road since 1752 and fought for it through limitless frustrations, visited it in 1774, even he was astounded to find people migrating over it "in shoals."

After Braddock's defeat, the advanced frontier settlements were temporarily abandoned or left to fight for their lives against the French and Indian raiding parties. Even Thomas Cresap had to withdraw for a time. The frontier deteriorated as far back as the Shenandoah valley, and the woods resounded to the crack of the Indian rifles, the scream of captured women and children and the tortured prisoners. But above these wilderness sounds was heard the muttering and grumbling of hard men who had been forced to abandon land and homes that represented the work of years.

This was the scene George Washington found when he arrived at Winchester in the Shenandoah valley, a place crammed with refugee settlers and their families. With such material he began to create a military force that could hold the Allegheny front and build a chain of forts around which the settlers could rally. Of split-log structures, arranged in a hollow square, and equipped with stout gates, well over a dozen were hastily erected. But as he went about his military duties, he wrote Governor Dinwiddie: "The supplicating tears of women and the moving petitions of the men melt me into such deadly sorrow." He listened carefully to the border talk. He learned why men fight.

The frontiersmen had long been disinterested in the British crown; now they became disaffected. A government that had demonstrated so vividly its inability to protect their lives and their property, that left them to shift for themselves, deserved little support. It was hardly better than no government. Even the exemplary Washington, struggling furiously to procure ammunition and shoes for his recruits, became so irritated that Governor Dinwiddie complained he was treated "with scant courtesy."

Over and over again the young officer listened to the pioneer refrain. Security from the Indians, access to free land—and, most of all, some voice in the decisions of their government. He sensed a population eager to grasp and mold its own future, and his spirit quickened as he understood what that future promised.

Great as the frontier education was for an officer who was to create the first American army, and to lead it through years of discouragement to eventual victory, these seething years contained an even greater education in political understanding. By the time Washington had been sent out to organize the frontier, he was already the most experienced man in Virginia's western affairs. He had commenced at the age of fifteen as a surveyor for the Fairfaxes in the South Branch valley. By his reading in the library at Belvoir, and his long talks at Greenway Court with the worldly and aging Lord Fairfax, he had profited largely. He had traveled widely, observed closely, and early been thrust into affairs of great importance. He had known the hard wilderness life, the subtle ways of the Indians, the treacherous characters of the frontier. But all these formal qualifications were now overshadowed, these experiences transformed, by an increasingly profound understanding of Americans in the making.

Perhaps it is still too easy to think of Washington as a country gentleman at Mount Vernon, sitting on the terrace of that famous mansion, looking across his well-manicured lawns with their deer parks and ha-has to the broad Potomac. This was not the environment that produced Washington the military genius, the political leader, the revolutionist. For the roots of that Washington we must look to the west.

The French and Indian War gave Washington a perspective of the strategic aspects of warfare on the continental scale. At Winchester he prepared himself to become a great leader of

American troops. But most important, during these years he gained a deep understanding of the motives of his fellow countrymen in the west, and through long residence and extensive landholdings he commenced to share those interests himself.

To know this Washington we must picture him not at Mount Vernon but at his Bullskin plantation at the foothills of the Blue Ridge in Frederick County, Virginia. We must think of him not as the central figure of some stiffly posed lithograph, surrounded by the tobacco panjandrums and adoring Negro slaves, but in a lively scene crowded by traders and land speculators, wheat farmers and men of the cowpens, backwoodsmen and the proprietors of sawmills, flour mills, and other frontier enterprises. Most of all we must steadily recall that his private fortune was not sunk in Tidewater tobacco land but carefully invested in vast tracts of wilderness acreage and held against rising values.

Washington's Bullskin plantation of 550 acres he acquired from Lord Fairfax in 1748 as compensation for his services as surveyor, then reckoned at a doubloon a day. Soon he added 15,000 acres of Ohio lands granted by the governor of Virginia as a military bounty to those who served in the French and Indian War. He doubled this holding by purchasing the bounty lands of fellow officers at a reduced price when it appeared that the grants would not be honored. From this transaction he ultimately realized close to half a million dollars' profit.

Constant buying and selling of lands of all sorts produced the first president's chief income. He owned not only plantations, piedmont farms, and wilderness tracts, but numerous lots in provincial towns. When informed that the national capital would be located in Philadelphia he made haste to buy a likely farm on the outskirts of that city in anticipation of rising values. But on the eve of the Revolution, in 1775, he was so land-poor that he replied to a friend who had requested a loan, "So far am I from having £200 to lend . . . I would gladly borrow that sum myself for a few months."

At his death it was authoritatively estimated that "General Washington is perhaps the greatest landholder in America." His inventory, aside from town lots, Mount Vernon and his wife's property, ran to more than 50,000 acres.

In addition to his participation in the Ohio Company, he was also a partner or shareholder in other great land companies: the Walpole Grant, the Mississippi Company, the Military Company of America, and in numerous corporations that were concerned with the construction of canals and internal improvements, and hence with the advancement of settlement and the raising of land values.

Washington was keenly aware of the art of developing real estate values. In addition to the quotation previously given, in 1767 he inquired how the greatest Virginia estates were made, and went on: "Was it not by taking up and purchasing at very low rates the rich back lands, which were thought nothing of in those days, but are now the most valuable land we possess?"

Such enterprise truly was much in the Potomac grain. The trade in land was a widespread form of enterprise. It flourished naturally in the traditions and habits of the place, stemming from the form of settlement and the lack of other satisfactory ways of holding capital and gaining a return on investment. Notably it grew from the landholding pattern of the lower Potomac and its natural extension to the interior.

One other characteristic went hand in glove with the traffic in land: the speculative instinct. Nourished by the spirit of the age, social customs, and the long habituation to the speculative nature of the tobacco crop also fitted it into the configuration of the region. Gambling with cards, betting on horses, and a tendency to turn anything into a wager were found everywhere. Land speculation fitted an age which speculated in everything: which bought shares in privateers, tickets at lotteries and raffles, and took fliers in bank stocks and canals. In England a deck of playing cards was published, each bearing a satire on one of the favorite speculations of the day. But there was little of the contemporary English attitude toward colonial land speculation in the Potomac valley, the attitude that moved the satirist to begin:

Come all ye Saints that would for little Buy,
Great Tracts of Land, and care not where they lye.

In the Potomac the counters were not the unreal paper shares of stock that could scarcely be distinguished from the gaming checks at Almack's and Boodle's. They were pieces of paper—grants, patents, deeds—that represented that most concrete of all things, the land itself.

The Potomac speculation was more like its cockfighting or horse racing, where careful training and handling balanced the odds. Just as English visitors in Virginia were astonished to find that the owners of celebrated race horses were accustomed to ride them in racing events, they found it distinctive that men with large speculative landholdings knew them in detail, and frequently had surveyed them. Councilor Carter with his theodolite, Washington with his surveyor's table and alidade, George Mason at his account books, and even the crusty Lord Fairfax deep in the heart of his Virginia grant at Greenway Court, knew what they were about when they bought and sold land; and they well understood what had to be done if they were to realize a profit from it.

His land interests aligned Washington not only with the frontier but with the dominant interest of his day. It did more. It committed him to the idea of an expanding America, a nation whose population would flow over mountains and into the great valley of the Mississippi, whose cities would grow, whose crops and industries would increase and diversify. In this effort he found himself at one with countless numbers of his fellow-countrymen, rich and poor, land baron and squatter, planter and farmer, trader and banker, all of whom found they were as heavily committed even if they did not own an acre. They were associated in a great variety of ventures, until at last they came to the greatest—the American Revolution.

Washington's trade in real estate was a marked development from the mere engrossing of estates as we have seen it in the time of the first Robert Carter. The effort to create land monopolies, of establishing a manorial system on the English model, could have had small part in the thinking of a man whose legs had carried him over the mountains and broad valley lands of eastern America, who knew its frontier peoples, and whose imagination ran ahead into the inland empire. The possibility of land monopoly may have existed in the minds of a few gouty Tidewater planters, but it faded quickly when the mountain barrier had been pierced and the limitless continental expanse spread out from the gateway at Pittsburgh.

The fate of manors was epitomized in a New World parable which illustrated the transformation of the pastime of fox hunting. The traveler John Bernard told it: "Hunting in Virginia, like every other social exotic, was a far different thing from its English original. The meaning of the latter is clear and explicit. A party of horsemen meet at an appointed spot and hour to turn up or turn out a deer or a fox, and pursue him to a standstill.

"Here," he continued, "a local peculiarity—the abundance of game—upsets all system. The practice seemed to be for the company to enter the wood, beat up the quarters of anything, from a stag to a snake, and take their chance for a chase. If the game went off well, and it was possible to pursue it through the thickets and morasses, ten to one that at every hundred yards sprung up so many rivals that horses and hunters were puzzled which to select, and every buck, if he chose, would have a deer to himself—an arrangement that I was told proved generally satisfactory, since it enabled the worst rider, when all was over, to talk about as many difficulties surmounted as the best."

The contrast grew. At home in England the enclosures accelerated greatly, and large estates grew larger, while the shrewd poet lamented:

> Ill fares the land, to hastening ills a prey,
> Where wealth accumulates, and men decay.

Yet in the Potomac valley the remains of the manorial system, the great tobacco estates, and the vast wilderness baronies were withering away. In their place a landscape of small farms was being created, peopled with a landowning class of sturdy farmers. Even the hired tobacco factors and trading-company employees were giving way before native business establishments both in the older shipping centers and on the fringe of settlement.

The New World stood out in clear profile, distinct from the Old.

SIX

CRADLE OF THE REPUBLIC

TOWARD THE END OF THE THIRD QUARTER OF THE EIGHTEENTH century the shape of the valley had become distinct. One man could know it well in all its aspects, could see in it a microcosm of the New World.

Traversing the valley section from the Potomac's mouth, the broad tidal bosom of the river was seen dotted with skipjacks and sloops carrying its petty commerce and the larger vessels of the tobacco fleet rocking with the current at plantation wharves. The great seines, drawn in by teams of horses on the river's bank, collected their fabulous hauls of herring and shad. In the fields that sloped to the river's edge, the year-round rhythm of tobacco cultivation progressed unceasingly, each plantation a miniature world with its strata of Negro slaves, indentured workers, clerks, and overseers. At the head of navigation lay the towns, where the merchants of Georgetown and Alexandria dealt in tobacco and wheat, Osnaburg and woolens; where the housewife's door opened each morning to a cobbled street alive with carts and peddlers, droves of cattle and turkeys from the country, or the carriages of gentlefolk. Beyond lay the wheatlands in the rolling foothills, and farther west in the valleys and ridges the Germans and Scotch tended their flocks and herds, a pastoral and idyllic society, stern and primitive in its self-willed isolation. In the mountain glades where the deer browsed, and on the rises, where the trees grew so thick they had to be felled from the top down so tangled were the branches with vines and creepers, the hunter and his dog found the deer, the bear, and the plenteous fur-bearing animals. The banks of mountain streams held sawmills busy with the virgin timber, and in the narrow defiles could be heard the sound not only of the woodsman's ax but of the miner's pick.

After nearly one hundred and fifty years of growth and change, the valley had filled up, an equilibrium had been achieved. It was a practical society these Potomac people had created. They were busy with agriculture, trade, and government. It was not a singing people destined to produce great poets, musicians, artists. Where the arts appeared they would be the practical arts of architecture and shipbuilding; the practical literature of law or politics; the practical music of work songs, devotion, or patriotism; even a practical form of painting, portraiture. Yet in its variety and balance a Potomac civilization had been achieved, delicately poised in its harmonies and conflicts, solidly planted in its future course. Like the transit of some glowing planet, its star had risen in the east over the London tobacco warehouses, but now hung in the west over the virgin lands of the Ohio. That vision now gripped the minds of its people.

Through the valley flowed an increasing commerce. Along the roads could be seen the hunter's pack train loaded with saddles of venison and bear, skins and furs. Or the annual caravans from the mountain settlements taking their ginseng, peltries, and bear's grease to market, and returning with salt, iron, utensils, and implements. The heavy ox-drawn wagons of the valley farmers, laden with wheat, came down to the shipping ports and milling centers, returning with loads of smoked river shad. Travelers multiplied, and the postrider was to be seen on the country roads and at the numerous ferries.

On the river itself increasing trade was observed. Below the falls the naval offices inspected and registered the shipments of tobacco and superintended the development of lighthouses, the services of pilots, and other aids to navigation. Above the falls, especially during the spring freshets, river commerce was growing. Like the Indians and the early traders, Thomas Cresap had been accustomed to come down the river in a stout boat, his gentleman's costume carefully wrapped in a waterproof cover against the day it would be worn in provincial towns. But now the river was becoming crowded during the high waters each spring and fall with strong, narrow barges of shallow draft from as far up as Cumberland, and even high up such tributaries as the Shenandoah, carrying hemp, wheat, hides, and other cargo. Timber was rafted down to the port cities. Everyone knew that Sir John St. Clair, Braddock's energetic quartermaster, who had voyaged down the river to Alexandria from Fort Cumberland with Governor Horatio Sharpe, had pronounced navigation in flat-bottomed boats feasible. The Maryland businessman, Thomas Johnson, agreed. Already plans were afoot—even the legislation had been passed in the Virginia House of Burgesses—to open the navigation of the upper Potomac, and the aspiring John Ballendine of Fairfax County had organized a company to undertake the work. The dreams of the Ohio Company, their cherished vision of a Potomac route to the west, still kindled the imagination of men in Virginia and Maryland.

All these dreams were of the future. To realize them within the framework of the colonial system seemed increasingly difficult. Even the present was troublesome enough.

2

As the tides floated the great ships with their cargoes of tobacco down the river, the planter may well have watched them go with an uneasy feeling. Trade advanced and receded with increasing swings of prosperity and depression. But when he turned to other forms of agriculture and enterprise, the planter felt the shackles of the British law steadily tighten, insistently directing him into the grand design of empire. In the imperial plan the role of the Potomac was not open to question. It was to grow tobacco for the British home market, and to supply the London merchants who sold tobacco to the world. Nine-tenths of the Potomac's tobacco was re-exported from England to the Continent. It was clear to anyone in Westminster that, should the colonies cease to grow tobacco and turn to other crops, Britain must buy from Spain, Portugal, or France. The assured market in the colonies for British goods would disappear. Equally self-evident was the proposition that, if the colonies were allowed to sell their tobacco directly to the European countries, the London merchants could engage only in a profitless competition.

In this system the tobacco planters held a wavering position. When the quantity and price of the tobacco crop were satisfactory, it often seemed a desirable state of affairs. The hard-won reforms of nearly a century appeared to have eased the planter's insecurity. Yet when the droughts caused his tobacco to fail, or sudden heavy rains made a coarse and abundant leaf of poor quality, he cursed the crop and the system that obliged him to keep to it. He thought himself "a species of property" of the British tobacco merchants, who steadily made from 100 to 200 per cent profit from the trade. Through good times and bad, planters in general found themselves deeper and deeper in debts which, Governor Fauquier reported, made them "uneasy, peevish and ready to murmur at every occurrance."

As their attention turned from agricultural problems, they contemplated the system of tobacco inspection. Along the river at each important shipping point the tobacco was inspected and graded—but the inspectors were provincial officers, appointees of the crown or the proprietary, and the fees they exacted deducted heavily from the planter's profit. The value of the system was beyond dispute. The intervening years only reinforced the argument advanced in Maryland by Daniel Dulany in 1743 that without "sound effectual regulations" of the tobacco trade the growers would be obliged to desert their plantations. (With a synoptic view of the effect of such a catastrophe on his western settlements, Dulany had gloomily added, "and very few, if any, will come into a country which is on the brink of ruin.") But often it appeared that the purpose of inspection was not to help the tobacco planter but to provide jobs in a corrupt government for a growing bureaucracy that fattened on the inspection fees.

The years before 1776 saw a gradual slipping away of the

loyalties of the tobacco planters. The final convulsion came variously in Maryland and Virginia. Significantly, as Professor Charles Albro Barker has shown, in Maryland it came from a governmental obstruction: in 1771 the two houses of the Assembly were unable to agree on the continuation of the tobacco inspection law, and the system was permitted to lapse. It became evident to all that Maryland's curious form of government had become unable to deal with the essential types of legislation. Everyone could see that a law upon which virtually the entire population depended had been smothered in a dispute over the proprietary.

When the "effective regulation" Dulany had demanded thus lapsed, spontaneous committees arose to continue the inspection work. In Port Tobacco twenty-five buyers set up their own system, and were shortly followed by similar groups in the other shipping centers. The pattern was much like the earlier spontaneous government of the frontier, when militia companies were organized to fill the gap left by a government incapable of supplying military protection.

But such exercises in self-government could not be left to the caprice of voluntary efforts. Small groups administering a tobacco-inspection program could only lead to the abuse of power, to unsound and self-interested restraints of trade. Voluntary militia companies too easily were controlled by factions, or led to irresponsible mob rule. Unrelated to some general form of government, they could only serve as the briefest expedients. To abandon the crown government or the proprietary was not enough, and the fact was appreciated by thinking men from the beginning. Their minds tuned to the words of John Locke, their imaginations did not hesitate to consider more sweeping changes.

On the wheat farms of the Piedmont a restless spirit was also noted, even among the placid Germans. Often farmers were insecure even in their lands, waiting with apprehension for decisions of contested titles in faraway courts. They chafed under a burden of quitrents, and resented the tithes exacted to support an established church to which their conscience did not permit them to subscribe.

CANAL CROSSING THE MONOCACY

In the eyes of many a Tidewater planter they were fortunate in their crops. Jefferson expressed this view when he contrasted wheat with tobacco. "The culture of wheat is the reverse in every circumstance. Besides clothing the earth with herbage, and preserving its fertility, it feeds the laborers plentifully, requires from them only moderate toil, except in the season of harvest, raises great numbers of animals for food and service, and diffuses plenty and happiness among the whole. We find it easier to make a hundred bushels of wheat than a thousand weight of tobacco, and they are worth more when made."

Yet a wheat inspection to standardize quality and secure premium prices had already been demanded in Virginia before the Revolution, and improvements in transportation were everywhere needed to get the grain to market. The frontier settlements were at a disadvantage to secure such changes, for despite the rapid increase in western population their representatives in the provincial assemblies were still few in number. In their view colonial government was not only oppressive and corrupt; it was unsympathetic and unresponsive to their repeatedly expressed needs.

A British traveler found organized bands of Germans patrolling everywhere as he made his way from Frederick to Hagerstown in 1775. When apprehended, he reported their remarks in such of their dialect as he could recollect. "One said, 'Got tamn you, how darsht you make an exschkape from this honorable committish?' 'Fer flucht der dyvel,' cried another, 'how can you shtand so shtyff for King Shorsh akainst dish Koontery?' 'Sacramenter,' roars out another, 'dish committish will make Shorsh know how to behave himself;' and the butcher exclaimed, 'I would kill all de English tieves as soon as Ich would kill van ox or van cow.' "

3

The "back lands," as the Tidewater planters persisted in referring to the upriver settlements, contained far more than German wheat farmers, and its population was moved by many factors that had little to do with making a living. Probably half the white people in the Potomac lived above the falls on the eve of the Revolution, and their grievances were many. In their poverty and isolation, religion assumed a large part in their lives, and the region was swept with revivals and filled with dissent. Unwillingness to support the established church resulted in such famous episodes as "the Parson's Cause" and underlined the popular belief in the avarice of the clergy. Complaints of congregations against drunken and dissolute clergy became general in all the major denominations. Even within the Anglican Church, a stoutly independent spirit was noted, presaging the day when the members of that church themselves would lead in its disestablishment and reform.

Many episodes revealed growing criticism of the Anglican Church and the movement for an independent church and clergy, but perhaps the outstanding one was the struggle in Maryland over the proprietary appointment of the clergy illustrated in the Bennet Allen affair in Frederick. As the aborting of the tobacco inspection law showed the inability of the provincial government to deal with essential regulation, the Allen affair tore away the curtain of church patronage and revealed the evils of a system under which appointments were made frivolously and the proprietary governor was hopelessly torn between the wishes of the proprietor and those of the people.

A friend of the Lord Baltimore—some said he had written a pamphlet in defense of the young lord, accused of raping a Miss Woodcock—Allen arrived in Maryland late in 1766, carrying a letter to the proprietary government with instructions that he was to be given "the best living available" in the province. Almost immediately it became clear that the worn and bald clergyman was loud and boastful, pretentious and a liar.

Governor Horatio Sharpe thought the matter distasteful and Allen pushing, but he attempted faithfully to comply with Calvert's instructions, only to find Allen even greedier than he thought. Allen suggested that he be given two parishes, one of which he would let out to a curate, and supported his case with a windy legalistic brief in defense of such pluralism. He accepted one church in Annapolis, so that he might ingratiate himself and keep an eye on possible other advantages from the provincial government. There he promptly became involved in politics, and engaged in a bitter and profitless controversy with the powerful Dulanys in the columns of the *Maryland Gazette.*

When in May, 1768, a vacancy was about to occur at All Saints' in Frederick, by far the richest parish in the province became available. It also lay in the sphere of influence of Allen's chosen enemies, the Dulanys. Allen had long had his eye on All Saints', and immediately commenced importuning the governor for the appointment. In his successful pursuit of this appointment, he gave up his ideas of pluralism. Situated in a wealthy and rapidly growing community, it had long been recognized that this parish was too large and should be divided, but the step had been postponed by the parish during the lifetime of its rector.

Allen realized that he must act swiftly before the petition for division reached Annapolis. While the old rector lay dying, he secured the appointment and rode out on a fast horse to Frederick. The dutiful Governor Sharpe managed to block a bill introduced into the Assembly calling for a division of the parish, having been assured by Allen that Lord Baltimore "would not suffer it to be divided." Allen arrived in Frederick on the day of the old rector's funeral and immediately sensed growing opposition, which he attributed to his quarrel with the Dulanys. The plot, he decided, must be nipped.

By a stratagem he secured the keys to the church from the sexton's servingmaid. He went into the church on Saturday and "read prayers, the 39 articles & my Induction" before a congregation of two, one of whom was "a tippling old barrister." Legally his induction was completed.

The following day, however, the infuriated vestry bolted the church door, only to be outwitted by the resourceful Allen who had arisen at four o'clock and entered through a church window by means of a ladder. When the vestry and the congregation arrived, there were some hard words that terminated, in Allen's later account to Sharpe, when he "leap't into the Desk & made my Apology & began the Service. The Congregation was call'd out. I proceeded as if nothing had happened till the Second Lesson. I heard some Commotions from without which gave me a little Alarm & I provided luckily against it or I must have been maim'd if not murder'd. They call'd a number of their Bravest, that is to say their largest Men to pull me out of the Desk. I let the Captain come to within two paces of me & clapt my Pistol to his Head. What Consternation! They accuse me of swearing by God I wo^d shoot him, & I believe I did swear, wch was better than praying just then.

"They retired and I proceeded, but the Doors & Windows flying open & Stones beginning to rattle, my Aid de Camp Mr Deakin advised me to retreat, the Fort being no longer tenable."

Allen appointed a curate popular in the locality and, implying that he might consent to a future division of the parish, retreated to Philadelphia. From that neutral corner he carried on the war with long, fulminatory letters to Sharpe, accusing Daniel Dulany of refusing to divide the parish because he wanted it left intact for Walter Dulany's sons. The younger Dulany indignantly denied this charge, asking, "To what was ye Tumult owing? Most certainly to ye Pistole—to ye Oath and ye violent & vindictive Spirit of ye Parson discover'd upon ye Occasion."

Soon Allen reappeared in Annapolis, and upon his first encounter with Walter Dulany the two commenced beating each other over the head with canes in a street row, summed up in the verdict of the Dulany faction: "Ye Booby acquitted himself, as was expected, like a Poltroon."

Men did not have to reflect long, as this juicy tale unfolded in taverns and churchyards, to see what was to blame. Those who did not belong to the established church they were taxed to support, were confirmed in their suspicions of its venality.

Those who did belong became more convinced than ever that until the church could be divorced from patronage appointments and be responsible to an American diocese with its own bishop its professions of faith would be empty indeed.

At the western fringe of settlement families had been abruptly halted in their progress to the Ohio after the French and Indian War by a royal proclamation forbidding land grants west of the headwaters of the streams flowing into the Atlantic. Beyond this proclamation line the lands were reserved to the Indians. Virginia's impressive western land claims were thus nullified. Well-intentioned as this decision may have been, it displayed an unfathomable ignorance of the dynamics of western life. To enforce it was a hopeless task, and neither the Maryland nor the Virginia authorities succeeded. The 1763 proclamation remained an irritant, and no man who ventured west of the line could expect security in his lands, the common services of law and government, or protection against the Indians. Yet before the Revolution thousands migrated in defiance of the proclamation, and the drift westward continued steadily all during the war. Indeed, the events of the Revolution stimulated the western movements, as the creation of new counties and the growth of settlements showed.

In the towns a broader interest in government arose. At the courthouse steps in Alexandria the ubiquitous tobacco that had been collected as taxes was auctioned. But the market square on which the courthouse faced was used as a drill ground, and by the eve of the Revolution already had its traditions. Here Washington had recruited and drilled his hastily improvised force that later surrendered to the French at Fort Necessity. Here the band had played "The Girl I Left Behind Me" when Braddock's regulars marched out toward Fort Duquesne. Rolling down from Winchester and the Shenandoah valley came the wheat-laden wagons, emblems of a new era in the town's life. From farther west came the pack-horse trains of peltries from the wilderness.

All parts of the Potomac region brought their goods to its metropolitan center and chief port. Standing at the head of Tidewater, the town was still new. George Washington had helped to survey it with John West in 1749. It occupied a balancing position between the lower and upper river civilizations, uniting and harmonizing them. At the Alexandria balls, between the stately country dances of the lower river they played the "everlasting jigs" of the back country.

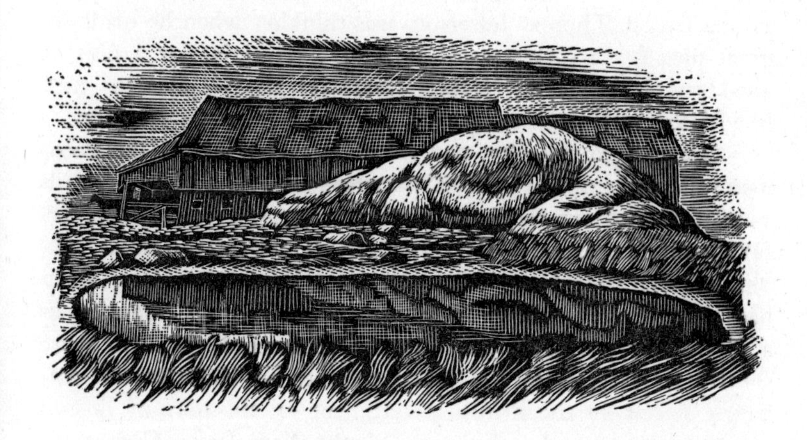

"The genius of the land was for the plantation rather than the town," it has been said. Alone, this little capital of a virtually unbroken rural area represented the urban qualities of Potomac life. In this theater the valley's conflicting ideas rubbed against one another, became smoother and more tolerant, and led to action. Here the political arts were cultivated.

Of the many interesting and talented figures whom Alexandria came to know, who visited its Indian Queen Tavern and drank their toddy at the Bunch of Grapes, none is more appealing than George Mason. He came there often from his home at Gunston Hall, a dozen miles down the river. A true provincial figure, Mason was at home in Alexandria and Fairfax County. He was miserable when as far away as Williamsburg, and his only appearance in national gatherings was in Philadelphia at the Constitutional Convention. Mason's life was narrowly oriented to his home, where he personally attended to the management of his plantation, and the care of his motherless children, and to the three principal circles of life in Alexandria. He was a diligent vestryman, concerned with the program of his parish. As a county justice, he participated for years in local administration. In his capacity as a trustee of Alexandria, for more than two decades he contributed largely to the development of the city. Perhaps it was of such a civic career that his young friend, Thomas Jefferson, was thinking when he made his great plea for a strong ward system in Virginia, for a unit of local government that would command the services of men like Mason and assure a vigorous democracy at a popular level.

Oriented to this city, and to the larger region of the Potomac valley, Mason nevertheless became a spokesman whose words carried throughout the colonies. On July 18, 1774, at the Fairfax County courthouse in Alexandria, he presented a set of resolutions that soon became known as the Fairfax Resolves. They had been drafted, like those of the other Virginia counties, as a means of instructing the Virginia representatives who were shortly to meet with the other colonies to consider the embargo placed by the British government on the port of Boston in retaliation for the dumping of the East India Company's cargoes of tea. In their essentials they formed the basis for the action of the Virginia delegation.

The tone of the Fairfax Resolves was conciliatory and moderate. They did not go to the extremes of similar documents prepared by Patrick Henry of Hanover, or Thomas Jefferson in Albemarle county, or reflect the popular tendency to magnify and multiply grievances to justify the inevitable course of action. Nor did they descend to the conservative admissions of the tobacco planters downriver that England had the right to regulate colonial commerce. Yet they are wholly lacking in equivocation.

The principal question at issue in the Resolves was the legality of the Navigation Acts, and the large body of legislation that carried through the details of the mercantilist scheme, upon which depended Britain's right to regulate colonial commerce and to tax colonial trade; the legislation, in short, by means of which the British government imposed the tobacco trade upon the Potomac and stifled new developments in manufacturing and trade. On this subject the Resolves offer no compromise. Indeed, their whole philosophical structure revolves around the treatment of this issue.

What began as a reasonable extension of the power to regulate commerce, and was tolerated by our ancestors despite abuses, wrote Mason, has now become tyrannical and unjust. This power can only be exercised by the provincial assemblies. Questions of the regulation of commerce and of revenue are inseparable. The right to tax can only be lodged in the people's own representatives. "Taxation and representation are in their nature inseparable." These syllogisms were not developed from reading Locke and other political philosophers; they derived from the fabric of the English common law in support of the dynamics of life on the Potomac.

Should Parliament, as it had increasingly since 1764, continue legislating the details of colonial trade and revenue, Mason continued, the powers of provincial legislatures will become meaningless. Such regulation can only result in tyranny.

At this point the Fairfax Resolves passed to practical matters, of actions to be taken to help the port of Boston in its struggle against the crown, recommending a "firm union" of the colonies through a Congress of colonial representatives, specific measures of economic retaliation, and of preparation for a long embargo on British trade by increasing local production of wheat and wool.

The following year Mason drafted a scheme for a county militia supported by local taxes, a measure that resulted in the creation of the first such body in the colonies. Shortly afterward he addressed this "Fairfax Independent Company," whose buff and blue uniform its commander, George Washington, later adopted as that of the Continental Army.

"We came equals into this world," Mason told his military audience, "and equals shall we go out of it. All men are by nature born equally free and independent." Just as men give up some of their natural rights to create governments capable of ensuring other rights, so it becomes necessary to give some individuals the power to command others. How, he inquired, can these militia officers be prevented from becoming despots? He answered: "The most effectual means that human wisdom hath ever been able to devise, is frequently appealing to the body of the people, to those constituent members from whom authority originated, for their approbation or dissent."

The officers of the Fairfax militia, he concluded, should be selected annually by the members of the company, rotation in office should be encouraged. He observed with satisfaction that the method proposed would "in a few years breed a number of officers."

If a single instance among many were needed to demonstrate how deep Mason's political beliefs were rooted in Potomac soil, his little speech to the Fairfax Independent Company would be chosen. The same spirit animated other declarations of great national importance.

To the themes of the Fairfax Resolves and his speech to the Fairfax Independent Company Mason returned in 1776, when designated to prepare the Virginia Declaration of Rights. By correspondence the colonies had come to agreement on the question of independence. But when faced with the necessity of creating a government of their own they quailed before this responsibility. After believing for some time that the Continental Congress would issue for their guidance "a well digested form of government," they received with some forebodings the Congressional resolution John Adams had introduced, recommending that the colonies adopt such forms of government as they thought expedient. But to Mason, ever the constitutionalist, the news was electrifying. "We are now going upon the most important of all subjects—government!" he wrote May 11, 1776. "We shall, in all probability, have a thousand ridiculous and impracticable proposals, and of course a plan formed of heterogeneous, jarring and unintelligible ingredients." Yet his misgivings were ill-founded. The Virginia Convention proceeded to appoint a drafting committee, which accepted with hardly a change as the basis of the new government the Declaration of Rights drafted by Mason in his library at Gunston.

Commencing in the spirit of the Fairfax Resolves, the Virginia Bill of Rights describes man's natural rights, the creation and dissolution of governments, and the nature of public service. It then describes the division of legislative, executive and judicial powers, observing that members of the first two should "at fixed periods, be reduced to a private station, returned into that body from which they were originally taken." The right of suffrage should extend to "all men having sufficient evidence of permanent, common interest with an attachment to the community." Certain general provisions relating to the courts are detailed, including the right of trial by jury. Freedom of the press is upheld. A preference is expressed for a militia to defend the state, since standing armies in time of peace are dangerous to liberty. Then follows a paragraph on religious freedom, drafted by Patrick Henry, and a conclusion which warned that "no free government, or the blessings of liberty, can be preserved to any people, but by a firm adherence to justice, moderation, temperance, frugality and virtue, and by frequent recurrence to fundamental principles."

With the Virginia Bill of Rights in his hand, the following month in his Philadelphia lodging house, Thomas Jefferson, with his "happy talent of composition," was writing: "We hold these truths to be self-evident, that all men are created equal, that they are endowed by their Creator with certain inalienable Rights, that among these are Life, Liberty and the pursuit of Happiness. That to secure these rights, Governments are instituted among Men, deriving their just powers from the consent of the governed. That whenever any Form of Government becomes destructive of these ends, it is the Right of the People to alter or abolish it, and to institute new Government . . . ." The language was that of Jefferson, but the ideas are those which Mason had written into the Virginia Bill of Rights, the thoughts that stem from a county resolution and a speech to a local militia company. They were the ideas of the Potomac region, of its planters, its merchants, and its frontiersmen.

Mason's genius was his power to generalize. He extracted the golden universal truths from the common clay of particular experiences. His gift was tempered by a profound knowledge and respect of the common law, and the long history of Virginia's colonial government; and by his diligent reading of the political philosophers of the seventeenth-century revolutions in England and the works of John Locke. Drenched, like all of his generation, in classical culture, he was yet alive to the burning issues of the present.

Why he did not play a more prominent part in the Revolutionary drama, his biographers have not clearly explained. He was not at his best in large assemblies. Jefferson found his elocution "neither flowing nor smooth," but acknowledged that "his language was strong, his manner most impressive, and strengthened by a dash of biting cynicism when provocation made it seasonable." His literary style was often nervous and sometimes abrupt, but with a singular precision in statement. His thoughts were arranged with system and care, giving a concise, logical quality to his writings.

This was the man who best represented the Potomac civilization in the Revolutionary period, through whom the region spoke to the new nation and to the world. Surrounded by more illustrious figures, his were the decisive formulations of the Revolutionary struggle.

Occupying a middle ground between a detached philosophical outlook and a deep immersion in practical affairs, George Mason fitted well into the Revolutionary pattern of the Potomac. The essential principle of that carefully balanced movement was suggested many years later by one of its leaders: "Sensible, however, of the importance of unanimity among our constituents, although we often wished to have gone faster, we slackened our pace, that our less ardent colleagues might keep up with us." The impressive measure of unanimity Mason achieved for his public statements, the absence of compromise or amendment, best evidences his skill. It was not dissipated in advancing extreme proposals, and then struggling to secure the adoption of some mangled portion of them, but in a carefully formulated middle course, supported by well-marshaled evidence

IRON FURNACE, ANTIETAM

and irrefutable logic, which had to be accepted or rejected as a whole.

The nicely balanced Potomac region was unequaled as a school for leadership, and for a time Alexandria was its classroom. Nowhere in the colonies was the political art in all its aspects more diligently cultivated than there. Its principal men were eager to lead, and prepared themselves carefully for their chosen roles. When they unleashed a rebellious continent from the mother country, it was commonly accepted that strong men would be required to lead the new republic. Yet Mason, in a retrospective letter to a cousin overseas, was to rail against the overstatement that "this great Revolution had been the work of a faction, of a junto of ambitious men, against the sense of the people of America. On the contrary, nothing has been done without the approbation of the people, who have indeed outrun their leaders, so that no capital measure hath been adopted until they called loudly for it."

5

The events of the Revolution bore down swiftly. But from the time of Lord Dunmore's revengeful expedition up the Potomac, which opened hostilities, until the final surrender on the peninsula at Yorktown, the river was never a theater of war. Occasional raids were attempted in the Potomac and a state of naval defense was maintained. The main events of the war took place far to the north or south of the river.

Driven from his capital at Williamsburg in 1775, Lord Dunmore evolved an ambitious plan that embraced Indian raids on the western settlements of Virginia and Maryland, an insurrection of the slaves, and a pincers attack up and down the Potomac. His western forces, under the command of John Connolly of Pennsylvania, John Smith of Maryland, and Allen Cameron of Virginia, were to march eastward down the river and meet the governor's fleet at Alexandria. He laid plans with General Gage thus to divide the northern and southern colonies along the line of the Potomac, incidentally pillaging the bulging tobacco barns of the lower river and devastating the plantations of Washington, Mason, and the principal rebels. A strongly fortified line was to be established along the river, with headquarters at Alexandria.

The vigilance of John Hanson, who had recently removed up the Potomac from Mulberry Grove to Frederick, and his American Committee of Observation turned the indispensable Connolly and his associates from their objective to a jail in Philadelphia, and Lord Dunmore's ships met such armed resistance everywhere that they soon left the Potomac, contenting themselves with raids in the Chesapeake.

The valley contributed richly with troops. A friendly rivalry distinguished the marches to Boston of Daniel Morgan's riflemen "from the right bank of the Potomac" and of Michael Cresap's company from western Maryland. These were prompt responses of the valley to General Washington's appeal. It does not diminish such exploits, however, to know that the contribution of the valley in feeding and clothing the Continental Army was perhaps of greater importance. That "granary of the Revolution," the protected interior valleys of the upper Potomac, was the source of supply upon which the army could rely, whatever adverse conditions of military interference, failing crops, or governmental bankruptcy were encountered elsewhere.

Perhaps the Revolutionary campaign that was best understood and aroused the greatest enthusiasm in the Potomac country was the expedition of the young redheaded colonel, George Rogers Clark, to clear the British from the Northwest Territory and firmly establish Virginia's claim to those western lands. The famous February march from Kaskaskia to Vincennes, four days of wading waist-deep through the flooded Wabash swamps, that led to the surprise and smothering defeat of the British garrison was the kind of individual exploit that stood out sharply in the muddled course of the war. Another old score had been settled. The fruits of Clark's exploit were tangible, and the victory sweeter because so long cherished.

With the peace, tired veterans straggled home, and soon

the old tendencies were resumed. Again the long procession of Conestoga wagons wound through the upper Potomac to the western gateways. But the war itself had given new opportunities.

6

The Revolution had begun with Virginia and Pennsylvania virtually at war with each other over their conflicting western claims. During the war these differences were compromised, and the boundary question settled. Some agreements had been reached between Maryland and Virginia on their mutual boundary, the Potomac, by which the Virginians secured the right of free navigation on the river, and the Marylanders the fishing and oystering grounds.

The possibility of relative unity among the states in the Confederation could hardly be said to exist, so unequal were their resources. Notably was this true of Maryland and Virginia, the former having no western land reserves and the latter claiming her 400-mile wide swath of land from the Atlantic to the Pacific. As a condition of entering the Confederation, Maryland spoke for all the smaller states and insisted that all western land claims be ceded to the Confederation. Gradually, New York leading, the other states came round to this position, and Virginia finally consented to give up her claims to the western lands north of the Ohio. Maryland accepted this compromise, thus leaving in the possession of the United States a public domain, "a common stock," in Maryland's phrase, "to be parcelled out by Congress into free, convenient, and independent Governments."

Among the other states such disputes were less amicably settled. Jealousy and distrust prevailed. The individual states showed little success in coming to agreement on matters of mutual concern. The collective behavior of their representatives in the Continental Congress was little better. Feebly constituted as the Confederation of States was, it was further weakened by the inability of its members to agree upon the proper course to pursue in the most fundamental national concerns. The state of the continent's defense was absurd; the enormous debt resulting from the war was everywhere admitted a collective responsibility of all the states, yet none of them wished to contribute toward discharging it, and the thought of allowing the central government to raise funds by direct taxes was anathema; a nation seeking to establish an independent economy faced the prospect of not one but a series of independent state economies, each quarreling with the others. Men of vision in many states could clearly see the Confederation rotting away; it was only a question of a short time until disintegration led to dissolution.

Even the wartime agreement between Maryland and Virginia on the regulation of the Potomac left much to be desired. In the years between 1778 and 1784 this had become plain. In the Virginia Assembly, on June 28, 1784, James Madison moved to deal with the matter by appointing four commissioners to meet with representatives of the state of Maryland to provide "concerted regulations between this State and the State of Maryland, touching the jurisdiction and navigation of the river Potomac."

The Virginia commissioners designated were George Mason, who had represented the state in the 1778 negotiations with Maryland; Virginia's attorney general, Edmund Randolph; the young and rising James Madison; and Mason's friend and neighbor, Alexander Henderson of Fairfax County. To meet them Maryland sent its three veteran delegates to the Continental Congress, Thomas Stone, Samuel Chase, and Washington's friend and neighbor, Thomas Johnson; and Major Daniel of St. Thomas, Jenifer, who was living at Retreat and was Maryland's "Intendant of the Revenue," or state treasurer. Of these eight representatives, six were later to be designated to represent their states at the Constitutional Convention.

State business in Virginia appears to have been casually organized. No copy of the legislation appointing and instructing the Virginia commissioners, or even the news of his appointment, had reached George Mason. His first intimation of the conference came a bare two days before the meeting, when he received a letter from the Maryland delegation, saying that they proposed crossing the Potomac south of Gunston Hall and would be pleased to have his company on their journey north to Alexandria where the meeting was to take place. Maryland had set the time and place for the meeting, and had notified the state of Virginia, but no one appears to have thought to notify the Virginia delegates concerned.

In this predicament Mason discovered that his neighbor, General Washington, had a copy of a later resolution of the Virginia Assembly concerning Pennsylvania, from which a general idea of the initial resolution could be patched together. He also discovered that another neighbor, Alexander Henderson, had been appointed a Virginia delegate, and that the resolution authorized "any two" of the commissioners to act. With this hazy background, the Virginia commissioners sat down with the Maryland delegation—Thomas Johnson being absent—in Alexandria on March 20, 1785, in the midst of a raging late-winter storm.

At the end of four days they accepted with alacrity General Washington's invitation to adjourn their meeting to the more comfortable surroundings of Mount Vernon, where another four days concluded their work.

The Maryland commissioners, Mason found, "brought with them the most amicable dispositions, and expressed the greatest desire of forming such a fair and liberal compact, as might prove a lasting cement of friendship between the two States; which we were convinced it is their mutual interest to cultivate." Their first concern, he noticed, was to secure the abandonment of tolls on Maryland vessels passing through the Virginia capes into Chesapeake Bay. Once assured of this, they were ready and willing to make liberal concessions on other points.

Perhaps it was fortunate that the Virginia delegates had not seen the legislation authorizing their activities, for uninhibited by specific directions they proceeded to consider Potomac matters in the broadest terms. They disposed of lighthouses, buoys, and other aids to navigation; of piracies and related questions. The matter of tolls was passed over without controversy. Then the discussion moved on to purely commercial questions.

It was agreed that the two states should adopt a uniform valuation of foreign currencies, and that interest on foreign and domestic bills of exchange should be the same. Mindful of the damage caused during the Revolution along the river and the earlier colonial convoy system, a naval protection agreement was proposed. Then the question of duties on imports and exports, differences in which had already led to extensive smuggling across the river, was taken up and agreement reached on uniform schedules of tariffs.

The fisheries of the Potomac were becoming of considerable commercial importance, and an even more significant source of income to the waning tobacco plantations along the river. At the Alexandria conference it was observed that "The idea of the right of fishing on both shores of the Potomac river is one the Marylanders are not fond of parting with." Since the legal boundary of the two states was not the middle but the high-water mark on the south side of the river, Maryland's rights to the entire river were obvious. The concession on fishing rights in the Potomac was lightly regarded by the Virginia commissioners, seeking Maryland's support to a longer-range objective.

The two states divided by the Potomac had much in common; but they also had divergent interests. Roughly, the river lay in a diagonal position between the parallels of latitude that bounded Virginia on the south and Maryland on the north. Along the Atlantic, Virginia's coast line was small; but toward the west her boundaries were immense. The reverse was true with Maryland, which had a negligible western frontier and no western land interests. Under the influence of such men as Washington and Mason, Virginia's interest in the upper river, and the gateway to her western domain, was paramount.

A suggestion of this interest is found in Virginia's belated addition of Pennsylvania to the states concerned with the development of the Potomac route to the west. In its resolution the Virginia Assembly also specified action on the removal of obstructions to navigation on the upper Potomac and the construction of a road "from the head of navigation to the waters running into the Ohio." Considering George Washington's lasting interest in this route across the mountains, there can be little doubt that this complex question was deeply considered at the Mount Vernon Convention—as the meeting became known.

However great the measure of unanimity between the two states that bordered the Potomac—and in 1785 this unanimity was impressive—the questions to be settled affected other states as well. Stemming from the Mount Vernon Convention, the commissioners' report recommended further meetings with Pennsylvania and Delaware, implying the notion of a regional bloc of states bordering Chesapeake Bay, and its tributaries and the Delaware. Maryland issued an invitation to these states to meet at Annapolis the following year.

But the muttering thunder of events on the national scene made it clear that not even regional co-operation would suffice. Upon receiving Maryland's invitation, Virginia sent out a call to all the states "to meet such commissioners as may be appointed by the other States of the Union . . . to take into consideration the trade of the United States; to examine the relative situation and trade of the said States; to consider how far a uniform system of commercial regulations may be necessary to their common interests and their permanent harmony. . . ."

The preoccupations of the "river of traders" were becoming of national importance. Yet only five states sent representatives, and even Maryland, which had conceived the idea of the conference, was unable officially to participate in the Annapolis meeting because its Senate had failed to consent to the enabling resolution. After three days, the Annapolis Convention had drafted a brief but sufficient report, recommending that "commissioners meet at Philadelphia, on the second Monday in May next, to take into consideration the situation of the United States, to devise such further provisions as shall appear adequate to them necessary to render the constitution of the Federal Government adequate to the exigencies of the Union." Upon receiving this report, Virginia hesitated no longer. She issued the invitation to the states whose representatives met as the Constitutional Convention.

The delegates who convened at Philadelphia saw the political art long cultivated along the Potomac at its prime. The issues at Philadelphia were old Potomac conflicts. A region renowned for its liberalism—they called it Whiggery—it was also a region that longed for sound national credit, solid defenses, a wide trade in staples, public roads and canals—and a steady increase in land values. Proud and jealous of its regional invididuality, it also wished to be part of a larger world.

To give the government of the United States the authority to raise money from direct taxes and to permit it to regulate commerce and trade—these were the indispensable concessions if a national government was to be created. But how to hedge these delegations of authority to preserve the freedom of the individual states was the seemingly hopeless question the Convention had to face. The twelve delegates from Maryland and Virginia could be forgiven some measure of optimism. Experience had shown them that agreement was possible. They were as ready to compromise their demands with other sections as they had been to compromise their own conflicts. As that most distinguished student of the Constitutional Convention, Professor Max Farrand, has pointed out, whatever the delegates' views might be from study or classical precepts, "when it came to concrete action they relied almost entirely on what they themselves had seen and done." The experience of living along the Potomac was of a high order when the great compromises of 1787 were being considered.

In a convention taken so lightly by its members that almost one-fourth failed entirely to appear, and larger numbers were absent for most of its deliberations, the Virginia and Maryland delegations were notable in the regularity of their attendance. Of the twenty-nine who opened the meeting, Virginia delegates alone numbered nearly one-fourth. The Virginia delegation actually arrived well in advance of the opening, and met together each afternoon to discuss the issues that would arise in order to present a closely reasoned and consistent front in the Convention. The fruit of this preparation was the so-called "Virginia plan," a 14-point draft, offered with the explanation that since the Convention had come together at Virginia's invitation, they felt obliged to advance a fuller proposal. It was accepted on the opening day as the working basis for subsequent discussions.

Through the summer months of debates and the tortuous compromises, the delegations from Maryland and Virginia, led by the men of Potomac—by Washington, Mason, Madison, Jenifer, and the others—showed amply their determination to create a union of states. Over and again, when the Convention seemed hopelessly deadlocked, they came up with formulas and compromises that made it possible to continue. Their struggles with the other delegates were duplicates of earlier struggles among themselves. From their ancient experiences they resurrected devices of government that would work, writing home to ask for copies of laws or contracts, with directions to look for them in this desk drawer or that pigeonhole. Their arguments were convincing. With broad strokes they hewed out the foundation beams for the new republic.

They were not successful in all cases. They were not satisfied in all cases. Indeed, the later addition of the first ten amendments to the Constitution, the Bill of Rights, showed that the nation rectified earlier errors of the Convention. But when the delegates on September 17, 1787, signed the engrossed copy of the Constitution, and repaired for dinner to the City Tavern in Philadelphia, they could recognize the imprimatur of "the middle states" on their four months' work.

In Alexandria the citizens were overjoyed, but their George Mason was sunk in gloom and had refused to sign the great charter. The document still lacked the bill of rights he considered essential, and in the failure to safeguard adequately the power to regulate commerce he believed he could detect a means by which "a few rich merchants in Philadelphia, New York and Boston" could monopolize staple crops "and reduce their value by as much as fifty per cent." He doubly abhorred the compromise that had led to this provision, by which the importation of slaves was to continue for twenty years. Events along the Potomac were to show how prescient his judgment was.

What Mason deduced from a long life, in good part devoted to the public service, he summarized in his will: "I recommend it to my sons from my own experience in life, to prefer the happiness of independence and a private station to the troubles and vexation of publick business, but if either their own inclinations or the necessity of the times should engage them in public affairs, I charge them on a father's blessing never to let the motives of private interests or ambition induce them to betray, nor the terrors of poverty and disgrace, or the fear of danger or of death, deter them from asserting the liberty of their country and endeavoring to transmit to their posterity those sacred rights to which themselves were born." Thus the old revolutionist.

7

The Revolution along the Potomac, in both Maryland and Virginia, had seen no violent overthrow of established government or political officers. Governor Eden in Maryland was allowed to depart in peace; and even in Virginia the very inferior Lord Dunmore boarded his British warship unmolested. In this homogeneous region, with its remarkable stability, the dead skin of the crown and proprietary governments slipped easily away, leaving beneath it the firm and healthy body that had been growing since the first settlers arrived in St. Mary's City and the Northern Neck. Save for the royal governors and a handful of their followers, or an occasional stray loyalist, there were few changes in the familiar faces at the seats of government. Even violence against the Tories was infrequent.

The natural leaders of the valley did not have to revolt to "take hold"—in Professor Barker's descriptive phrase—of their

PIEDMONT LANDSCAPE

FROM SUGARLOAF MOUNTAIN

governments. The Potomac saw no great mass uprising, no violent disruption. Little penetration from outside occurred. Only one year before the Declaration of Independence the very idea of a separate nation would have been found strange and intolerable in Tidewater. The pattern of change was found in other institutions. The recodification of provincial law was carried through by lawyers who had risen to prominence under the proprietary governments before the Revolution. The reform of the land question was in the hands of the large landowners. Even the reform of the established church came from within and was controlled by the church leaders themselves. Institutions were not overthrown; they were subtly modified by the people who knew them best, who had lived longest with them.

When the time came to build new governments in the states of Maryland and Virginia, a marked conservatism was noticed. Here, as in other states, consonant with the idea of voters having a substantial interest in the community, the franchise was carefully restricted to men of property. In Maryland the fifteen members of the state Senate were chosen by an electoral college of forty members for a five-year term, almost without regard to residence, making it "the most aristocratic body known to the Union." Membership in the lower house was more circumscribed with property qualifications than in any other state. In Virginia "the fencing of society by the institution of government" took place on somewhat more moderate lines. Yet the suffrage was limited to freeholders who held title to 100 acres of unimproved or 25 acres of improved land. A system of indirect election of senators was invoked. And an inequality of representation among the sections of the state was adopted hardly more liberal than that which had existed in colonial days.

For all the brave talk about the rights of man, some things remained to be done before they were given full effect.

SEVEN

"SMOOTH BLUE HORIZONS"

OVERLOOKING HARPERS FERRY FROM THE FAMOUS ROCK THAT still bears his name, Thomas Jefferson gazed eastward through the gap in the Blue Ridge. The description of this scene prepared for his *Notes on the State of Virginia* has passed into literature. During the years when there was a taste for the picturesque, it even created a small literature of its own. For a time it was memorized and recited by school children.

"The passage of the Potomac through the Blue Ridge," Mr. Jefferson wrote, "is perhaps one of the most stupendous scenes in nature. You stand on a very high point of land. On your right comes up the Shenandoah, having ranged along the foot of the mountain a hundred miles to seek a vent. On your left approaches the Potomac, in quest of a passage also. In the moment of their junction they rush together against the mountain, rend it asunder, and pass off to the sea.

"The first glance of this scene hurries our senses into the opinion, that this earth had been created in time, that the mountains were formed first, that the rivers began to flow afterwards, that in this place particularly they have been dammed up by the Blue Ridge of mountains, and have formed an ocean which filled the whole valley; that continuing to rise they have at length broken over at this spot, and have torn the mountain down from its summit to its base. The piles of rock on each hand, but particularly on the Shenandoah, the evident marks of the disrupture and avulsion from their beds by the most powerful agents of nature, corroborate the impression.

"But the distant finishing which nature has given to the picture, is of a very different character. It is a true contrast to the foreground. It is as placid and delightful, as that is wild and tremendous. For the mountain being cloven asunder, she presents to your eye, through the cleft, a small catch of smooth blue horizon, at an infinite distance in the plain country, inviting you, as it were from the riot and tumult roaring around, to pass through the breach and participate of the calm below. Here the eye ultimately composes itself; and that way too the road happens actually to lead.

"You cross the Potomac above the junction, pass along its side through the base of the mountain for three miles, its terrible precipices hanging in fragments over you, and within about twenty miles reach Fredericktown, and the fine country round that. This scene is worth a voyage across the Atlantic."

Shrewd as was Jefferson's inspection, and bold as his conjectures were, he saw only a small fragment of the geologic picture. His belief in cataclysms may have been erroneous, but his taste for scenery was above reproach. The tangle of raw nature at Harpers Ferry and the luscious wheatlands in the valleys beyond still form a startling contrast. Today we know both scenes to be part of the related drama of this restless earth. More important, the decisions made by nature in ages past, its teeming life and its lifeless rocks, remain controlling in the lives of the Potomac people. The resources of the earth, and the shape taken by the land over geologic time, have framed the life of the Potomac above the falls.

In the upper valley, the limestone, the iron ore, and the coal, each individually proved its usefulness, but it was their presence in a small area that made them together important to the iron furnaces and the later industries that rose upon this foundation. As it turned out, the iron deposits were of relatively low quality and quantity. The coal, too, was gradually diminished. These resources of the valley were unable to sustain the development they had originally encouraged. As competition from richer producing areas appeared, they waned. They did, however, help distinguish the life of the upper valley from that of Tidewater, and for nearly a century supported the larger share of its prosperity.

The shape of the land had its influence, too, not least the drainage system. Although the geological history of the Potomac region covers a long period, even as geologists reckon time, the final shape of the river is of relatively recent origin. As the leisurely young Potomac began to flow eastward to the ocean, meandering across a gently sloping peneplain that stretched from the mountains to the sea, it cut its channel in the distinctive sweeping curls which can be seen today as the river crosses the mountains and the Great Valley. As an uplifting process gradually took place, the river cut its way deeper and deeper into its originally formed channel. As Dr. Ernst Cloos of the Johns Hopkins University geology department succinctly puts it, "When the land rose, the river cut down into its bed slowly, and then couldn't get out any more."

As matters stand today, the Potomac *flows across the mountains* that have risen to their present height since the river was first established in its bed! A further paradox, as it crosses the Great Valley, is that the valley floor is hundreds of feet *higher* than the level of the Potomac river.

The Potomac is the oldest thing one sees in the landscape. Much of the eloquence of the exposed river bed as it is seen in the immense gray rocks at Great Falls expresses this.

Of the greatest practical importance was the circumstance that the river had created a natural waterlevel route deep into the heart of the mountains that could easily be followed by men in their search for a route to the west.

A further consequence of the work of nature was the honeycombed landscape of the upper Potomac that lent itself to the development of numerous widely different communities, that could exist without friction in close proximity to each other. Within these mountain valleys the streams run north and south, tributaries of the few like the Potomac that persevered in making their way eastward to the ocean. The upland tributaries are fed from the ridges by smaller creeks that run in at right angles, and give a trellised effect on the map. To either side of the Potomac, as it threads its way among the mountains, deep in the channel it cut long ago, are long, narrow tributary valleys, some eight or ten times as long as they are wide, isolated in themselves, and within each of which travel is still difficult. In its upper reaches, the Potomac itself, at least its North Branch, turns south to parallel its tributaries.

As the earth rose, the river narrowed and cut deeper into its channel. High above the river can be seen today the gravel-covered terraces where once lay its older beds. Remnants of these descending plateaus prominently mark the Potomac valley from Washington to Cumberland. Below, the river flows in the lower channel it has cut over the years.

The Potomac commences at the Fairfax Stone in the valley-and-ridge country of the Appalachian highlands in western Maryland and northern West Virginia, making its way north and west to Massanutten Mountain. Here it crosses the Valley —variously called the Cumberland in Maryland and the Shenandoah in Virginia—to the Blue Ridge and the minor valleys that lie eastward in its course. At Point of Rocks the river penetrates the last mountain barrier, the Catoctins, and emerges to the rolling Piedmont country. Just above Washington it pours turbulently over Great and Little Falls, and feels the tidal pulse of the ocean.

The transit of the Potomac through these four distinct landscapes: corrugated valley-and-ridge country, the Great Valley, the Piedmont, and Tidewater, gives it variety. The river itself gives it unity. In each province a special life exists, miniature civilizations more closely allied with other similar civilizations north and south of it than with the dissimilar civilizations above or below it on the river. As the river flows through the four provinces it changes its personality: now shrill and clear as it threads the narrow upper valleys, then more sedately, entrenched in old meanders, as it crosses the Great Valley, more fully and yet more slowly as it cuts through the rolling Piedmont floor, and then with barely perceptible movement after it falls to join the brackish tidal waters that seem but the greatest arm of that "sunken meadow," Chesapeake Bay.

2

From his Piedmont farm near Leesburg, on the land Thomas Lee of the Ohio Company had purchased in Loudoun County, came John Binns, traveling down the wheat road to Alexandria. The Revolution was just over in that year, 1784, and a yeasty atmosphere filled the Potomac with new possibilities. Sloping from the Hog Back on his right, Binns noted the thrifty farms of the Quakers and the Pennsylvania Germans with their tight fences of limestone slabs, their big barns, and their houses with the wooden sundial on the south wall lest they forget the fleeting hours. Farther along the road as it fell off to the south he observed that the landscape changed. Here one saw porticoed houses and slave quarters, rail fences, and the beginnings of tobacco fields. With the eye of a practical farmer, he contrasted the land. It was the same land, but treated differently. He noted the muddy streams with their red charge of silt, and traced their color to the clay in bare cornfields. His thoughts went back to his own hilly acres where every spring the soil, cultivated to a depth of two or three inches with a single-horse "trowel hoe," washed steadily away in the beating rains each April and May.

The insistent cropping to wheat was exhausting the soil even more quickly than had the monoculture of tobacco in the lower river. Part of the country was still thickly wooded, but already the farms had a shaggy look. In the fields the grey stone shelved from the sod. Everyone had a little lot where they made a cash crop of coarse, cheap tobacco, and what little manure came from the few cattle was spread there or in the kitchen garden. Fields with old stumps still standing were either left untended to grow up in sedge and briers, or so many cattle were turned in that they ate the heart out of the grass. The idea of rotating crops was considered of trifling importance, and it was found convenient to plant the same fields continually to corn or wheat. As in all the years since the Potomac was settled, the ax and the hoe remained the favorite implements; plows were just beginning to appear.

Every fall saw gangs of men in the wheat patches, cutting the grain with reaphooks. They carted it to the threshing floor, where it was beaten with flails or trodden out by the feet of animals. The chaff was separated by throwing it against barn doors on windy days, or by using a hand screen. Only the high prices that prevailed after the Revolution could long support an agriculture so primitive, so wasteful, and so destructive.

Binns was a young farmer in his middle twenties, and he had been born in Loudoun. All his family were there, and his father had already served thirty years as clerk of the county. He had become a lieutenant in the county militia during the Revolutionary War, almost as soon as he was old enough to hold the commission. His roots and his ties were all in Loudoun County. But how could a man stay in a place where the people were steadily abandoning their farms and moving west to richer lands? The wasted state of his own farm and its steadily decreasing yields and values seemed to be reflected in the farms he passed along the road. What was it the sundial said?

> The noiseless foot of TIME steals softly by,
> And ere we think of MANHOOD age draws nigh.

As he mused on the problem, he recollected what people were saying about Israel Janney, a Quaker farmer in the neighborhood. An energetic and bustling family, the Janneys had come down from Pennsylvania and settled in Loudoun. On one of his trips back to Chester County, Israel Janney had secured some crushed limestone from a friend and brought it back home in his saddlebags. He spread it on some oats, and they did remarkably well. But then Binns remembered that Janney was an eccentric farmer who planted clover and left much of his land to cattle. He pondered the matter further as he cantered along.

In Alexandria, Binns fell in with the captain of an English vessel whose hold was ballasted with chalk stone. He may have been told something of the handful of farmers in Britain who were spreading chalk on their tired fields. Like Israel Janney, he loaded his saddlebags with "two small stones, weighing about 15 lbs." and brought them back to Loudoun County. A man had to try something or he would certainly lose his farm. "The noiseless foot of time" would not wait. Binns pounded his chalk to powder with a sledge hammer, sifted it through a hair sieve, and giving it to a tenant instructed him to scatter it on four or five hundred hills of Indian corn.

Yet his mind was of an experimental bent. He kept records, and compared the growth of that cornfield with others that had not been treated, and he was gratified with the prompt increase in yield. He tried the experiment again on the succulent native blue grass—they called it spear grass then—and on wheat, rye and barley. He got samples of all the different kinds of limestone coming to Alexandria. Some came from as far off as Nova Scotia. He tried pounding up the native limestone. He crushed it coarse and then fine, and studied the results. He even bought a ton of Israel Janney's lime.

All this was interesting and productive enough. His farm lands were steadily improving, and each year as he studied his records he knew that some new thing had been learned. But the thing that impressed him most was the response of clover to the lime. The clover seed that Israel Janney was selling to the farmers roundabout, a quart at a time, met with an indifferent reception. People tried it, the way they tried poultices or water cures or anything else for a serious illness. But on the lime-treated fields of the Binns farm, the clover was superb. He was so proud of it that he named the farm "Clover Hill."

After eight years of constant experimenting, his soil had steadily improved. The yields were phenomenal. He raised twice the corn and four times the wheat his neighbors succeeded in growing. Binns noted that "lands which at present are so light and subject to be washed in gullies, will become stiff and prevent washing away of the soil," when lime was applied. People began to visit Clover Hill to see with their own eyes what had been accomplished.

In 1792 Binns moved to another farm, worn out like most of the farms in the county, and recommenced his program of soil improvement. By now he felt he knew the important things: limestone, clover, and cattle—and deep plowing. Gradually the worn-out lands he had acquired were brought back to a state of high cultivation.

This convinced him. In 1797 he began to buy other farms, to rehabilitate them, and to sell them to the steady stream of customers who passed through Loudoun County on their way west. It is easy to imagine with what enthusiasm these immigrants, recently arrived in the port of Alexandria, or regretfully leaving their Tidewater farms, abandoned their reluctant departure for the hard life of the frontier to settle in thriving Loudoun County. Binns became more and more concerned with the profitable traffic in farms, but not to the extent that he forgot his interest in the soil.

He built a mill where the limestone could be ground by water power, and gave away free samples. He redoubled his efforts to spread the gospel of lime, clover, and grazing. The tired droves of cattle and hogs by the thousand on their way from the Ohio country to Baltimore, Philadelphia, and the eastern markets picked up when they came to Loudoun County, and fattened quickly on the succulent grasses. Such Piedmont feeder lots became an enterprise. The people of Loudoun had stopped emigrating, and John Binns was a rich man.

In 1803 his missionary zeal led him to write a book, published across the river in Frederick by John Colvin, fiery editor of the *Republican Advocate. A Treatise on Practical Farming* he called it. It sold for fifty cents, and told people everything they wanted to know about "the Loudoun system." Binns's pamphlet, he apologized in the preface, was "not written in a scholastic stile." It was forthright talk to dirt farmers like himself, to men who were murdering the land because they knew no better.

There was not much generalizing in Binns's book, although

he added some essays on agriculture in a second edition. It was a simple, autobiographical account of his experience. In his anecdotes you could fairly smell the Potomac country, and its references were to the hillsides of Loudoun, the Catoctin Mountains, the islands in the Potomac near Leesburg; to the German farmers north of Clover Hill, and the Quakers to the west; to Alexandria, Frederick, Leesburg, and the other towns of the valley.

Binns's publisher, a crusading Jeffersonian in the heart of the Monocacy wheat country, well understood the importance of what Binns was saying. It was fundamental to Jeffersonianism. He may even have helped him say it. Colvin pushed the book, printing as many copies as he could in the face of the limited supply of paper from the local mills along Tuscarora Creek. He filled the columns of the *Advocate* with testimonials to "the Loudoun system." He printed letters signed with distinguished Potomac names: Brent, Spotswood, Washington—descendants of men who had settled the river farms, whose stake was in the valley. He sent off complimentary copies to Thomas Jefferson, and the president in Washington took time from Napoleon and the state of affairs in Europe to send copies to members of the English Board of Agriculture.

To Sir John Sinclair, Jefferson wrote that he was sending, "The enclosed pamphlet on gypsum by a Mr. Binns, a plain farmer, who understands handling his plough better than his pen. He is somewhat of an enthusiast in the use of this manure; but he had a right to be so. The result of his husbandry proves his confidence in it will be found, for from being poor it has made him rich. The county of Loudoun in which he lives, exhausted & wasted by bad husbandry, has, from his example, become the most productive one in Virginia; and its lands, from being the lowest, sell at the highest prices. These facts speak more strongly for his pamphlet than a better arrangement & more polished phrases would have done."

The president wrote William Strickland in the same vein, adding that before the introduction of "the Loudoun system" people had been "going Southwardly in quest of better lands," but that "Binns's success has stopped that emigration."

The "Loudoun system" of lime, deep plowing, clover, and grazing became widely adopted. For nearly a century it was the principal system of agricultural reform. Across the river in Maryland, some handy-andy invented a lime spreader. Lime was sold by the ton, clover by the sack, and the valley filled with cattle. It saved the wheat country of the Piedmont. Within a few years after its inauguration, Binns reported, "there are some in the settlement that will be obliged to desist from threshing, being unable to find room in the mills, or yet deposit any more in their granaries." The mills were flooded with wheat; the granaries of the Potomac bulged with the grain. The itinerant painter of tavern signs and coach panels could be forgiven his enthusiasm for such scenes as "The Peaceful Kingdom" and "The Land of Plenty."

The buying and selling of farms in Loudoun County went on briskly and the farm work itself continued, until at his death in 1813 Binns was able to free all his slaves and provide handsomely for his widow. Ten years later his rival, Janney, died. Between the two of them they had saved Loudoun County, and put the wheatlands of the Piedmont on a solid basis that lasted a century, until new crops took the place of grain. It would be profitless to divide the credit between them, and neither was fully responsible for the great revolution in Piedmont agriculture. Hazy experiments in the use of lime were going on in many places. In Leipzig the German farmers were trying it; in England experiments were under way; in Pennsylvania Judge Richard Peters and Jacob Barge were using lime; even in the Tidewater farms of the Potomac, as in the Hudson valley, the imported chalk from Britain was spread on the fields, along with the closely allied marl.

The accomplishment of the Potomac was the "Loudoun system," and the practical experience that made it plainly workable, and Binns's pamphlet and his energy in demonstrating and publicizing the new methods. Hardly "a village Liebig," as has been suggested, this agricultural publicist was perhaps more an American edition of his English contemporary, Robert Bakewell, the livestock breeder whose celebrated farmhouse was visited by royalty to examine the pickled legs of mutton on the kitchen wall, that startlingly illustrated the size of bone and quality of fat Bakewell was endeavoring to breed. Binns's demonstration was the three-field system of crop rotation, where the lime worked wonders with the clover, the clover fed the manure-producing cattle and fixed the nitrogen of the soil, and the long-rooted wheat flourished with its specific need supplied. So great were the increases in yield that, with all the difficulties of the Hessian fly and the inefficiency of an agriculture restricted in horsepower and with primitive tools, the wheat culture flourished. Into this happy system the Potomac farmers plunged, secure for once in a balanced agriculture, and for fifty years or more wheat was king in the region.

3

The German peasants with their small grain culture had gradually broken down the exclusive domination in the Potomac of the pioneer corn culture, taken over almost intact from the Indians by the first settlers. The method of killing forest trees by girdling to make cornfields, the corn itself, the planting in hills and cultivating with wooden implements, the husking bees, the primitive wooden mortar and pestle for grinding, the hominy and the johnny cake (even the very words)—all these wasteful and primitive parts of the Indian maize culture were adopted intact by the English settlers of the lower river. Somewhat improved over fifteen decades, in essence it remained the same. Now this system was challenged by a new one. The maize culture, adapted to the nomadic life of the Indians, passed westward with the white nomads and hunters of the frontier. The coming of wheat, and a diversified agriculture that included cattle, immeasurably multiplied the attachment to a fixed place. The social world of the farmer was replacing that of the hunter and speculator.

Along with the cultivation of small grains, the peasants brought with them an habituation to another new world crop, greatly improved in Europe since its first introduction there—

the white potato. With them also came hardy orchard trees, for the first time supplying fruit for the table of the modest settler who could not afford the luxury of a greenhouse or a gardener, and linked with it the characteristic Piedmont neighborhood event of the apple-butter boiling. Flax and hemp also were part of the diversified and self-sufficient peasant agriculture, supplying the needed fibers for clothing, sacks, and other essential purposes. Oats and rye were supplementary grains.

But increasingly wheat dominated this scene as the cash crop. When the first quarter of the nineteenth century had passed, Maryland and Virginia were growing more than half the wheat produced on this continent, and their ascendancy continued to the eve of the Civil War.

Drawn into the orbit of the upper river's expanding wheat boom, even the tobacco plantations grew wheat. On what had once been the "lusty soyle" of Tidewater, led on by the example of Councilor Carter and the early agrarian reformers, the grain occupied an increasingly large position. Yet the hour was late, and the "galled and butchered lands" made only poor crops. At Nanjemoy, in 1774, the traveler Nicholas Cresswell "went to see them reap Wheat. The greatest slovens I ever saw, believe that one-fourth part is left in the Field uncut." He reckoned the grain indifferent and the crop light, only seven bushels to the acre. The initiative of Tidewater had waned, and their response to the new agriculture was unsteady.

With frantic experimentation, stolid indifference, or stoic grace that region steadily spiraled downward. At Hooe's Ferry, a generation after the Revolution, Isaac Weld sketched a dreary landscape of worn-out tobacco fields growing up in broomsedge, pine, and cedars, empty houses and cabins in ruins. His contemporary, the Duc de La Rochefoucauld-Liancourt, journeyed through the entire Potomac valley, and his dismal observations on Tidewater amply confirmed Weld's impression. By 1834 one could read in the *Farmer's Register* the report of a visit to the once-model plantation of Mount Vernon, describing the "ruins of capacious barns, and long extended hedges," a scene the writer summed up with the remark, "a more widespread and perfect agricultural ruin could not be imagined."

During the years of decline there was no lack of hardy individuals, of clear vision, and of firmly expressed opinions, like those enthusiasts of deep plowing, Robert Russell, of Loudoun, and Thomas Moore, of Montgomery County. The new agricultural press was crammed to the rules with their writings, advocating the Loudoun system, terracing, contour plowing, manure, rotation, and marl. In 1803 John Taylor of Caroline, in a series of letters to a Georgetown paper, worked out the rationale of agricultural reform he was to publish a decade later as the vastly influential *Arator*, a book that exercised the widest influence throughout the plantation world, and led ultimately to political views of even greater importance. Yet even Taylor's influence was to be least in the lower Potomac country.

By the time Taylor's *Arator* had appeared, however, it was clear that the seat of progressive agriculture was to be found in the Piedmont and valley districts. In Maryland it was the valleys of the Monocacy and the Conococheague, and in Virginia the counties of Loudoun and Fauquier, that delighted the eyes of foreign travelers. Here, in the Piedmont and the valleys, ringed with mountains so blue with haze that only the bold profiles marked the horizon, the most seasoned observers felt they could see into the future of American democracy.

The Potomac civilization bent itself toward wheat; the crop colored its thinking and its activities. From flax it made fine linen sacks for flour, some of which have lasted for seventy-five years, with their gay blue borders and proudly stenciled names. The hemp made coarser sacks for grain and the indispensable twine for binding the sheaves. From the forest came white oak to fashion flour barrels, and the Conestoga wagons and canalboats that hauled the wheat to mill and the flour to market. Taverns were named the Sheaf of Wheat or the Plow. Merchants traded at the Sign of the Golden Sheaf. Towns acquired characteristic names like Wheatlands. Benjamin Latrobe, who knew the Piedmont as well as he did the Tidewater, used wheat as a motif in his embellishment of the Capitol in Washington. The needs and possibilities of the new crop filled the imaginations of men.

The growing of wheat brought a profound change in the national diet. Even in far-off New England valleys the old brown breads, the rye loaves, and the Indian meal breads began to disappear. The poorest families, even on farms, insisted on white loaves of wheat bread as their staple. The self-sufficiency of the New England towns began to crumble, and her farmers moved increasingly from their rugged climate and glaciated soils to the west or to the growing centers of manufacturing. But in the wheat regions, in the Genesee, in Pennsylvania, and in the Potomac, stability and prosperity were the rule.

The old common white wheat long continued to be grown, for small as it was it made a fine flour in demand for pastry. But minor strains and varieties appeared. A forward white wheat, the Sicilian, was introduced by the end of the eighteenth century. Well before the tumultuous repeal of the corn laws in England, greatly expanding the world market for wheat, the soft red winter wheat was introduced, to be followed closely by the hard red spring and hard red winter wheats, both suited to bread flour. The significance of the introduction and development of new seeds is beyond question, but the origin of the new varieties is clouded in obscurity. Here and there a clear patch provides a glimpse of the children of Mennonite farmers in the Crimea hand-picking the red grains of Russian wheat to send to cousins in Pennsylvania and Maryland. Then the clouds close in again.

The improvement of wheat in all its complexity, from the selection of the seed to the cultivation of the grain, is suggested by the steadily reducing amounts of it needed to make a crop. On the eve of the Revolution, it was estimated that a farmer generally needed two bushels of wheat to sow an acre. But by the end of the century, within hardly a generation, this had been reduced to half a bushel. A few wealthy planters, who generally led in the adoption of new machinery, imported the wheat drill that Jethro Tull had invented, but for many years the seed was sown by hand. Until the coming of a better agricultural press, and the agricultural societies and fairs, the improvement of wheat was limited to a small circle of neighbors, each collecting his own seed, and a few exchanging it with other farmers. It was the mark of a thrifty Negro farm hand in Frederick County that he carefully picked the best of the wheat left uncut in the corners of fields, and hand-sieved it to sell for seed. Steadily, in large ways and small, anonymously, the wheat improved. The work of generations, and of thousands of people laboring like ants, did it.

The history of improved wheat resembled the slow improvement of cattle, which all during the colonial period had been scrubby little black or dun cows that gave little better than a quart of milk daily. One who seems to have done better than the rest was named "Five Pints." They were casually fed, often browsed in the woods the twelve months around, and many farmers would not milk cows during the winter because they thought it would weaken and kill them. Or it was like the improvement of the scrubby pack horses of the pioneer days to stout draft animals and fleet riding horses. Yet gradually and obscurely the livestock improved, partly from imported stock and from native selection and breeding, but perhaps more through better understanding and care.

The trace of agricultural invention is clearer. Iron implements came on swiftly in the atmosphere of freedom that followed independence. Until then plows were generally made of wood, and in a single day a skillful plowmaker could produce one, fashioning the beam, post, handles, moldboard, and landside from a well-selected tree. Yet the light wooden plow barely scratched the surface of the land. The crude harrow, also made of wood, and such hand tools as scythes for hay, sickles for grain, and hoes completed the farmer's equipment.

Now came the golden age of the plow. The English agriculturist, Arthur Young, sent Washington an iron plow—which disappointed because it would not cut a furrow six inches deep, as Young had claimed. Jefferson, while in Paris, and doubtless under the stimulation of new Dutch and English plows he had seen, worked out the designs for a moldboard plow that could break the heavy soil of the Piedmont. These indications heralded the great change, which came around the turn of the century. The Newbold plow, for years a standard design, was patented in 1797. Jefferson published his scientific views on plowing in 1798. The Small plow, based partly on Jefferson's ideas, came in 1802. All these plows were of cast iron, and as the small iron mills increased at the forest's edge along the foothills, they were locally made. A farmer could go to a casting shed with a new idea, and come away that evening with a plow that met his specifications. Experimentation grew, and progress became steady. The Potomac's share was large, and its leadership was marked.

The cast-iron plows, numerous and specialized as they were, were still clumsy in design, and farmers disliked them because they were brittle. When a plow broke, it had to be discarded and a new one purchased, often at the most pressing season of the year. Rolled-iron plows, from local smithies, superseded them. The advance of industry went hand in hand with agriculture. By 1819 Jethro Wood, in the Genesee country, had produced his three-piece plow, with standardized, interchangeable parts, and within twenty years the steel plows of Lane and Deere were being made.

To all this progress the Piedmont farmer brought a conservative skepticism; he changed slowly. Many claimed the improved wooden moldboard plow did more work than the newer cast-iron plows—and they were probably right. The trowel hoe and the fluke hoe persisted, limited in their depth of cultivation to a bare three inches. Even the new plows were drawn by only one horse, and cut but little deeper.

In their farm workshop in the Shenandoah country, the Scotch-Irish John McCormick and his son Cyrus were busily making plows. The 1816 McCormick plow was so great an improvement over the commonly used Newbold plow that the floods of orders caused the McCormicks to manufacture it in Leesburg and Alexandria, as well as at their factory in Auburn. The French Academy examined the McCormick plow, and pronounced it unequaled.

Their minds filled with the panorama of the wheat country, the McCormicks invented new types of plows, a threshing machine, a water-driven hemp breaker, and finally a mechanical reaper that cut vertically but was not wholly successful. In the stream of new contraptions that engaged the wheat farmer's fancy, the reaper had a special place. It was all very well to have horse-drawn mechanical seeders, horse rakes, threshing machines, straw cutters, and other machinery; but when the heavy season of the year came, and the wheat was golden ripe, it could not be left standing in the fields. To cut it with cradles was only a little improvement over the scythe; a good man could not cut an acre a day. Before the crop was cut and the sheaves gathered, rain and storms were likely to come and spoil the crop. Manpower was scarce, and especially so at harvesttime when everyone was busy with his own crop.

To break this dilemma continued to intrigue Cyrus McCormick, as it did many other men, long after his father's death. Finally, in 1831, he invented a successful reaper that cut horizontally, and commenced to manufacture it in Rockbridge

County. McCormick's reaper fought it out with Obed Hussey's, and for a time the choice between the two was very close. Competitive demonstrations were held on many valley farms. Either would cut as much grain as five men with cradles, and five more behind them with rakes, and could harvest fifteen or twenty acres a day.

4

The advance of wheat cultivation was followed hard by the construction of gristmills and flour mills. Even the old hominy block, with its wooden pestle, taken from the Indians, was supplanted by stone hand mills, and finally by the water-driven tub mill. Along every possible stream where water power could be tapped, a swarm of small mills arose, as alike as grains of wheat in a sack.

The stream was dammed to create a millpond that could be tapped when the mill was used. The gate opened, water poured into the millrace, and turned the big wheel. As the water wheel went round, gears spun an axle to which the millstones, in their boxed enclosure, were attached. Wheat went in at the top and came out ground at the bottom. Then the bran was sorted from the flour. It was as simple as that. Any man who owned a millsite, and had the stone or wood to build the mill, could become a miller.

But the miller's art and the complexity of the wheat seed itself resulted in a bewildering profusion of flours, each with a different color, usefulness, and even a different nutritional value. Soon there were more different kinds of flour than there were different kinds of wheat. To safeguard future sales, brand names arose from the confusion, and the markets of the world from Europe to South America came to know *Monoquacy* and the other brands of Potomac flour, their names burned on the heads of barrels made of good Piedmont white oak. Like the earlier tobacco planter, the miller and the wheat farmer now acquired a world perspective.

With its primary source of power, the mill was easily converted to other work. Charles Carroll established a mill on Carrollton Manor that was used not only for grinding wheat, barley and corn, but for sawing lumber and grinding cement. The Delaplaine gristmill on the Monocacy was one of the first to grind limestone. At the mills, the farmers' hay and straw were baled for shipping and storage. In the amorphous countryside, an unrelieved landscape of farms, the mill became an important center and the miller an outstanding figure. His business led to wide connections. People found it easier and safer to leave the money for their crops at the mill than to take it home. The mills expanded and became banks, sources of credit; they issued scrip which frequently had currency beyond their neighborhood. When mills failed, it was a calamity to the entire community.

Millowners became men of enterprise, and from small beginnings crushing stone, making cement, or sawing timber expanded into the construction of roads and buildings. Throughout the year they kept their mill wheels turning at an astonishing variety of jobs. Many millers kept stores, as had the tobacco factors a century before. The mills even reached across and joined hands with the ironworks. It often seemed that with water power a man could accomplish anything. In a single tributary to the Potomac you could find paper mills, powder mills, nail mills, flour mills, wooden mills, linseed-oil mills, a flax mill, sawmills, and fulling mills, as well as the omnipresent distillery. Early nineteenth-century maps, like Charles Varlé's map of Frederick and Washington counties in 1808, were sprinkled wth symbols of mills of all sorts.

Around these centers, each the living nucleus of a tiny world of its own, towns arose. Before the middle of the nineteenth century, when railroads had not completed the liquidation of provincial barriers, the upper Potomac was divided into tiny communities, each self-sufficient and each almost identical with the others. Like the geographer Thünen's theoretical city on a plain, you could plot their size and location, their types and their distribution, from the smallest of the crossroads towns to the largest and most diversified centers of the wheat country, like Frederick and Leesburg, Hagerstown and Winchester.

Each community achieved its little urban nucleus, usually

around a mill, but sometimes at a quarry, an iron furnace, or even a ferry. Here the crops went to market, to the gristmill, the woolen mill, the tannery, the distillery. Here the farmer obtained his needs: flour from the mill, plows and other ironware from the smithy, credit from the mill or bank, information from the village press or the post office. Here his church, the local fair, the village school, the tavern. The faded relics of some of these Piedmont and valley towns can still be seen, but the enlargement of trade areas with improved roads, and the ribbanded development along the railroads, have left them moldering with only a general store and post office, a church, and a few houses, faded indeed from their busy appearance of a hundred years ago. Often a walk up the stream will bring you to the vine-covered ruins of the old mills, and, infrequently, to one where the mossy wheel still turns.

In each of the self-contained cells that honeycombed the rolling Piedmont could be found self-sufficient farms and a balanced and uniform urban center. The Tidewater plantation had been a world in itself, but always with a direct trading line to London and Liverpool, a means of maintaining some connection with the world of trade, of politics, and of fashion. With the exception of Alexandria and Georgetown, both created as late as the middle of the eighteenth century, there were no towns in Tidewater worth the name, despite the labor and exhortation of the crown for more than a century, and the appeals of many individuals like Francis Makemie. So long as the urban focus of the lower Potomac was in England, the planters had no need for other towns. When this orientation changed, they could look increasingly to Baltimore, Norfolk, Richmond, and other established American cities to supply their needs.

In the upper Potomac, between the falls of the river and the mountains, towns arose out of agrarian necessity. You could measure the growth of the Piedmont towns—and that of many metropolitan centers as well—by the wheat-laden wagons that found their destination there. Long after good transportation came, wheat was almost the only article originating in the Piedmont that moved in any quantity beyond the borders of the country neighborhood. The farmer called the towns into being to serve his needs. To mill his wheat they were located first along the streams; to ship his wheat they moved closer to the turnpike, the canal, the railroad. To make the farmers' plows and implements, forests were leveled and the iron mines opened, furnaces built and labor assembled. So visibly set on their agricultural underpinnings, these flour-sack towns were permeated with an unmistakable atmosphere of rurality. The wool-hatted farmer, smelling of the barn, was the ultimate customer of every enterprise. Even those who lived in the towns were only a few years removed from the wheatfields and often held to them as an anchor in a venturesome age; many a merchant or tradesman closed his business every summer to help harvest the wheat.

In towns of scarcely a thousand inhabitants, around 1800, you could find an astounding variety of trades and professions. In addition to doctors, lawyers, dentists, midwives, teachers and ministers, there were mechanics by the score, often with the most exotic callings. In Cumberland one of the earliest streets, Mechanic's Street paralleling Will's Creek, was full of them. To clothe both farmer and townsman there was the hatter with his beaver pelts, the tailor, the boot- and shoemaker, the stocking weaver, the milliner and mantuamaker, and beneath the sign of the short-clothes with a wild deer courant the man who made buckskin breeches and gloves—all backed up by the woolen mill, the fulling mill, the tannery, dyers, and others. To house the inhabitants there was a real-estate dealer who sold town lots and Kentucky lands, a conveyancer, builders of all sorts, and house painters, brickmakers, nailmakers, glaziers, joiners, and cabinetmakers, blue dyers, and coverlet makers. To equip the farmhouse of the period, the maker of spinning wheels, the artisan of the curled hair mattress, the coppersmith with his apple-butter kettles and stills, the maker of stoves and fireplace backs, the apothecary "at the Sign of the Golden Mortar," the potter and earthenware maker, the distiller and the watchmaker all labored. The farmer came into town to trade with the maker of bridle bits, stirrup irons, saddle trees, harness of all sorts, the printer and the publisher at the "Hot

Pressed Bible Works" where you had a choice of bindings. There were also merchants by the dozen, tavernkeepers, and numerous others, even down to the comb maker and the man who sold drums, and the fashionable "Francis Pie, of Georgetown, Potomac," who came to the Sign of the Indian King, a tavern, for periods of three days to sell an assortment of millinery and jewelry.

This was hardly all, for the millers bought their imported French buhr-millstones from the stonecutter, the hunter came to the gunsmith, the drover to the saddler, the teamster to the wagoner, and those with the means and a fine taste patronized the gold- and silversmith. Just on the edge of town, where water power turned the wheels of numerous mills, you could buy plaster of Paris, nails, or take the cloth the women of the family had woven to the fulling mill, your wool could be carded, your clover seed cleaned, and your barrels made by the cooper, to say nothing of countless places that ground corn and wheat.

"I recollect," an old Cumberland valley resident wrote, referring to Hagerstown in the early part of the nineteenth century, "seeing every winter, when there was much snow on the ground, numbers of two-horse sleds coming into town from quite a distance, bringing wheat, rye, oats, green and dried apples, feathers, and beeswax, honey, venison, hay, &c. This was a harvest for the merchants of the town, who all kept extensive apartments of every kind of merchandise, drygoods, groceries, queensware, shoes and boots."

An integrated society had been quickly achieved, where the farmer, the miller, the smith, and the wagoner played the major roles. The townsmen, recruited from the countryside, were scarcely distinguishable from the farmers, and the towns themselves, loose country neighborhoods, scarcely distinguishable from the farms. Charles Town, with its eighty half-acre lots, possibly half of them actually built upon, well indicated the urban ideal. Or the Quaker settlement of Sandy Spring, where the village seems actually composed of farms. Every proper home had a kitchen garden, liberally fertilized with stable manure, and most families in towns kept a cow. The countryside itself was never far distant. The townsman liked to remind himself of it: in one town, an outcry arose when it was proposed to erect a high building at the end of the main street, blocking a handsome rural view.

Such small communities dotted the Piedmont. Each had a flavor and a lively individuality of its own, much alike as they may have appeared on the surface. We have a pleasant view of one of them, perhaps idealized a little, in the mind of an old resident. He recollected the slave market at Licksville, near Noland's Ferry across the Potomac, where Quakers from Pennsylvania and southern planters from as far as Georgia came to buy or sell; and the neighboring race track; the cockpit in Dry Branch Hollow; the dogfights at George Kephart's; the elephant kept as a curiosity by Samuel Stover; the bearfights at Nathaniel Kidwiller's; or the Negro, Horace Brewer, who could outrun a man on horseback a hundred yards and return, and who always won. His mind turned fondly back to Wesley McAbee, who made shoes, to Peter Leaply, the blacksmith, and Conrad Buckheimer, the cooper of countless flour barrels. He recalled Thomas Norris, the Negro who grew incomparable strawberries, and other truckers and market gardeners who were celebrated for their early vegetables and succulent cantaloupes, and the miller who kept eels in the millpond. It was the kind of neighborhood where people were famed for their specialties, their cider, or their tansy bitters. In the general store the row of hogsheads and barrels contained Potomac River herring, whisky, and molasses. On the shelves would be found snuffboxes and clay pipes, needles and dry goods, as well as the sovereign remedies of laudanum and castor oil. You stood outside the store in the rain to get your mail from the surly postmaster. With all its narrowness, its traditions, and the circumscribed routine of its citizens, life in the agricultural village was spiced with eventfulness.

In the sections settled from the lower Potomac, a sporting note prevailed, with horse races of all kinds, old-fashioned tournaments, cockfighting, the coon hunt, and the generous traditions of Potomac hospitality and the groaning board. Among the placid Germans the social events were of a more practical turn, with hog killings, corn huskings (where you kissed your girl if you found a red ear), apple-butter boilings, house and barn raisings. Before the middle of the nineteenth century the two strains of people were commencing to intermingle, to intermarry. The traditional community dinners that punctuated the year reflected the change: at Easter, the ham dinner, with sauerkraut, mince pie, and sherry wine; in the winter, the cabbage dinner with boiled middling, pumpkin pie, and hard cider; at Christmas, the turkey dinner, with plum pudding, brandy sauce, and eggnog with cake.

The frontier had rolled west over the mountains. Occasionally a family would pursue it to Kentucky or beyond. Rarely a party, like that led by Captain Swearingen of Shepherdstown, would migrate farther west to become the first settlers of a new city like Chicago; a Nathaniel Rochester would migrate from the Cumberland valley to found a new city on Lake Ontario; or a new Barnesville in Ohio would arise to resemble the old Barnesville in Montgomery County. In the main, the upper Potomac was stable and prosperous, but on its fringe the frontier lasted many years, and the frontier type was found in urban forms as well. The wheat miller and migrant, Daniel Boone, who appeared briefly on Seneca Creek, was an example. An avatar of the frontiersman appeared in such a character as Samuel Rinehart.

Leaving his home in Pennsylvania in 1828, pack on his back, to strike out for himself, Rinehart journeyed southward a little and came to the Point of Rocks, where he was given a job hauling logs out of the Potomac. Four months later he was working as a clerk in the local store. Six months after he moved up the river a few miles and became superintendent of a lumbering firm at Harpers Ferry. After nearly two years the Chesapeake and Ohio Canal came through, and Rinehart went to work as a division superintendent a dozen miles farther up the river at Sharpsburg. Three years exhausted that, and he became a clerk at the Antietam Iron Works, but he soon returned to the canal as a section boss, and migrated farther up the river to Hancock, where the canal enters the tunnel through Sideling Mountain. Another six months and he moved to Cumberland and became a contractor's superintendent. A year and a half later he retraced his steps to Hancock and opened a store. This was not to his liking, so two years later he moved his store across the river to the mouth of Cacapon Creek, where he did business for the next year until washed out by a depression resulting from the suspension of work on the canal. He moved back to Hancock, got a line of goods on credit, and opened another store. To improve his fortune he also invested in a Pennsylvania tannery a few miles north of Hancock. Paying off his bills must have kept him in Hancock, for he remained there until 1860, when he moved to St. Joseph, Missouri, and opened a bank. He sold the bank the following year and returned to Hancock, where he reopened his store. After the Civil War, prompted by his experience in the tannery, he built a sumac and citron-bark mill, and shortly added to it a sawmill. From the timber he commenced building canalboats. That led him into the operation of a line of boats, and canal contracting, and soon into local politics.

Such men frequently found that they had lived a dozen lives in one. But at the end they often found themselves rich or with nothing for reasons they could never quite discover. Restless and bemused, they belonged to the age of Micawber, and shared his pathetic faith in progress. Expansion and change were everywhere in the Potomac in the nineteenth century, and if a man looked sharp his fortune might quickly and easily be made.

5

Through the Piedmont the Potomac is a wavering and untrustworthy frontier, separating two civilizations. From place to place, as from time to time, its precise boundaries have changed. The river remained in its immemorial course, but the swirling currents of culture were often affected more by soil, climate, or settlement than by the river's bed. As the Tidewater plantations of the seventeenth and eighteenth centuries were strongly united by the river, and an identical civilization was created on either bank; or as the civilization of the Great Valley marches from north to south with magistral indifference to the river; in the Piedmont dim lines were established by the various strains of settlement from the north and east. These changed in emphasis over the years, the real cultural frontier running generally a little north of the river itself.

In the lower Monocacy valley the settlement of southern Carrollton Manor was heavily impregnated by Tidewater families, but above Frederick it became predominantly German. In the Middletown valley, from the Potomac to Burkittsville, the plantation form of Tidewater agriculture prevailed; north of that point, the diversified agriculture of the Germans commenced. In the Cumberland valley the plantation influence became even weaker, limited to scattered plantations like Montpelier and Stafford Hall, built by descendants of George Mason, set high in a landscape dominated by settlers from the north. Even on the Virginia side of the river, Tidewater settlers traditionally stopped at the Opequon; west of that the northern settlement began. In Loudoun County it is still possible to draw a line separating the area of settlement originating in the lower river from that of the Germans and Quakers coming from the north.

Perhaps the significance of this division is most spectacularly seen in Loudoun. The Virginia farmers of northern Loudoun County raised a regiment that fought in the Union Army, yet in the southern part of Loudoun the young Mosby recruited his renowned cavalry regiment of guerrillas. The difference persists today in Loudoun between the Democrats of the South and the Republicans of the North; between the survivors of the plantation system with their traditions of hospitality, an outdoor sporting life, and the social and political ascendancy of freeholders, as contrasted with the frugal husbandry and co-operative spirit of the northern yeomanry. Today the situation is perhaps a little confused with the heavy immigration of Northerners into the hunt country, but there are some who can still laugh at the parody by Ogden Nash satirizing their new neighbors that begins:

Carry me back to Ole Virginny,
And there I'll meet a lot of people from New York,
There the Ole Marsa of the Hounds is from Smithtown, or
Peapack, or Millbrook,
And the mocking bird makes music in the sunshine accompanied by the rattling shaker and the popping cork.

The Tidewater island in Loudoun County is one part of an archipelago of such Tidewater settlements that run through the valley, contributing a distinctive strain with distant root and mysterious vitality. Nowhere was the common character of these settlements more explicit than in the land that is no longer Virginia, the eastern panhandle of West Virginia west and south of Harpers Ferry. Here along the Bullskin Creek, the terrain that had so impressed George Washington—open country, with buffalo grass as high as a man's head, and groves of sugar trees—became thickly planted with other Washingtons. Here the knife of Tidewater cut deep.

As early streams of settlement swirled around the headwaters of the Big and Little Bullskin, a confusing pattern of land grants resulted. Like other Potomac barons, Lawrence Washington secured a sizable grant from Lord Fairfax. Joist Hite, Fairfax's implacable enemy, claimed a part of the land. Crown patents from 1730 covered another part. On these lands could be found early settlers from Tidewater, Pennsylvania German, and Quakers, as well as a few Scotch-Irish like the McCormicks.

Increasingly, however, the Tidewater settlers prevailed. The McCormicks moved up the Shenandoah valley. Quakers like Isaiah Pemperton sold their lands and moved away; or, like Colonel Robert Worthington, took on the predominant Tidewater hue, joined the army, and became members of St. George's Chapel, an establishment so Anglican that even the high-backed pews were imported from England.

When George Washington first came to the Bullskin as a tall freckled surveyor in the spring of 1752, he found settlers from his native Westmoreland had preceded him. He boarded with the Stephenson family, who owned over 800 acres of fertile Bullskin land adoining the tract he and his brothers had inherited from Lawrence Washington. One of Mrs. Stephenson's sons helped him with his chain and compass, and later another led one of the first militia companies from the Potomac to him at Cambridge. There can be no doubt that Washington fell in love with this country. He purchased additional tracts of land, and commended it highly to his brothers and their children, who settled there.

In 1756 his eldest brother, Samuel Washington, moved to the Bullskin. For him the architect of the remodeled Belvoir and the reconstruction of Mount Vernon, John Ariss, built Harewood, a stately limestone mansion of the mature Georgian period. The nearby house of Fairfield, which Ariss later built for himself, he sold to Warner Washington in 1784. The remaining work in this vicinity that has been traced to Ariss's drafting table is the architect's own Locust Hill, a tiny but charming house in his characteristic manner, built shortly after the sale of Fairfield on a rocky hillside overlooking the Bullskin tract the architect had leased from George Washington.

In the architecture of John Ariss the increasingly conservative spirit of the lower Potomac is readily seen, whether expressed in his Tidewater houses or those in the Bullskin Creek neighborhood. The tendency to lag behind the prevailing taste in England, commented on by Latrobe, had already marked such works as Buckland's Gunston Hall. In one of his earliest buildings, Kenmore, Ariss perpetuated the traditional building form but altered and enriched it by the addition of ornamental plaster and surface decoration. His bow to the taste of the Potomac, however, was one of recognition and respect, for the houses the architect built for himself show no deviation from that taste.

The tendency to build spare, undecorated exteriors, of simple but dignified proportions, with well-paneled and richly decorated interiors, was in keeping with the desire to conserve the major architectural effort for the rooms that carried the greatest social emphasis. Perhaps it was a persistence of the old colonial pattern that looked to England for its styles and imported its mantels and paneling for entire rooms like the pieces of a prefabricated house. For all its typical manner, life on the Potomac was not essentially luxurious, however much it wisted in that direction. It was fond of display, but inconsistencies betrayed this to be an affectation. Travelers noted the absence of real luxury. They pointed out the lack of such personal rooms as libraries, studies, music rooms, or even small dining rooms. They contrasted these deficiencies, found even in the largest mansions, with the sumptuously appointed drawing rooms and dining rooms suited to the entertainment of large companies. It appeared to many that the houses along the Potomac were designed less for agreeable family living than for entertaining visitors and for show.

Ariss's design for Harewood expressed this characteristic blend: instinctively retardataire, formalistic to the point of ostentation, but necessarily meager in execution. Behind the large and imposing limestone walls, devoid of major decoration but correct in their adherence to the engraved plates of Ariss's source books, the visitor came upon one room of superlative charm and finish—the large Palladian drawing room, fully paneled from floor to ceiling. Each window opening and doorway of this room was framed with Doric pilasters, and the green marble mantelpiece with its richly carved egg-and-dart pattern was well calculated to strike the visitor with its imported splendor. Yet aside from this one room and the stair in the center hall, the remainder of the house was quiet and unimpressive, and wholly lacking in decoration.

The work of John Ariss, covering a period of nearly forty years, was remarkably static. The plans of his houses were largely based, as the research of Mr. Thomas Waterman has demonstrated, on *Vitruvius Scotius*, a book of measured drawings by the Edinburgh architect William Adams, showing houses modest enough in their size to be suited to the planters' requirements. Ariss found a further decorative source in James Gibbs's *Book of Architecture*. While the Virginia architect freely adapted the work of Adams and Gibbs, the buildings identified with him show no development or direction over the period of his career. In this, as in other ways, his art was sympathetic to the spirit of his age and place. It was at home in the Potomac Tidewater, and the extension of that civilisation in western Virginia.

Charles Washington had arrived in the neighborhood of the Bullskin plantations shortly after the end of the Revolutionary War, and commenced trading in land. In the fall of 1786 he was granted permission to lay out Charles Town, in eighty lots of one-half acre each. At the southern edge of the town he built his own home, calling it Happy Retreat. The Tidewater settlers might have given the name to the whole neighborhood, clear back to the Apple Pie Ridge where the prim Quaker ladies with their picnic baskets gathered for meeting on Sundays. Not far off, at the headspring of the Little Bullskin, Bushrod Corbin

Washington later built a celebrated house, Claymont. Around the Washington plantations the countryside became thickly planted with Pages, Lees, and other families bearing Tidewater names.

In the Bullskin country Tidewater society, planting itself afresh, was invigorated by the rich lands and virgin resources, and the steady yield of wheat. But it did not change or develop. Rather it continued to revolve slowly, a quiet pool where debris gradually collected, away from the main stream. As the plantations of the lower Potomac declined and were abandoned, and emigration to Kentucky or the southern states took place, the original homes of plantation life were deserted. But in the upper Potomac valley along the Bullskin the old ways of life lasted another generation or more. Washington Irving, visiting the scene to collect material for his life of George Washington, was entranced with it. Thackeray found it so fascinating he was moved to write *The Virginians.*

The great and the near-great passed time agreeably in this community. Prince Louis Philippe and his brothers visited here. Dolly Madison was married in the splendid paneled drawing room at Harewood. It was a gay and interesting society, yet almost from the beginning it had a faded quality. If it would not change, the alternative was a gradual senescence.

As early as 1781 George Washington was writing of the builder of Harewood: "In God's name how did my brother Samuel get himself so enormously in debt?" In a few decades the brilliant legal talent of Bushrod Washington, which had so delighted his illustrious uncle, was sinking into snuff-covered desuetude on the Supreme Court. The ultimate fatuity of Colonel Lewis Washington, great-grandnephew of the first president, when he was captured as a hostage of John Brown during the raid on Harpers Ferry, marked the end of an era.

Embalmed, the isolated plantation life of the Charles Town neighborhood persisted uninterrupted until nearly the eve of the Civil War. But the retrospective mood of the early nineteenth century was growing. People read Sir Walter Scott, and built their new Episcopal churches in the Gothic style.

Even Charles Washington's Happy Retreat was renamed Mordington. From the moldering tobacco barns, and the mansions with their crumbling stucco, the paint peeling from the porticoes, even the deserted ruins of St. George's Chapel, issued a nostalgia for an earlier and happier day.

A small but impressive collection of writers appeared in the families of Cooke and Pendleton, and a few poets tried their wings, rising like May flies in the late afternoon sun. As they read Scott they recalled their own *Chronicles of Border Warfare,* inscribed by Alexander Scott Withers from old frontier legends, or examined the *Notes on the Settlement and Indian Wars* by the Reverend Joseph Doddridge. They considered the tales they told their children, of the Wizard's Clip where the scissors were never still until exorcised, or other fantasies straight out of the Pennsylvania German superstitions. Attitude kept them from such resources ready at hand; from the Negro tales of the plantation, the lore of turnpike and tavern, the episodes of canal and railroad. It kept them from using even the Indian stories that were still being handed down around the hearth, of the Lover's Leaps, the battles around the flaming stockade forts, the tales of derring-do on the old frontier. Nothing in the valley experience equipped them to understand those who were unlike themselves. Much as they admired what Washington Irving had done with similar materials found in the Hudson valley, they could not find the way to use this ore as the raw material of a living literature.

The great shadow of Irving hung over most of the Charles Town school. The lighthearted essays of David Hunter Strother, published in *Virginia Illustrated*, with his own drawings, were entertaining enough efforts in the Knickerbocker manner, but the effect was strained and the residue insignificant. Strother found his later career as an artist-writer in the pages of *Harper's Illustrated* under the name Porte Crayon. A Union officer in the Civil War, his *Personal Recollections* are balanced and amusing.

The Blackwater Chronicle, a Narrative of an Expedition into the Land of Canaan, showed the characteristics of Strother, but although it contains his illustrations it was written by his cousin, Philip Pendleton Kennedy.

Another cousin, John Esten Cooke, made an impressive effort. In better than a dozen novels of the region he tried to master the forms of creative writing, but when he had finished what remained was hardly literature. His facile writing had grace and humor, but it was distinctly of a vintage when chapters were headed, "In Which . . ." After a brilliant career in the war, he settled down to write *Surry of Eagles Nest*, *Mohun*, and a series of other war novels, thinly autobiographical and in places more accurate than most of the partisan histories of the period. Cooke's view was one common among the veterans of defeated armies: not the soldiers, he insisted, but the Richmond civilians had lost the war for the South. Some of the drama of the expiring moments of the Confederacy breathed in his pages, but his novels shortly became unreadable. A graceful history of Virginia was his most enduring work.

The most significant figure who deserves a place in this circle was undoubtedly John Pendleton Kennedy, whom Parrington has called "one of the most attractive figures of his generation." His novels in the historiated Charles Town style, *Swallow Barn*, *Horse-Shoe Robinson*, and *Rob of the Bowl*, showed a sensitive observer, endowed with magnificent descriptive powers. *Swallow Barn* is loosely organized, rather on the model of the Sir Roger de Coverley papers, with a plot you could write on a postal card, but gives perhaps a more discriminating picture of the declining plantation life of Tidewater than does the work of any other writer. More swiftly moving, and with less digression, *Rob of the Bowl* is set in the forgotten seventeenth-century world of St. Mary's City, the first capital of Maryland on the lower Potomac, and was the author's masterpiece.

John Pendleton Kennedy's life shows perhaps one of the great difficulties of the Charles Town circle: the unwillingness to concentrate seriously on the pursuit of letters. He gave one decade of his life to it in the 1830's, writing his three novels, editing a highly valuable literary magazine in Baltimore, and befriending Edgar Allan Poe. He moved in literary circles, knew Irving well, and gave Thackeray so much material for *The Virginians* that it was believed for years he had written a chapter of that book. Yet he was a lawyer, a railroad magnate, for six years one of Maryland's representatives in Congress, and secretary of the navy. His life, even to the provision in his will leaving $5,000 to cover the republication of his works, evidenced the prevailing belief that literature was an ornament, a token of gentility, but hardly to be considered as a vocation.

A little way up the Shenandoah Valley, in New Market, in a house little more than a log cabin, the poet Joseph Salyards worked in solitude at a battered walnut desk by the light of a lamp contrived from an old lard tin. A classicist born out of his age, a great linguist, a diligent scholar, he dedicated his life to poetry. It was a day of disbelief, but in this backwater Salyards was writing an epic poem on a religious theme, "The Day of Doom," and exercising his muse with a somber concoction called "Idothea, or the Divine Image." One of the few authentic Potomac figures that belong in the romantic movement, Salyards was not fluttering his wings to break the bars of his cage. He found life agreeable, and enjoyed his surroundings. In "Idothea" he described the valley sunrises he saw:

> The orient lines are misty now no more;
> The golden reins are flashing at the door;
> The gate unfolds—Time's ancient songs begin;
> The king of glory and of day comes in.

The highly conventionalized literary circle at Charles Town were not prepared for such ethereal splendor. They regarded Salyards as an eccentric. At the other extreme they could shudder at the career of Edgar Allan Poe, whom John Pendleton Kennedy had known in Baltimore, but who must have shaken them up with his reviews in the *Southern Literary Messenger*, especially his scathing dissection of the popular novel *Norman Leslie*. They wanted to remain amateurs, to follow the genteel tradition, and usually ended as dilettantes.

CHURCHYARD AT CHARLES TOWN

Good conversation, good fellowship, good living dominated their lives, and it was of them that Kennedy was thinking when he wrote of "the mellow, bland, and sunny luxuriance" of the old-time society. Theirs was the literature of the dining table, when the cloth is removed and the ladies have retired, the ripe tales of reminiscence and retrospection. They took their cue from Irving's lack of contempt for regional themes, but were unable to raise their treatment of them above the level of provincialism. Border people, in an age that was driving apart planter and industrialist, North and South, they maintained their poise but refused to believe anything very seriously. From the high level of Kennedy's novels, they sank gradually to watered-down romances of analine luridity, to sentimentalism.

Kennedy was fortunate in living most of his life in Baltimore, by then connected by railroad with Harpers Ferry and Charles Town. But his cousins, deeply rooted in their plantation ways of living, were bewildered by the choice of a vocation. They, too, could write their sketches and essays, novels and a few poems, but the environment was lethargic. The endless routine of plantation life had grown sluggish. Already people were taking the old gristmills apart, quarrying them for building stone, signaling the end of an era. Even the law had become a limp profession, now that most of the conflicting land titles and boundary disputes had been settled. Life at the bar, or even more on the bench, left much time for literary exercises but offered little stimulus.

Judge Daniel Bedinger Lucas, who lived just north of Charles Town on the road to Harpers Ferry, became known as "the poet of the Shenandoah." His life span coincided with that of Swinburne but, compared to that robust Englishman, his work was pallid. In his poem on the meeting of the Shenandoah and the Potomac at Harpers Ferry, he voiced the contemporary interest in the picturesque:

> What pinnacles sublime seem formed by Nature's hand
> To mock the puny works of human art!

But he was happier in oratorical forms, as later in the same verse:

> A thousand echoes will from cliff to cliff respond,
> A thousand ripples break from shore to shore beneath,
> A thousand breezes bear on high the rushing sound,
> As far the white foam, flashing, flings its crystal wreath.

His first collected poems appeared in 1869, with the characteristic title *The Wreath of Eglantine*, and included his most famous, a patriotic comment on the Confederacy: "The Land Where We Were Dreaming." But despite the title of this poem, and stern reminders in the ruined landscape left by the war, there was little evidence that anyone was as yet awake.

EIGHT

THE MARCH OF INDUSTRY

AT THE END OF THE REVOLUTIONARY WAR, WHEN THE MEN OF Potomac considered what to do with their independence, one thing was clear. The political bonds broken, it remained to sever those economic ties which had kept the planter in thrall to the British merchant. To accomplish this, a wide-ranging native industry must be developed to supply the goods formerly imported. Heed must be paid the century-old words of George Alsop, the Maryland poet who had written:

> Trafique is earth's atlas that supports
> The pay of armies and the heights of courts
> And makes mechanics live that else would die
> Meer starving martyrs in their penury.
> None but the merchant of this can boast
> He, like the bee, comes laden from each coast
> And all to Kingdoms as within a hive
> Shows up those riches that doth make them thrive . . .

The iron furnaces and their auxiliaries fitted into this plan. So did the gristmills, the snuff factories, and the other industries reared on the agricultural foundations that had been laid in the Potomac country. But more was needed.

To slake the thirst for domestic manufactures after the Revolution, men were ready with plans for mills and factories of all sorts. The manufacture of finished goods formerly imported from England was everywhere hailed as a national urgency. The demand for American products at home seemed limitless, to say nothing of remoter possibilities abroad. In the Potomac country, as everywhere, small mills were expanding and new ones were begun. The progress of local manufacturing was a subject of the warmest interest in every newspaper and public discussion.

To this world ripe with new opportunities came John Frederick Amelung of Bremen, lured by the stories he had heard from the wayfaring merchant, Benjamin Crockett of Baltimore. The connection between the Baltimore merchants and wheat exporters and the north German ports had become intimate, and Amelung must have kindled to Crockett's report of the rapidly expanding American city on Chesapeake Bay where every other man you met on the streets could speak the German language. He found no difficulty in convincing himself and others in Bremen of the truly magnificent possibilities in Maryland of manufacturing glass.

By 1784 Amelung had found subscribers in Bremen for nearly £15,000 and commenced to recruit a company of skilled glassworkers from all parts of Germany, especially Bohemia. These he amplified by other craftsmen needed to make up a self-sufficient community. He purchased tools and machinery, and chartered his own ship.

All this activity greatly excited the British trading community in Bremen, and soon Amelung found his preparations handicapped not only by the inordinately hard winter but by mysterious and increasing difficulties in getting exit permits for his workmen, and securing their tools and equipment. With only sixty-eight workmen, part of a much larger force of three hundred he had hoped to bring, and leaving a lieutenant to smuggle out an additional fourteen hands later by way of Holland, he set sail in May. After a difficult passage of sixteen weeks the party arrived in Baltimore the last of August.

An admiring account by a descendant describes the procession of John Frederick Amelung, with his wife, three daughters, three sons, a sister-in-law, their English, French, and German tutors and governesses, and domestics, from their landing to the Indian Queen in Baltimore. They were cordially received.

Still more tangible evidence shows that Amelung was hardly an artisan in glass, as he has often been described, but undoubtedly a prosperous Bremen merchant. His retinue of butchers and bakers, physicians and teachers, shoemakers and blacksmiths, also makes it clear that, while the manufacture of glass was his principal aspiration, he was prepared to move in other directions should circumstances in the New World warrant.

In Baltimore, Amelung conferred with his agents, Keener and Mercer, and with Charles Carroll of Carrollton, Thomas Johnson, William Paca, and other Maryland leaders, to whom he presented letters of introduction, and before long was persuaded to settle his company on a "tract of land on Patowmack, not far from the mouth of the Monocacy" in Carrollton Manor. After determining that the supply of fuel and potash were satisfactory, he purchased 2,100 acres, probably from Charles Carroll. When his company had raised another seven or eight thousand pounds from American subscribers he later added a thousand acres of mountain woodland. The site of his colony he named New Bremen. He purchased shares in the Patowmack Company and looked forward to the day when his factory would be connected by water with the principal markets of the nation, even the world.

In 1785 these facts were stated for the information of prospective investors, with the additional information that "I have made a beginning at glass making." He had built one "glass house" equipped to manufacture bottles, window and flint glass, and was beginning to construct a second. For his workmen he had built "dwelling houses for one hundred and thirty-five now living souls." Overlooking the glassworks, the prospectus omitted to state, he had also constructed Mountvina, a 22-room mansion, complete with a large wine cellar, a ballroom, and all the other necessary apparatus for good living and lavish entertainment, surrounded by acres of terraced gardens and well-tended orchards and vines. The prospectus also announced the establishment at New Bremen of a German school "that will educate children in English, German and French languages, writing, ciphering, music, to play on the harp, harpsichord, flute and violin; I have the masters for this purpose already here." The employees at the glassworks were organized into a full orchestra, and performed not only for balls, but on every ceremonial occasion. A half century afterward, in the spacious attic of Mountvina, a bored little girl accompanying her mother on a visit, found four golden harps.

For all the grand manner and the entertaining asides, the brimming bowl and fireworks displays for distinguished visitors, there is little question that Amelung was capable and aggressive. He did produce glass to the value of £10,000 a year, large quantities of it, incomparably better by all evidence than any other being made in America. Flat glass for windows, a flood of bottles, practical glass for jellies and household ware came from his glass houses. He began to make mirrors. But of all the products of New Bremen, the finest were the beautifully clear, slightly smoke-toned presentation pieces he seems to have made for the express purpose of winning friends for the new enterprise.

So lost was the trace of this fine work that when, in 1928, a covered goblet inscribed "New Bremen Glassmanufactory—1788—North America, State of Maryland," and on the reverse "Old Bremen Success and the New Progress," was found in Bremen and deposited in the Metropolitan Museum of Art, it was seriously contended that no American glassworks could possibly have made such a piece at that date, and it must have been produced in Germany. In the twenty years since, the great series of Amelung presentation glass has been unearthed, authenticated, and described, firmly establishing the New Bremen works as the most significant episode in the history of early American glass.

Amelung's glass was like the architecture of Mountvina, where despite the conventions of colonial building and a solidity of interior trim that evidenced the taste of Northern Europe, an unmistakable native air prevailed. Hardly a mere copy of the contemporary European forms, from the very beginning, and with increasing boldness, it struck out in new directions. It established an affinity with the austere art forms of the Pennsylvania Germans, and blended with their cultural landscape.

Amelung presented to President Washington "two capacious goblets of flint glass exhibiting the general's coat of arms"; but he would hardly have appeared as the tongue-tied bumpkin, the stammering immigrant, of this much-told Mount Vernon story. To his neighbor, Baker Johnson, he gave a set of case bottles, duly inscribed and dated in a wreath of laurels ornamented by the traditional motif of two doves. Other glass was presented to influential members of the state and federal governments, and to the Masonic organizations that sprouted during this period. While these solicitous presents were gratefully received, and impressive testimonials returned, their deeper motive soon appeared.

The difficulties of a young enterprise, with overambitious leadership, could not be concealed. Foreign glass was flooding the free American market, glass in familiar English and Irish patterns, which found a ready sale to customers long accustomed to its standard patterns and quality. Despite the fact that he had completed two glass houses, and raised nearly £8,000 from American subscribers, Amelung found he had overextended his business. He lacked working capital. When he wanted to pay bills incurred by his physician or to buy a violin, he had to persuade the Frederick merchant, John Shewell, to accept payment in glass.

In 1788 Amelung petitioned the Maryland Assembly for a loan of £1,000, and his good friends in Annapolis, Carroll, Johnson, and the others, soon arranged matters satisfactorily. The following year when Congress met for the first time, another friend and supporter, Charles Carroll of Carrollton, introduced the first tariff bill, providing a 10 per cent ad valorem duty on imported glass.

Passed July 4, 1789, this measure does not seem to have given much immediate relief, for in 1790 Amelung petitioned the House of Representatives to grant him financial assistance. In urging the tariff the faithful Charles Carroll described Amelung's efforts to establish a new enterprise, and confessed that "he now finds himself gravely embarrassed in prosecuting the business." He referred to the rise in the price of grain as a contributing difficulty. Yet he was confident that if the House granted the requested loan of three or four thousand pounds, this would "enable him to surmount every difficulty." Mr. Stone and Mr. Boudinot spoke up for the project with enthusiasm, and declared Amelung's glass unequaled. For a time it appeared the committee would sanction a loan of $8,000.

Yet the distant rumblings that were to characterize a long line of subsequent tariff debates were heard. Mr. Smith of South Carolina thought the loan would establish an unwise precedent; he asked was it not unconstitutional for Congress to lend money to a constituent? Mr. Sedgwick pointed out that many native manufacturers languished for such aid, and demanded to know why Congress should help foreigners out of their business difficulties. Mr. Jackson averred that Maryland would benefit more than other states, should the proposed loan be granted. The project faltered, and slowly came to a complete stop, the committee finally recommending that Amelung take his troubles to the Maryland Assembly.

The Congress, however, did recognize the weakness of the 1789 tariff, and shortly after its decision on Amelung's loan passed a new tariff raising the duty on glass to 12½ per cent, with an additional 10 per cent duty on goods imported in foreign vessels. But Amelung's enterprise at New Bremen was too far gone to receive much benefit from this.

He was able to secure a stay on the payment of Maryland's loan, and attempted various expedients. He sold a good part of his landholdings. Shortly before 1793 he took James Labes of Baltimore into partnership. In those gloomy days, with the vultures sweeping lower, he must have taken satisfaction in marrying off his three daughters, and completing the education of his sons.

The end of New Bremen came in 1795, when the remaining lands, the glass houses, and the dwellings of the workmen were sold. Even the names disappeared: New Bremen became Fleecy Dale, a woolen milling center, and still later and more prosaically

Park Mills. Mountvina decayed, the 8-room addition was removed, and cows were pastured in the gardens. When Rochefoucauld-Liancourt visited what had been New Bremen in 1796 he found it "near extinction," and wrote its epitaph in the candid observation, "This manufactory has shared the fate of almost all first establishments of this nature."

But the seed pod burst, and glassworks sprang up everywhere. Thomas Johnson purchased some of Amelung's machinery and built the new Aetna Glass Works on Bush Creek, hauling sand up from Ellicott City in the empty wheat wagons returning from Baltimore. He later built another glassworks on Tuscarora Creek, supplied from the Five Mile Sand Flat along the Catoctins. Both plants were manned by former Amelung workers. A foreman named Kolenberg bought Amelung's second glass house on Bear Branch, and operated it for years. Some efforts were even made to continue production at the original works. A party of New Bremen glassmakers set off on foot for the Ohio country, but when they encountered Albert Gallatin in Wheeling he persuaded them to start a glassworks on his own lands at New Geneva. Others later moved to Pittsburgh where they were absorbed in the firm of Bakewell and Page.

These new impulses, each with its nucleus of trained workmen, were perhaps of greater value to the nation than the New Bremen glass itself.

2

The pots and pans of the plantation smith and the trade in iron manufactures that the British had pinched off before the Revolution were the first struggling efforts to create something from the bog-iron deposits of the Tidewater. Long before the Revolution, the richer hillside outcrops of iron in the Piedmont were being worked with pick and crowbar, and the night was brightened by the hillside fires of the charcoal burner and the furnaces of the iron mills. Maryland iron alone counted for nearly half that exported from the colonies.

The wartime needs for cannon and guns greatly stimulated

the establishment of ironworks, and the embargo brought on by the war threw the Potomac valley largely on its own resources. A dozen gunshops were founded in Maryland alone, most of them along the foothills in Frederick County, and others were located in Virginia, each able to make a score or more of muskets a month. When independence was declared, there were iron furnaces already, and more were soon set in operation. When American cannon blazed at Yorktown, the value of these iron furnaces would be seen.

In 1776 the Catoctin Furnace, one of those owned by the Johnsons, received orders from the Council of Safety to furnish 20 three-pound cannon, 20 two-pound cannon, 40 swivel guns, and 200 iron pots of two to four gallons for the use of the army. These made, other orders followed.

The operation of the eighteenth-century ironworks was highly standardized. Against the hillside stood the blast furnace, where it could be readily charged from the top with wagonloads of ore, charcoal, and lime. Huge bellows, often driven by water power, built up the necessary heat. In front, on the lower level, was the casting house where the molten iron was poured. Here pig iron, pots, kettles, stove plates, firelocks, and similar articles were cast. Often that was all. The big house of the ironmaster and the smaller cottages of the workers completed the picture. Such industry remained close to its agricultural base, and nearly everyone had a tiny farm of his own. At Catoctin Furnace and near the mouth of Antietam Creek illuminating ruins of such old furnaces still survive.

In favorable situations, some ironworks grew more complex. Then the pig iron was heated again until it became soft, and beaten out by water-driven trip hammers to make a more refined bar iron for blacksmiths. Often the smiths would set up close to the furnace, and make their tools, locks, implements, and other articles. A further development was the bloomery forge, where more heating and hammering made wrought iron, or the rolling and slitting mills, where iron plates and nails were manufactured.

The iron furnaces multiplied, and each spring their products were floated down the tributaries to the Potomac in increasing quantity. Enterprising men like Thomas Johnson and his brothers, close friends of George Washington, owned a chain of furnaces along the foothills. Their operations spread out to include the ownership of cheap mountain lands where the charcoal was made, lime kilns, or harnessing their water power to sawmills

and gristmills to secure a balanced seasonal work load. They acquired an interest in river navigation and hard roads that surpassed even that of the wheat farmer.

A cross section of the industry was seen at the mouth of Antietam Creek, draining the Cumberland valley north of the Potomac, where an ironworks was begun in 1765, located virtually on top of a rich outcropping of ore. A couple of years earlier the site had been offered for sale, with the pertinent observation that iron ore had been found close to a stream to work the furnace. The hillsides grew abundant hardwoods for charcoal, and iron deposits were later found on the Virginia side of the Potomac as well. The entire valley was underlaid with limestone for fluxing. The pig iron could be shipped out by river barges or ferried across the Potomac to reach the blacksmiths of Winchester and the Shenandoah valley. It was a magnificent combination, as the Antietam Iron Works proved. One of the earliest ironmasters of the Antietam was the Huguenot refugee, Launcelot Jacques, who had commenced as a partner of Thomas Johnson in 1766, but within a decade struck out for himself. Presently there was not one but several ironworks, closely situated along an 8-mile stretch of the Antietam, the Fort Frederick Furnace, with its related works, the Frederick Furnace; the Mount Aetna Furnace; the Great Rock Forge; and others. By the time of the Revolution they were casting cannon balls and shells, forging cannon, and manufacturing guns and small arms—and prospering mightily.

With the end of the war, bar iron "of the finest quality" was advertised for "all kinds of nails, brads & sprigs," and smithies sprang up to consume it. Nail factories appeared. When James Rumsey of Shepherdstown was building his pioneer steamboat, the neighboring Antietam Iron Works made many of the parts he required.

Presently the Black Rock Forge of Daniel and Samuel Hughes began to move ahead of the others. They already had "two hammers, four fires, a substantial dam and a considerable head of water," and during the war they had made cannon. In 1796 they purchased the Mount Aetna Works, and four years later expanded it. They added Great Rocks Furnace and the Old Antietam Iron Works south of Sharpsburg. At their Antietam Forge they made bar iron, fireplace backs, stoves, and other castings. The earlier ironworkers who survived became more interested in such specialties as nails, or turned themselves to ropewalks where they made "mill and well ropes, bed cords, leading lines" and other articles of hemp, leaving the Antietam Iron Works to dominate the stream and supply the local trade.

The water power of the Antietam, which ran the furnace bellows and raised the heavy trip hammers of the forges, ran many other mills, not only those related to iron like the slitting mills, but sawmills, gristmills, limestone crushing mills, woolen mills, spinning mills, hemp mills, fulling mills, and a swarm of others.

3

As the iron furnaces produced increasingly, small industries sprang up to work the pigs into finished goods. Every blacksmith became a manufacturer, seeking to expand his market and his line of products. Into the valley the government itself came, to establish an armory and arsenal at the point where the river could drift the pig iron down to Harpers Ferry, and where water power was abundant.

Only a little time after the Congress, impressed with the defenseless state of the nation in a world seething with armed powers, and the uncertainty of relying upon private manufacturers of ordnance, had authorized the establishment of three or four armories where guns could be made and stored, President Washington was writing to the secretary of war calling his attention to the water power at Harpers Ferry, and recommending it as a site for one of the new gun works. He stated it was "the most eligible spot on the river from my point of view." Washington's recommendation undoubtedly embraced his knowledge of the water power, the growing iron industry of the upper Potomac, the water routes from the furnaces to Harpers Ferry, and some useful military considerations, as well as his constant interest in the Potomac route to the west.

In 1796 the government purchased its first land at Harpers Ferry, to be added to as time went on. By late that year muskets were being produced in small quantities, probably in the buildings shown in the famous view of Harpers Ferry sketched in 1802. Additional tracts of land were purchased from time to time, principally from the Shannondale Bloomery Mills, which had been started about 1760, and as the armory expanded it was found necessary to have extensive tracts of mountain land for charcoal, and an adjacent supply of iron ore as well.

The early production of the armory was limited to muskets and some cannon, but the invention, in 1811, by John H. Hall of the first satisfactory breech-loading rifle profoundly altered things at Harpers Ferry. In 1817 Hall was appointed assistant armorer, and instructed to begin manufacturing his flintlock

rifle, which took its name from the place of manufacture. Additional buildings were commenced for this purpose, and Hall invented much of the machinery employed in rifle manufacture. By 1827 an inventory showed thirteen buildings owned by the government, four of which were devoted to Hall's operations. About 10,000 rifles were made to the time of Hall's departure in 1840, and an examination board's report showed the Harpers Ferry rifle had a rate of fire of eight rounds per minute as contrasted with two rounds per minute for the ordinary musket. Presently new designs were prepared, and the famous Minié rifle—the standard Civil War rifle—came into production. Many other pieces were made, including the 54-calibre Harpers

Ferry pistol, the standard side arm in what must have been an army of heroes.

With the coming of the railroad, the town boomed. Perhaps as many as five hundred workers were employed at the armory and at the arsenal where the guns were stored. By the middle of the century they had helped raise the population of Harpers Ferry to nearly three thousand making it the most considerable industrial town on the Potomac. Mostly recruited in the vicinity, the workers transplanted the sporting blood of the Potomac from its traditional rural setting of cockfights and horse races to an urban stage, where armory workmen in beaver hats and long coats commanded such spectacles as the international prize fight between Yankee Sullivan and the Englishman, Ben Caunt. Urban difficulties arose, too. Cholera broke out about 1850. The housing shortage made it necessary for the government to build some homes for its employees. And during the superintendency of Colonel Lucas, when the spit and polish of military life became the rule at the armory, unrest broke out among the workers, who finally sent a delegation to Washington to petition President Tyler for relief of their grievances. Recollecting their trade, Tyler brightly advised the delegation "they must all go home and hammer out their salvation."

In the iron trade, the race was to the swift. When it dawned, furnaces were built everywhere, but soon the narrowing supply of good ore and the cunning of the iron masters left but few survivors in each of the economic provinces into which the upper Potomac region had divided itself. These merchant furnaces remained, each to dominate its own locality, until all were swept out of existence—the last ones in the years following the Civil War—by the coming of steel and the concentration of the industry in Pittsburgh and a few other iron centers.

4

From the forest had come the beginnings of the industry that had failed to find a congenial foothold in the Tidewater country. The forest made the potash for the New Bremen glass factory, and the charcoal for the iron furnaces and lime kilns. In the Piedmont, once past the thin band of pines that skirted the fall line, the land was thickly grown up in hardwoods: virgin white oaks, red oaks, and large hickories grew on the hillsides, while along the streams yellow poplars raised their lofty crowns, mixed with black walnuts. The choicest of these trees were gradually and systematically cut, and for many years a steady exporting of white and black oak, hickory, black walnut, and even locust continued. To fill a special order for ship's timber, it was not considered extraordinary for an English firm to send representatives to the Potomac to superintend the selection and cutting of the wood.

The bulk of the timber found local markets. The black oaks were the foundation of a thriving shingle business. In the forest camps, the trees were cut down close to the ground, and sawed in shingle lengths. Split in blocks, the shingles were rived out by hand, sawed and shaved to proper size, and then piled in sections to cure. The timber cutting was so gradual that a man could spend a lifetime in such a trade. Masters of ax and adz, with rhythm and strength to split a dry 40-foot chestnut log in one blow, they not only hewed the beams for houses and mills but graduated to many specialized tasks. They became boatbuilders to fashion flatboats and keelboats, those floating lumberyards of early river navigation that went down the river never to return, and shaped the commodious canalboats of a later day. Houses, barns, churches, mills, and other buildings showed the woodsman's art. Even the crude houses of logs, in the Scandinavian manner, in their hands became tightly built cabins, notched and pegged to stand for centuries.

As late as the Civil War, much of the Potomac valley was still in virgin timber. The really spectacular ravishing of forest resources that occurred elsewhere did not happen in the Potomac Piedmont. Yet the process of deforestation was relentless, and where the cutover land was not farmed it covered itself with scrubby pines of small value.

The forests disappeared into flour barrels of white oak,

shingles, building lumber, and other uses. The wheelwrights and wagonmakers, the ship- and boatbuilders, the coopers and carpenters took their share. Tanneries were everywhere, with their tanbark walks spreading in all directions, and the chestnut oak and white oak melted away, often only the bark taken, the logs left to rot in the woods. In the little, self-contained worlds into which the Piedmont divided itself, no one thought twice about using the abundant timber. In the extravaganza mills and even bridges were built with all their beams of black walnut.

Farther west in the mountains a more rapacious attitude was found, a true frontier spirit of depredation that later lent itself to fantastic forms of waste when the railroad and the steam-power saw had greatly increased the ability to destroy natural resources. In a region where whole mountain ridges were repeatedly burned off by children, merely because it made the huckleberries more plentiful, the idiocies of their elders were even more notable. At the very end of the nineteenth century it was still possible to send "woodhicks" by the hundred into the virgin forest that framed the North Branch of the Potomac and strip it naked of hardwood for a distance of almost fifty miles.

Nearly a century would pass before men could understand the consequences of cutting down trees. In the Piedmont, timber was cut by slower methods and with more discrimination—but only for the characteristics of the wood as fuel. The fast- and slow-burning woods were carefully blended to fuel the lime kilns, and the manufacture of charcoal was the indispensable accompaniment to the iron furnaces for the first hundred years of their existence. When Benjamin Latrobe ascended Sugarloaf Mountain in 1810, he sketched a typical Negro family who lived in a bark shelter while engaged in charcoal burning. The charcoal pits were everywhere in the hills. The first railroads and steamboats were long powered by wood, as indeed were those preposterous early agricultural machines, the steam plow and the reaper. The licking flames of a new industry finally laid waste the forests, which it was nobody's responsibility to replant.

As the charcoal forests began to disappear, men took more interest in the coal laid down by the older, prehistoric Allegheny forests. For almost a century before it was mined commercially, it had been known that the upper Potomac contained coal. Braddock's map noted a coal mine, and the Potomac coal was used as early as 1804. But the distance of the coal banks, as they were called, from the large markets hampered its use on any considerable scale. Every farmer still had plenty of timber of his own. Few great cities yet demanded the fuel. The blacksmith preferred the steady-burning charcoal. The low-browed Vulcans of the iron furnaces, long habituated to charcoal, swore that good iron could not be made with coal or coke. The woodsmen and charcoal burners opposed the use of the new fuel. The first toddling steps of industrialism were not based on coal, but on wood. Temporary construction crews of railroad and canal lived in "Slabtown." Even the first plants in the Potomac to make illuminating gas employed the rosin gas patent. The first coal-mining operations were undertaken not so much for the coal alone as for its local use in producing iron; the early companies almost invariably joined the words "coal" and "iron" in their names. The use of coal advanced slowly. As late as 1841 the Baltimore *Sun* could observe, "The coal region is at present one of the most unproductive regions in the state."

Yet on the bounty lands west of Cumberland, every winter in the slack season on the farms, a short-lived trade erupted, with part-time coal miners and teamsters digging the coal from exposed seams high up on the hillsides and hauling it in wagons from the coal banks along valley roads to the head of navigation on the Potomac at Westernport or Cumberland. To make the square-ended flatboats, or the later keelboats, that carried the coal downriver, other seasonal trades came to life during the winter months. From the coal piles accumulated at Cumberland during the winter, the loaded barges with their crews at the sweeps would float from their river harbors during the spring freshets, and commence the hazardous trip down the Potomac.

With two men at the sweep oars, an oarsman at the bow, and a helmsman at the stern, the coal barges made their way over foaming shoals and rocks, following the channels that had been sketchily marked by the buoys of the Patowmack Company. If the state of the river had been correctly estimated, and the flat-bottomed boats with 25 or 30 tons of coal were not overturned in the rapids or broken on rocks in the stream, they arrived at Harpers Ferry or Georgetown in two or three days. The boat and its cargo sold, the grimy crews commenced the long walk back to Cumberland or Georges Creek to begin the next trip. In the short season of navigation the pace was furious, and instances were reported of crews walking as far as 60 miles a day on the return trip. At the peak of the river traffic, as many as fifty boats a week left Cumberland, loaded with flour and other farm products as well as with coal. Until the railroad reached Cumberland, this was the only way of getting the upper Potomac coal to market.

In those early days there was little system in mining. There was little idea even of ownership, and the itinerant seasonal miners worked the coal almost where they pleased, with no other tools than could be found lying in any barn. On the Sheets farm coal had been dug as early as 1816 and hauled into Cumberland, where it was used in a glassworks. The patrician Captain John Porter, on his farm "Rose Meadows," worked an outcrop of coal following his return from glamorous exploits in the War of 1812. The coal lay close to the surface, often so exposed that it could be mined with a crowbar and pick.

Into these primitive scenes near Cumberland, among the miners in their shallow cuts and the brawling drovers of the National Road, the cowboys in the glades and the farmers in the narrow valleys, came a new figure, that of the Maryland engineer and cartographer, John H. Alexander. Experiences as a railroad surveyor had convinced Alexander that it was idle to pretend to accurate engineering while a precise and generally accepted system of base points was lacking, to which individual surveys could be tied. The down was scarcely grown on his cheeks in 1833 when he had persuaded the governor of Maryland to secure funds from the legislature for this purpose. When he was twenty-one, the governor had placed him in charge of the preparation of the state's first official map. Under Alexander's careful promotion the map expanded to include a coast survey, harbor studies, a continuing geological survey, and many other useful projects, not least the pioneer geological survey of western Maryland.

The sparks always flew from Alexander. He had graduated brilliantly in the classical studies at St. John's College in Annapolis, and had dutifully read law for four years before turning to science. When he became the first state engineer his mind already displayed the inclusive habits that later led him through a myriad of scholarly enterprises and achievements. Greatly skilled in mathematics, he delighted in all its branches, and was never happier than when developing new methods of computation or working with his scientific colleagues of the Coast and Geodetic Survey. Proficient in languages, he acquired a mastery of seven and a smattering of more. While sufficiently concerned with philology to write a dictionary of the Delaware Indian language, and a 10-volume dictionary of English surnames, he used his languages more as a powerful device to enlarge his store of information on all subjects. His highly practical works were liberally seasoned with the most recondite, yet pertinent, references to the technical literature of the world. At his death, his library numbered more than five thousand volumes, and contained works in thirty-one different languages, as well as numerous maps, charts, and other items.

Like his New England contemporary, George Perkins Marsh —the two careers are strikingly parallel in many respects—he moved in the worlds of practical and political life as well as in those of scholarship. To learn the process of brickmaking, he worked a month in a brickyard. He prepared an official report on the *Manufacture of Iron* in Maryland, in 1840, covering all the principal furnaces in the state, with detailed notes on their methods of operation and statistics of production, and containing as an incidental feature a more extensive historical sketch of the colonial ironworks than any subsequently to appear. He was an active participant in railroad matters, and brought his great knowledge to bear on the many-sided question of railroad location. Hardly had California been secured than he issued a pamphlet calling for a transcontinental railroad line; while premature, its arguments in favor of a southern route might well have been considered twenty years later. He was consulted by the citizens of Wheeling in their successful effort to make that Ohio port city the western terminus of the Baltimore and Ohio Railroad. He concerned himself with that seminal question of his age, the standards of weights and measures, and

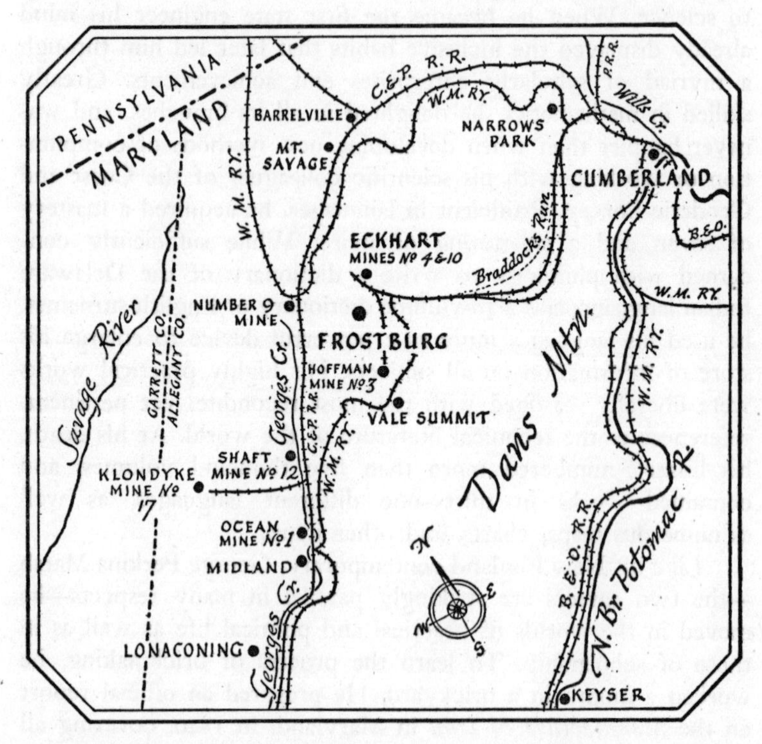

prepared both a well-received history of English units of measurement, republished in England a decade later, and the official Maryland report of 1845 on weights and measures.

A good year before he was engaged as state engineer, a career that lasted from 1834 to 1841, Alexander became interested in the coal resources of the Georges Creek Basin. In 1835 he secured a charter to incorporate the Georges Creek Mining Company, and the following year, with his associate, the state geologist, Philip T. Tyson, he issued an elaborate prospectus of the corporation.

Among its detailed maps and geological data, Alexander's prospectus contained a contoured map of the coal basin, "exhibiting, as nearly as possible, the configuration of the ground, by means of horizontal curves, and also the principal coal beds." While the pioneer cartographer, in explaining this first use of contour lines, stated that the system had been used in the "trigonometrical survey of Maryland," a parsimonious legislature had not yet provided funds to publish that map. The 1836 prospectus of the Georges Creek Company thus contains the first contoured map printed in the United States, an application of a method of presentation theretofore used only in Europe for hydrographical charts. Even the use of the term "contour lines" did not appear in England until 1844. Perhaps Alexander was correct in placing the location of the "principal coal beds" second in importance to his new mapmaking technique.

In making plans for the corporation, Alexander could scarcely have been unaware of the unrest in labor, growing from a desire to better its conditions of work and living. Every construction project of any scale had labor difficulties. Inability to control his Irish workmen had cost James Rumsey his position with the Patowmack Company. The Baltimore and Ohio Railroad was repeatedly in trouble when the Irish laborers employed by its contractors rioted or refused to work, and the militia was frequently called out to quell gatherings of workmen. The National Road nearby was harassed with similar difficulties. An ugly pattern of revolt and repression was being created. Perhaps Alexander was more immediately motivated by the meeting held in Cumberland early in 1835, where the independent-minded mechanics aired their grievances. Whatever the motivation, the incorporation of the Georges Creek Company was notable for the provisions the state gave the new concern in controlling the habits and morals of its employees.

The "Lonaconing Residency"—as the act creating the company town was called—gave the company paternalistic powers over its workers broad enough to prohibit drinking (save possibly on special occasions, as in deference to the Scotch mode of treating the first visitor to set foot in the house following the New Year); to permit a compulsory inspection of employees' homes to ensure their neat appearance; the exclusion of all dogs; compulsory school attendance; an obligatory subscription of fifty cents deducted from the monthly wage account to support a system of medical care; and a variety of company stores, and other philanthropic and disciplinary measures. Under these powers the Georges Creek Military Company was formed, to patrol the district and make arrests. Inevitably, as in other company-owned industrial towns, such feudal powers in corporate hands led to their abuse and revocation, but before then a measure of benevolence often tempered their application. Who can say how much of the white-painted cleanliness and verdant charm of Lonaconing today, its settled appearance so unlike the transient squalor of other older coal-mining settlements, derives from the firm policy of another day?

In 1839 the company was still engaged in experimental projects, awaiting an outcome of efforts to secure cheaper transportation to the principal coal markets. They commenced the manufacture of iron with coke, "almost for the first time in America," as Alexander pointed out. A furnace had been built at Lonaconing with a 50-foot stack, where over two hundred hands were engaged in producing 75 tons of iron a week. Yet the corporation was not thriving.

The difficulty was not in coal mining. Any industrious person could go and quarry the coal, and load it into wagons. The important thing was to get it moved from the mines to some place where it could be marketed. This led inescapably to the formation of many small companies, but the ability to control transportation factors determined the outcome of competition among the mines. The value of coal lands was negligible unless transportation was available, and all early enterprise centered on this essential.

The road laid out by Nemacolin and Thomas Cresap, over Will's Mountain, later followed by Washington and Braddock, led fortunately to the Georges Creek basin, where the phenomenally rich 14-foot vein of coal had been uncovered at several places along the narrow, winding valley. These mines were located within reach of the growing town of Cumberland. The difficult wagon haul over the mountain was not long a satisfactory solution of the transportation problem. Inevitably, as in other countries, men began to wonder why the railroad in the mine should stop when it reached the surface.

Even before the Baltimore and Ohio had laid a rail, the iron and coal companies of western Maryland were obtaining charters from the state legislature permitting them to construct railroads connecting their mining properties with Cumberland, where, the recent appropriation of one million dollars by the legislature gave assurance, the Chesapeake and Ohio Canal would eventually terminate. The half dozen companies that received charters were soon incorporated, and little bits of railroad were being built, often only of planks, along which tiny wagons drawn by horses moved the coal. Still, the Baltimore and Ohio was not to reach Cumberland until 1842; and the canal would not come until 1850. The small, weakly financed mining companies faced a long period of waiting.

One of the largest of them, the Maryland Mining Company, was representative. Its charter of 1828 permitted it to engage in mining, to build railroads, and to operate a bank. Not for a decade did the company make any tangible move; then they opened the bank in Cumberland. Gradually they began to buy out the earlier small operators, but before long the slow-witted company found itself in financial difficulties, and sold its coal interests to the larger and more aggressive Cumberland Coal and Iron Company.

In this situation Alexander's Georges Creek Company made valiant efforts. Boosted by English investors, they built a tramroad from the mine at Lonaconing to Cumberland, and tried to weather the long period of waiting for better transportation by making iron. While it was thus expanding slowly, what remained of the Maryland Mining Company was taken over, and reorganized with an ominous name—Maryland and New York Iron and Coal Company. Alexander must have watched with amusement as this firm of Northerners began buying up lands which he knew contained little or no coal, often paying fabulous prices in the belief, then general, that coal underlay every hill in the region. Yet he seems to have overlooked the essential thing the New Yorkers saw: if a railroad could be built through the Georges Creek basin, to be ready when the Baltimore and Ohio reached Cumberland, a rich revenue would be assured from the coal traffic.

The New York company made a tremendous effort to build its railroad, one facet of which led it into rolling, in 1844, at its Mount Savage ironworks, the first solid track rail made in the United States—an achievement that was recognized when the Franklin Institute awarded the company a silver medal. By 1843 the New Yorkers had the ironworks in operation, and the company's railroad reached there two years later. Before it could be pushed up the valley to Frostburg, however, disaster overtook the company. Its resources exhausted, save for the production of rails at Mount Savage, a series of reorganizations eventually made it necessary to set up the railroad as a separate corporation, the Cumberland and Pennsylvania Railroad, in which the reorganized Mount Savage Iron Company held by far the largest share of stock. Under this new impetus, the railroad was pushed up Jenning's Run to Frostburg, standing at the headwaters of Georges Creek.

At Lonaconing, hardly ten miles distant from Frostburg, the Georges Creek Company had not been inactive in its own railroad-building effort. During the Civil War the company finally completed its connection down the valley to the Baltimore and Ohio at Piedmont. But it was too late. Almost at once its new line was swallowed up by the rapidly expanding Cumberland and Pennsylvania, which in turn was immediately devoured by a still bigger fish—the Consolidation Coal Company.

A creature of Boston and New York financiers, the Consolidation Coal Company was chartered by the state of Maryland in 1860. No more significant gesture, on the eve of the Civil War, could have revealed the border state's essential sympathy with northern capital and its disbelief in the agrarian course of the southern Confederacy. Uniting three smaller coal companies, with the acquisition of the railroad through Georges Creek the Consolidation Company moved into a dominating position. Before long its principal remaining competitor, the Cumberland Coal and Iron Company, was maneuvering for a merger.

While the Cumberland Company was trying to fight the Baltimore and Ohio's monopoly of rail transportation in the upper Potomac by making loans to "canawlers," it was becoming recognized that the coal of western Maryland had superior properties. A committee of naval officers had found it the world's best, ton for ton, and in the increasing steam-coal trade it was in great favor to fuel ocean vessels as well as for domestic purposes. The export market for bunker coal began to seem attractive. A shipload of Georges Creek coal even left Baltimore for Newcastle!

The Consolidation Company was liberally supplied with New York capital, and its backers included such names as Corning, Aspinwall, and Varnum. Soon others were added, Warren Delano becoming a director in 1864, and James Roosevelt in 1868. Little hesitancy but solid realism marked Consolidation's approach to the coal business. During the Civil War the company surveyed a route from Cumberland to Hagerstown, looking to a junction with the Western Maryland Railroad that might break the strangle hold of the Baltimore and Ohio on the coal traffic at Cumberland. It bought thirty canalboats. Then the Pennsylvania railroad man, James Milholland, became president of Consolidation. In the two years after the war, from 1865 to 1867, the company better than doubled the amount of coal it shipped.

In 1870, after the abortive effort to secure a merger of the two companies, Consolidation bought out the Cumberland Coal and Iron Company, and emerged the unchallenged giant of the Cumberland coal trade. From this deal it acquired another railroad, more canalboats, wharves, barges in New York, as well as 7,000 acres of coal lands. Consolidation now owned about five-sixths of all the Big Vein coal in the Potomac, and could ship it out to the canal at Cumberland over either of its two railroads, or to the Baltimore and Ohio at the same point, or to Piedmont,

HARPERS

FERRY

where it also met the Baltimore and Ohio. Its superior transportation made the cost of its coal, laid down in any likely market, cheaper than the coal of any of its smaller competitors. It owned its own wharves on the river at Cumberland, Georgetown, and Alexandria, and in Baltimore; and it had a fleet of schooners to carry the coal to any of the seaboard cities. In 1871 the company clinched its position of dominance over the remaining independent mineowners, by securing authorization to build a two-mile spur connecting the company-owned Cumberland and Pennsylvania with the Pennsylvania Railroad, providing a third outlet. The Baltimore and Ohio Railroad might well grow apprehensive.

The 213,000 tons of coal shipped by Consolidation in the banner year of 1867 dwindled when compared with the 614,000 tons shipped in the year of the panic of 1873. Yet the difficulties of the depressions of 1873 and 1877 showed an inherent weakness in the company that dominated the region and was so largely controlled by capitalists in remote cities. Miners and boatmen, perhaps prodded a little by the Baltimore and Ohio Railroad, grumbled loudly at the wage cuts that came with the panic. But it was soon to appear that absenteeism was not the only issue.

To improve its position with the dissident elements in Maryland, Consolidation stockholders debated whether to move the company's offices from New York to Cumberland, but matters soon quieted down when substantial blocks of stock were purchased by the Baltimore and Ohio Railroad, and one of its rising vice-presidents, Robert Garrett, was added to the Consolidation's board of directors. Maryland feelings were assuaged still further the following year when a Baltimorean became president of Consolidation. In the careful words of the Consolidation Company's official historian, "Thus Baltimore financiers secured control of the largest bituminous coal producing company in the country, and the Baltimore and Ohio Railroad was automatically secured the large revenues from the coal trade of Alleghany County." No one was surprised when, in 1888, the president of the Consolidation Coal Company became president of the Baltimore and Ohio Railroad as well.

From this time until 1906, when a perspicacious Congress decided that railroads could not deal in the commodities they hauled, the Baltimore and Ohio cleansed itself by transferring its controlling interest in Consolidation to a carefully built Baltimore syndicate. Thus pure in law, it was difficult at the time to see what difference the new legislation made, either to the railroad or to Consolidation. The far-flung holdings of the company had already spread beyond the Potomac valley, south and west to the Fairmont district of West Virginia, and north to the Somerset district of Pennsylvania, both deep in the heart of Baltimore and Ohio territory. But freed from railroad control, Consolidation spread quickly into Kentucky, and in 1927 boasted of being the largest commercial producer of bituminous coal in the nation. Its recent merger with the Pittsburgh Coal Company, to form Pittsburgh-Consolidation, removed the last great competitor in the Northern Appalachian field. A colossus had emerged, its feet solidly planted in the history of the Potomac valley.

In its gradual concentration the coal industry was reflecting a process that occurred everywhere as the small local industries that had prevailed during the first half of the nineteenth century were absorbed or eliminated. The age of trusts was at hand. Even where no natural monopoly was possible, as in coal, because of the very abundance and diversity of the fuel, something akin to that result could be artificially created by manipulating the means of transportation. It was against this background, against corporate as distinguished from individual ownership and management, that the miners attempted to work out their role in the industry through nation-wide agreements for fair wages and working conditions.

As late as the eighties mining was still the hard, slogging job that had changed little in the decades since it began. A mineworkers' leader, who knew the work well from long years in the mine, David J. Lewis, described it in miners' language thus: "We first undermined it; then cut it at the sides; sledged it out of the middle; scrutched it at the top, and then lifted bottoms." Drills and powder were not used until later. Ventilation was crude, often only a shaft with a fire kept burning at the bottom to form a rude chimney. As in the "adit mines" of the British Isles, drainage was often easy because of the sloping floor of the mine tunnel, and the orange-stained boulders of Georges Creek illustrate the resulting pollution of streams throughout the coal country from mine drainage. In the mines, horses and mules dragged the coal carts to the surface, along the wooden track.

The mine mules, some lanky and some "four-foot" because of the limitations of the height of the coal seam, were ever-present in the coal country, and celebrated in song and story. One of the miner's songs that has persisted is the comical and spirited "Whoa! Mule, Whoa!" The wry lyric commences with the miner singing:

> I am the owner of a mule, the finest thing you'll find.
> It has two ring bones on his shins, and both his eyes are blind.
> He can kick a flea off his left ear, he never will stand still.
> I'm going to buy a hat for him, if it costs a dollar bill.

And goes the repeated chorus:

> Whoa! mule, whoa! Don' you hear me holler?
> Tie a knot in his tail, or he'll slip his collar.
> Why don' yuh put him on the track?
> Why don' yuh let him go?
> Every time yuh stop that mule, it's whoa! mule, whoa!

Other songs, gay and ribald and some sad, were heard. The journey from the town to the mine portal, and then the long underground walk to the coal face itself, produced their functional music, bold with the resonance of hillside and tunnel, heartening music to men in dangerous work, music to shorten the weary miles. Up and down the Allegheny Mountains the songs traveled with the miners along with the development of the industry.

As the Piedmont wheat country and the Great Valley had crossed the Potomac at right angles, spreading out to north and south, so did the coal region. It recruited new hands from mines in Pennsylvania and sent out the parties that opened fresh seams to the south in West Virginia. It became known that the Big Vein was only part of a seam of coal that ran 900 miles from Pennsylvania into Alabama. Yet the early Potomac mines maintained their individuality.

The predominant accent of the men of the Georges Creek mines was Scotch. Large numbers of English and some Welsh, following old family occupations, also entered the mines. But the capital, the enterprise, and the leadership seemed for long firmly held in Scotch hands. In the dogged obstinacy of the Scotch owners, as they clashed with the Scotch leaders of the Scotch miners, you could almost hear the ring of claymores. Much of the mining art, the drainage and ventilation, the pillar and stall system, even the plank roads and barges, duplicated Scotch and English methods. In the mining towns themselves, with their whitewashed stone cottages, standing wind-blown against the sere hillsides, there was often more than a touch of Scotland in the landscape.

The Scotch stood together. One of the periodic causes of labor difficulties in the early days was the tendency to throw some "undeserving Irish papist" out of a job in the mines, so that the son of a Scotch family in the town could be taken on instead at his father's trade. All this paled, however, by the time of the first great strike in 1882, which lasted six months and saw the first importation of Germans, Swedes, and Slavs as strikebreakers. When the depression in 1894 brought on the second great strike, the Germans and the Swedes were already out of the mines and back on farms. By the third great strike in 1900, there was no question of using cheap immigrant labor to break strikes, but the Scotch influence was still there among the mine-owners and the labor leaders, perhaps thinning a bit under the pressure of the environment, shading off to admit some characteristics of the mountain people everywhere in Appalachia, tempering itself to more modern times, blending finally with the Welsh and English.

Wedged between the coal fields lying to the north and south, the Maryland miners of Georges Creek early developed a sense of statecraft. Their union organizers were forced to recognize inconsistent laws in Pennsylvania, Maryland, and West Virginia, states which in this area are a scant twenty miles apart. They noticed also the different policies of mining companies contending against each other, and the varying wage scales and working conditions. Naturally they reached out for broader organization. To match the growing monopolies in coal and railroads, and deal with absentee ownership, they abandoned the Knights of Labor with its mumbo-jumbo; they steadily widened the miners' union until it became national in scope.

The watershed divide in the development of the new national unionism was the great strike of 1900, a four-month struggle when the miners demanded 60 cents a ton, and which ended

with their demands unsatisfied. Eugene V. Debs, Samuel Gompers, John Mitchell, William B. Wilson, "Mother" Jones, and other celebrated labor leaders and agitators converged at Lonaconing to hold mass meetings so large that Moran's and Jackson's halls and the Maryland Hotel could hardly hold the throngs. The speakers went from place to place, giving their messages of encouragement to the strikers. Seven bands rallied the crowds, and a big parade was held. But nearly five thousand miners in Georges Creek could not make the strike effective so long as thousands of other miners elsewhere continued to work.

The earlier threat of immigrant strikebreakers had been countered; but there remained the problem how to make effective a strike against one company, or one small region, while other companies and other mining districts were unaffected. If any doubts of this lingered by 1900, they must have been dispelled when the miners of Georges Creek saw the Consolidation Coal Company continuing to fill its orders for coal during the four months the district was strikebound, by the simple expedient of supplying the coal from other districts.

In the effort to secure effective industry-wide bargaining power, the miners of the tri-state region in the upper Potomac took a leading part. Their union leaders included some who grew to become political figures of national stature, like David J. Lewis, who had a long and distinguished career in Congress, and William B. Wilson, who became the first secretary of labor. The lessons learned along the Potomac were applied increasingly on a national scale, with ultimate success.

5

As the midpoint of the century passed, time seemed to be whizzing by with uncomfortable velocity. Apprehensive passengers on a train moving at twenty miles an hour would take out their pencils and write consecutive sentences to prove their mental powers were unimpaired. The day when a fast horse could outpace Peter Cooper's locomotive was far distant, and the railroad had outrun the turnpike and the canal. Increasingly its net spread over the valley. Heavier trains, faster trains, more frequent trains came on. The countryside responded to the railroad, and towns bloomed or faded with their distance from the line and the life-giving coal it brought to fuel what local industry survived.

The railroad had found along the Potomac a valley organized into little self-sufficient cells, their centers a day's journey by team or oxcart from each other. As such communities were knit together by the railroad, the old life began to disappear. Gone was the self-sufficient household that made its own candles and soap, its colorful quilts and carpets, wise in the arts of the kitchen and the pantry. In the village store the shelves filled with bolts of cloth from Lowell and Fall River, and the spinning wheel was moved to the attic or barn. The traveling tailor settled down in Baltimore, and ready-made clothes appeared in stores. The curing of hides on the farm, or at the village tannery, and the making of shoes by the village cobbler became a thing of the past as the shoemaking machines of Brockton poured out their flood of cheap footwear. The local coachmakers in towns along the National Road, even before the end of stagecoach traffic, yielded to the $500 and $600 coaches from Troy and Concord. Even the saddlers and harness makers put their heavy tools away and contented themselves with selling and repairing the gear bought at a distance. The railroad created a turbulence that left nothing unchanged.

The world of wood and water power was vanishing, and the self-contained and decentralized life it produced, the life known by Jefferson and Jackson, was withering away. Vines grew over the cold blast furnaces, weeds covered the slag heaps, and the banks of idle ore pits caved with the elements. The moss-covered wheel of the old gristmill was silent, its millpond filled with silt, the limestone banks of the millrace crumbling. The pleasantly situated hill towns that had been located on the well-drained slopes where the roads were solid, close to the forest and the rushing millstream, were abandoned, and along the railroad, deep in the valley, new towns were created. Whole villages disappeared, leaving only a sparsely attended church and a moldering graveyard, and moved down to the railroad. Wild animals returned to their ancient haunts near the abandoned towns; deer were seen in the grass-covered street of Trammelstown and browsing among the tombstones.

In the new railroad towns the emphasis lay increasingly on trade rather than production. A few old valiants fought it out, trying to mill wheat by steam power, or to manufacture some local specialty, but in most cases the forces of concentration were too strong. Here and there a few succeeded, raising on the outskirts of towns enterprises more suited to the new times. A Hagerstown might put its skilled craftsmen to work making organs and furniture, but the valley town of the future was destined to be a railroad town, like Martinsburg, or a farmer's town, like Winchester or Frederick. Only with difficulty could an industry rise above the mere extraction of natural resources from the earth. Trading and wage earning dominated. Shrewd manufacturers migrated to more favorable locations, and the valley became silent again save for the chuff-chuff of the railroad.

In the rapidly changing times, the bolts were shaken loose from established institutions. Education, the family, the church, political parties found themselves caught in the swirl of events. New schools, new faiths, new political beliefs sprouted. A man could arise in the morning a good Democrat like his father, and go to bed converted to the tenets of the new Republican party. In a bewildering period, people clutched their old beliefs with fanaticism. Many were undone.

The career of Francis Thomas spanned these eventful years, and reflected the disturbing qualities of a localism spreading into a sectionalism, a sectionalism expanding into a dynamic system of national organization.

Thomas emerged in public life in 1822 at the early age of twenty-three when he was elected in Frederick County to the Maryland state assembly as a Democrat on a platform of reapportionment—an issue closely linked to the slave question. After four terms in Annapolis, he was sent to Congress, where he served eight years in the House. In his triumphant career he was president of the Chesapeake and Ohio Canal Company, and a successful lawyer. He was elected governor toward the end of the long depression that had begun in 1837. His administration showed outstanding courage in refusing to repudiate the large state debt inherited from the period of internal improvements, involving the property tax for the first time as a major source of state revenue. To this point it could be said that his political development, while outstanding, showed little to distinguish it from that of many another liberal Democrat in the mountain regions of the eastern states.

A personal tragedy of extraordinary magnitude caused his retirement from public life for almost twenty years, and then on the eve of the Civil War, an almost legendary figure out of the past, he left his isolated mountain farm near Frankville and stumped western Maryland to rally citizens to the Union cause. His hoary figure appeared in the streets of Cumberland to harangue the people. He raised three thousand volunteers to form a Union regiment. Most important, he re-entered political life as a Republican, and in Congress became closely allied with Thaddeus Stevens in the extreme radical wing that attempted to impeach President Johnson, and forced on the South the horrors of the reconstruction period. As the brilliant but unsteady Stevens was influenced profoundly by affairs in his private life, so too was Thomas.

During his early years in Congress, toward the close of the decade of the thirties, Thomas had lived in Washington at a political boardinghouse occupied by senators and congressmen, among whom was the powerful Senator Benton of Missouri, then nearing the peak of his thirty years in the Senate. There he met Sally McDowell, a 15-year-old miss from Lexington, in the valley of the Shenandoah, a daughter of the governor of Virginia, who had been sent to Washington to attend boarding school in Georgetown under the supervision of her aunt and uncle, the Bentons. She did not go to school, but absorbed an education of a sort in the older society around the boardinghouse table.

Presently Thomas—a bachelor nearing forty—became aware that the Bentons were regarding him with a speculative eye. Friends told him they had matchmaking in mind, and the flirtatious Sally McDowell announced she had her cap set for him. A leisurely Victorian courtship proceeded over the next five years, and when Sally McDowell was twenty and Thomas forty-two, in June of 1841, they were married at Cob Alto, her father's home near Lexington. This brilliant match linked two leading families in Piedmont democracy, and it had overtones of national importance. Thomas was advancing rapidly in political life: during the summer he was married he campaigned vigorously as candidate for governor of Maryland from his home, Montevue, on the outskirts of Petersville, a short distance from Frederick. Already he was regarded as a leading candidate for the Presidency in 1844.

From the very day of his wedding, Thomas was uneasy and

apprehensive, yet head over heels in love. He was disturbed by the Virginia customs, by gay parties, kissing cousins, and the easy life, and his strict view of the proprieties of wedded life was further complicated by the larger public responsibilities he was about to assume in the public eye in Annapolis.

Within a month after the newly married couple had returned to Frederick, Thomas discovered an affair between his wife and a young cousin of the McDowells' he had taken into his law firm. In attempting to remonstrate with his wife, he only succeeded in arousing her protests. She teased him by saying, "Mind, I have not yet broken my marriage vow." When riding with Thomas and her cousin she amused herself by poking fun at some flaw in her husband's dress.

The apprehensive and elderly husband began to fear he was cuckolded, and in his mind every incident was magnified to horrific shape. He suspected clandestine meetings, and when his suspicions were confirmed, was led to further imaginings more difficult of proof. His mind swarmed with old law cases he recollected of marital infidelity, and he was trapped into dealing with his personal problem in the manner of a prosecuting attorney. He cross-examined his wife at intervals, closely questioned her friends, and caused her to write a letter to her cousin demanding that he return to Virginia. He investigated her past and dredged up her old love affairs, especially a 5-year-old flirtation with Senator Linn of Missouri, which had led Mrs. Linn to write indignant letters of protest to Mrs. Benton. Still, the demands of political life were pressing, and he was somewhat pacified when Mrs. Thomas told him she was going to have a child.

On returning from an electioneering tour, he learned by chance from a Frederick physician that his wife's child had miscarried. All the old anxieties were renewed. The day following his election he returned to the cross-examination of his wife with new zeal. He demanded to know what had happened in Charles Town, where his wife had gone to visit in his absence, and forced from her what he thought was a confession of abortion. He pressed the case, and learned that in Charles Town she had procured a drug. A hysterical female of the household stammered out additional evidence. Yet it was hard to know whether his wife was mocking him or not. The cross-examination continued to pile up shreds of circumstantial evidence that Thomas's imagination wove into a pattern convincing to himself.

He appealed to his wife, to her parents, to Senator and Mrs. Benton, to return her to Lexington, but Mrs. Thomas refused to leave, and the couple were established in the Executive Mansion in Annapolis before any satisfactory solution was reached. When Sally McDowell Thomas was prevailed upon reluctantly to return to her home, she delayed her arrival by stopping over with the Bentons in Washington, where the affair was rehearsed with the senator, who pronounced it "a sad delusion." The Missourian —his eyes filled with political stardust—still hoped that something could be done to patch things up, and tried to convince both Thomas and the McDowells that this outcome was possible as well as desirable.

During the development of the Thomas affair, an incredibly large number of people had become involved in it, owing in part to Thomas's methods and his retinue of lawyers, physicians, political and other advisers; in part to his wife's intemperate chatter; and in part to the disgruntled Bentons and the McDowell clan. The scandal spread and grew, its flames fed by the prominence of the individuals concerned, and fanned by political animosities. Thomas was obsessed, and could talk of nothing else: he dragged it into his discussions on the floor of the Assembly and in Congress. He fell out with old friends who would not listen to his ranting.

The gossip spread. It was said that Governor Thomas habitually locked his wife in her bedchamber; that he took two pistols to his wedding bed—and by repetition, to this was added the unfortunate Thomas remark that "one of them will do for her father, the other for her brother." The Thomas faction countered by describing the meager wardrobe brought by the "portionless bride" to her new home. The scandal flew far beyond Frederick and Lexington. Sectional resentment flared. In Maryland the governor's behavior was defended, while in the Shenandoah valley the relatives and friends of the McDowells loudly proclaimed the virtue and chastity of the Virginia governor's daughter. The truth was impossible to discover even by such a close participant as Mrs. Benton.

When all efforts to effect a reconciliation failed, including a trip to Virginia, where his life was threatened by the McDowells, Thomas sought a divorce. Under the laws of Maryland, a divorce could be granted by the county courts sitting as courts of equity, or by the High Court of Chancery. The Maryland divorce laws had been changed in December, 1844, at Governor Thomas's instigation. But there was no precedent for a divorce when one of the parties was outside the state and would not appear as a witness. Both sides employed the most skilled counsel of the day to master this dilemma, and such legal pillars as Reverdy Johnson, Francis Scott Key, and Walter Jones later appeared to defend Thomas.

Unable to effect either a divorce or a reconciliation, and harassed by the spreading gossip that was embarrassing to his family and damaging to his political aspirations, the desperate Thomas took the extraordinary step of publishing a statement of his domestic grievances to make clear his version of the affair. He detailed the results of his own investigations, marshaled correspondance from many persons concerned, and much other evidence to support his view and actions. He took care that one copy of his 50-page statement was laid on the desk of every senator and representative in Congress. Published in March, 1845, after he left Annapolis, his elaborate rationalization commenced with this revealing sentence: "Conspicuous calumniators, who have labored to forestall and pervert public sentiment, have left me no alternative but this, to me exquisitely painful publication." This obtained a response which repeated efforts to secure divorce proceedings had not brought forth. Thomas was sued for libel in Washington, and Mrs. Thomas sued for divorce in Virginia.

The Thomas pamphlet struck sparks everywhere, but they were nowhere brighter than in Lexington, where a public committee of professors, doctors, clergymen, and other worthies was appointed and passed resolutions affirming the high moral character of Mrs. Thomas. The New York *Herald* was moved to comment, "This is certainly a new mode of settling connubial difficulties—by public meetings and resolutions," and recommended it to the attention of such observers of the American social scene as Alexis de Tocqueville. In the end, Thomas was found guilty of libel, Mrs. Thomas won her uncontested divorce in Richmond, and in Maryland the marriage was dissolved the following year.

The affair showed a Maryland drifting farther from Virginia and the southern colonies, and acquiring more of the complexion of a northern state. In Tidewater the rising commercial and financial center of Baltimore found its affinity with Philadelphia and New York. In the mountains, the new source of power in the coal mines was a token of an industrial order that steadily advanced. Throughout the state the rapidly spreading net of railroads heralded the dissolution of an older order, the coming of a new one. The ancient political alliances were breaking down, but new ones were slow in appearing. In the cities the chauvinistic "Know-Nothing" party terrorized the electorate by bringing cannon and swivel guns to the polls, and the needle-pointed shoemaker's awl was so freely used to intimidate political opponents that it became a party emblem; Locofocoism was rife in the back country; the old whigs were riven by sectionalism. Against this background some of the episodes of the Thomas case acquired a significance that transcends the personalities involved, or even the sectional dispute on either bank of the Potomac.

Senator Benton, who had carefully matched his own daughter with one presidential aspirant, John Frémont, fully understood the dynastic elements in the Thomas-McDowell marriage. He could remember the days of Maryland's golden-tongued William Pinkney, who had spoken for the slaveholding states during the debates on the Missouri Compromise. He saw in Thomas a likely exponent of the Benton brand of post-Jacksonian democracy, who might even become an exponent of the peculiar Benton ideas of expansion and empire. He was keenly aware that, given only reasonable opportunities, Thomas might become president in 1844. In Virginia, political leaders could congratulate themselves they had captured Maryland's foremost Jacksonian. When the domestic rupture came that blasted this fine dream, their first response was to oppose any upset in their plans; and when this was impossible, they tried to blackguard Thomas.

His harsh experiences unbalanced Thomas. After his divorce, when he tried to recapture his political career, he found himself defeated when a clever opponent pinned the abolitionist label on him. In the seclusion of his mountain farm, his venom against the Virginians simmered during almost twenty years of retirement. When he emerged, on the eve of the Civil War, it was as a radical Republican. On the national stage this attitude led to extreme consequences, and the symbols of personal psychiatry were acted out as laws and national policies. It was a troubled and uncertain time, and nowhere more so than in Congress. When railroad politics had finally jimmied the Pennsylvania Railroad into the national capital, the headstrong leadership of the day was amply shown by a legislative provision that the new station be designed as a replica of the Lancaster home of Thaddeus Stevens! A people much given to symbols and portents, his countrymen might have been impressed with Thomas's death, when struck by a locomotive on the line of the Baltimore and Ohio. The political corruption of the national administration, the railway politics and frauds, were prepared by such revengeful monomaniacs as Stevens and Thomas. Bold shadows of national shame were cast by such figures, shadows that reached from the Potomac to far corners of the nation.

6

The early streams of settlers journeying to the west had scarcely taken up more room than a squad of soldiers on the march. Braddock's 12-foot-wide trail over the mountains was not completed as a route for heavy traffic to the Ohio until 1818, and even then the National Road was rough traveling and often closed in by the heavy mountain snows of western Maryland. To either side of the road, where cabins were built, they occupied at best a narrow strip at the bottom of a mountain valley, between a rushing stream and the beginning of a steep ridge. "Goin' up, you can might' nigh stand up straight and bite the ground," the native report ran, and "Goin' down, a man wants hobnails in the seat of his pants." As in the neighboring country to the south, tall tales were told of men who fell out of cornfields while hoeing, or of cattle which fell out of hillside pastures where they grazed. Where the drainage basins interlaced, marshes were found in the glades at an altitude of half a mile. Although agriculture found but a slight foothold here, the isolated mountains, covered with an impenetrable tangle of brush, remained an undisturbed hunter's world until well after the coming of the railroad. An exact description of it, Meshach Browning's *Forty-four Years of the Life of a Hunter*, when published in 1859, was received with the incredulity of a campfire tale.

Browning's autobiography appeared at a time when few men remained who had known from their own experience the days when deer and bear were plentiful, but when everyone could still recall the old tales of frontier hunting. The leatherstocking tales of Fenimore Cooper, idealizing the hunter's life, had prepared a large audience for such an authentic document. Browning's book showing a quaint survival of an earlier life passed through many editions. Its luxuriant detail and vernacular style were eagerly absorbed by clerks, mechanics, and other restless people caught in city routines who longed to escape to the free world of the hunter and the woodsman. Even the sons of wealthy farmers, cramped more and more into a specialized agriculture, read with enthusiasm of the golden age. With its enticing descriptions of 20-inch native brook trout caught by the hundred, fat 400-pound black bears, and 12-point bucks, with 80-pound saddles, and Browning's careful instructions on the hunter's art, a man could feel that with a stout heart he too might do what the great hunter of Garrett County had done.

Meshach Browning had moved west from Frederick County with his widowed mother, and lived a pioneer life of great hardship until he commenced to earn his own living. Almost from the beginning he was strongly attracted to the hunter's life, and acquired a reputation as a marksman and a hunter. Browning impressed General Briggs, who was about to direct a survey of the state of Ohio, by taking his rifle and putting three shots into a piece of paper no larger than a half dollar at a distance of thirty steps. To continue his pursuit of Mary McMullin, later his wife, he remained in the Maryland mountains, and his narrative describes the alternating ease and hardship, the isolation and perils, of those little squatter communities at the turn of the century. The occupations and pastimes of the frontiersman's daily life crowd his pages, gently rubbing against such frontier legends as that of Betty Zane, and the increasingly eventful episodes of the hunt.

Browning's forty-four years ran from 1795 until 1839, in which time, he says, "I think I found out as much about the nature and habits of the wild animals of the Alleghenies as any other man, white, red, or black, who ever hunted in those regions." Nor was this idle brag. It has been authoritatively estimated that in his career he killed nearly 400 bears, 50 panthers, nearly 2,000 deer, and innumerable wild turkeys, otters, coon, and wolves (which carried a bounty of $8). A professional hunter, he not only kept his family in meat, but regularly carried out to the settlements saddles of venison and bear, skins and furs, some of which he took as far as Annapolis, Baltimore, and Washington. Of all his hunting, he liked best to hunt the black bear.

"The extraordinary success which I had in bear-hunting," he writes, "requires some explanation, which I will endeavor to give. I always kept two good dogs; one of which walked before me, and the other behind. The one in front would wind the bear, and lead me up to him on that side on which he could not smell me, for I would come on him unexpectedly. If, by chance, he found us coming on him, and ran, the dogs would overtake him before he would be out of sight. The moment I would see one run, I would send the dogs after him; and as I could run almost as fast as any bear could, when the fight began I was close up, and a shot was certain death. In many cases, however, I killed them with my knife; but only when the fight was so close that I was afraid to shoot, lest I should kill a dog; which has often been done. I never in my life shot a dog in a fight; for I always took a knife in a close contest."

His many decoys and stratagems in finding and following game were matched only by his coolness in closing in for the kill. The limited power and accuracy of the long rifle, even in the hands of an expert marksman, made it necessary to get very close to the game before shooting, and even then, at a distance of thirty yards, the piece did not have sufficient killing power to finish off a large animal. Most of Browning's hunts ended with the hunter's coup de grace delivered by his long knife. Intimacy with such close fighting gave him a boldness that terrified even contemporary hunters, and during the winter he frequently entered bears' dens, roused them and shot them by candlelight.

Browning's remarkable physical strength and endurance in the snow-pocketed hills where the Potomac and the Youghiogheny rise are evidenced not only by his own accounts but by the manifest difficulty of skinning and hanging up the fruit of the chase singlehanded. In its wilderness state the country was even more rugged than it appears today, with a carpet of laurel so thick you could walk on it, and the high winds and winter snows of the Alleghenies. His success as a hunter, however, was due more to his canny pursuit of the game. Repeatedly his superior craft enabled him to outwit other hunters. He knew the animals he hunted, their feeding habits at different seasons of the year and different hours of the day, their breeding customs. When on the trail of game, he craftily approached upwind. Often he followed in sight of game, maneuvering for a better shot, standing motionless when the shortsighted bear looked his way, moving through the dry leaves only when his quarry moved also, calling to his prey to make it turn or rear to present a better target. He developed the art of shooting deer along the streams by candlelight, masked by a bark screen set in his canoe.

With all his self-reliance and industry in the woods and glades, Browning was ill at ease in company. He had little of the inbred poise of Cooper's fictitious Natty Bumpo. When he appeared in eastern cities, proud in his white buckskin hunting shirt with its fringed cape, to sell his meat and skins, he was much admired, but he loathed the eastern people and the life he saw there. Although he loved his wife and children—he had 120 descendants living at the time of his death in 1859!—he was unable to face their problems. Affairs of property distressed him, and he repeatedly fell into business disaster. At one point he lost everything, and his wife and four children were reduced to living in a small house, 10 by 12 feet, while he carved out a new home with his ax in the virgin woods. Among the little hunters of rabbits, squirrels, and partridges in the eastern part of Maryland, his own stories, even when watered down, were scarcely believed.

Browning's philosophy in hunting, as in life, was: "If a man undertakes a dangerous enterprise, with a determination to succeed or to lose his life in the attempt, he will do many things with ease, and unharmed, which a smaller degree of energy never could or would have accomplished." "If I got into a dangerous scrape, the greater the danger appeared, the more anxious I was to win the fight." His hunting experiences reinforced this conviction, for "a hunter may steal on a bear, and shoot him through the lungs, when he will die in three minutes; but if the dogs attack him, and worry him, he will live and fight for fifteen minutes, and perhaps longer, though wounded in the same manner." He expressed an ethic of hunting that made it more than butchery, and helped form the hunter's code that emphasized the chase rather than the kill.

In the glades of the Appalachian highlands, their native meadow grass so high "a man could tie it in a knot over the withers of his horse," the deer long browsed in great numbers. Their gradual extermination, Browning believed, was due not to killing by hunters but to the steady depletion of the rich grass by cattle grazed there for Pennsylvania and Virginia growers. "From April to September they crowded the glades with hundreds and thousands of cattle, eating, tramping, and running over every place in the glade country." The herders shot the deer without discrimination, and the feeble laws were so hopeless that the mountineers finally took matters into their own hands and shot the invading cattle by night in large numbers. This violent action temporarily halted the interlopers, but the National Road had made the territory too easily accessible and soon grazing was resumed with greater intensity until the game finally disappeared. The deer are still hunted high up in Garrett County, in the laurel swamps and over the huckleberry ridges of Laurel Mountain that Meshach Browning once climbed; and an occasional black bear is found. But with Browning's death, an era had ended in the Potomac country.

NINE

THE POTOMAC ROUTE TO THE WEST

THE INEVITABILITY OF THE POTOMAC ROUTE TO THE WEST HAD impressed itself upon thinking men all through the eighteenth century. The rich potentialities of the Ohio country and the Mississippi lands beyond were already well known. Imaginations kindled at the possibilities that might be realized once cheap transportation linked the productivity of the west with the port cities of the eastern seaboard.

Since the days of the Ohio Company, the men of Potomac had been ready to show by maps and mileages how much closer than any other part of the Atlantic coast they were to the Ohio. This widely held conviction was hardly of the least importance among several factors that had led to the situation of the national capital on the Potomac.

The merchants of Baltimore viewed the matter from a somewhat different perspective. They contended that Baltimore was the salt-water port farthest inland, closest to the Ohio, and argued that it was the natural port for the great prospective trade with the interior. The abrupt growth of their city lent conviction to this view. Yet the issue was still conjectural, and a merchant casting about for a place to settle after the Revolution might well have thought, as Benjamin Stoddert did, that the Potomac trading centers were just beginning to grow.

In 1783 Stoddert carefully weighed the respective advantages of Baltimore and Alexandria before determining to settle in Georgetown. He gave some ingenious reasons for his choice. In Georgetown there was but one retail shop, and the town "had not a man who had ventured five pounds on any foreign voyage." In the six years of peace following the Revolution, Stoddert's decision seemed amply confirmed. During those years Baltimore experienced a recession from its wartime prosperity, while the exports of Georgetown in the two years after the peace had come to surpass those of Alexandria, and in another four years amounted to half the total exports of Baltimore. Stoddert found the principal shipment was wheat, and he reasoned that, since Georgetown was nearer the wheat-growing regions than either Baltimore or Alexandria, it was bound to attract more of the grain trade. Even the decline of trade in the Potomac when the French Revolution began in 1789 did not seriously disturb his reasoning.

Other traders must have thought similarly, because the years following Stoddert's choice saw a significant movement of merchants to Georgetown. The names of Peter, Forrest, Corcoran, Riggs, and Laird showed the character of the immigration. Yet the Scotch were in Baltimore as well, and the future of the Potomac as a route to the west would be decided between them.

An impressive display of unity along the Potomac had marked the years that followed the Revolution. The war itself, the events issuing from the Mount Vernon compact, early plans for the development of the river (from the days of the Ohio Company), had all shown the essential identity of interest of the Virginia and Maryland sides of the lower Potomac. Even in western Pennsylvania, where the settlements were disaffected from the Tidewater capital of Philadelphia, a substantial measure of agreement was to be found on the development of the Potomac route to the west.

Yet as the commissioners sat at Alexandria and Mount Vernon to consider ways and means of cementing the bonds of unity with the west, a little more cement might have been spared to point up the already crumbling structure of unity in the Potomac itself.

From the high point of the Revolutionary period, the states

of Maryland and Virginia were drawing gradually apart. Perhaps the very creation of the states had something to do with this process. Whatever the first cause, powerful contributing influences soon became apparent. The days when interstate compacts could easily be framed soon became hazy as the heat of sectional strife marked the beginning of the internal improvements program. Each time a road or a canal was to be undertaken, the localities directly affected were to be found snarling at each other. The natural advantages of the Potomac route to the west, which could have been realized if the states had continued to act in concert, were dissipated in conflicts and delays over the choice of routes. The states concerned could not act together. Indeed, they were scarcely able to overcome local struggles within their own jurisdictions. All three states concerned with the development of the Potomac route to the west were ultimately led into absurd efforts to build all-Maryland, all-Virginia, or all-Pennsylvania routes, and within each bitter factions were ready to spring at each other's throats.

What this disunity ultimately cost the Potomac may be seen by looking at the New York metropolitan region today. Had the logical Potomac route to the west been pushed through without quarreling and delays, the great port city of the Atlantic might well have been in Chesapeake Bay. Norfolk and Baltimore, Georgetown and Alexandria, would have been great metropolitan centers. The Virginia Assembly was not joking when late in the eighteenth century it passed a resolution urging the merchants of Alexandria to bestir themselves lest the port of New York surpass them.

2

The Mount Vernon compact had declared that the Potomac River was to be considered "as a common highway for the purposes of navigation and commerce to the citizens of Virginia and Maryland, and of the United States . . ." but the compact itself singularly lacked mention of any positive measure to open the river to navigation above the Great Falls. For two miles here the river stormed its way over huge, jagged rocks to drop nearly 75 feet. In the high waters of late winter it was not unusual for gray pinnacles of ancient rock, which in normal season stood 40 feet clear above the rushing water, to be lost to view in the flood. Even above Great Falls, major obstacles to navigation occurred at Seneca Falls, and in the vicinity of Harpers Ferry, and minor ones at a score of other places. Yet between these falls and rapids were long, smooth stretches of the river that must have seemed more alluring to the boatmen than even they are today: and however naïve, for many years it had been a commonly expressed and soberly entertained view that "once the rocks had been removed from the river" all would be well with upstream navigation.

The commissioners at Mount Vernon had no occasion to advance proposals of their own for the development of the Potomac: two auxiliary proposals were well under way that would fill this gap. One was the plan to establish a navigation company that would open the river by removing rocks, establishing channels, and cutting a canal around the falls. The other was a proposal to construct a road connecting the headwaters of the navigable tributaries of the Potomac and the Ohio. Together the two developments would connect the Potomac Tidewater with the waterways of the Mississippi valley.

Both of these complementary schemes were well launched by the time the Mount Vernon conference took place, and were known in detail to the participants at that meeting. They had been the subject of large gatherings of Maryland and Virginia citizens, at both Alexandria and Annapolis, only a few months before the official commissioners of the two states met. Indeed, the legislatures of both states had already appropriated funds for the construction of the road; and the Patowmack Company, which was to undertake making the Potomac navigable, had been established and was successfully advertising for subscribers.

The aspirations of the Patowmack Company differed little from the old pre-Revolutionary navigation scheme of the promoter, John Ballendine, but they were pursued with great determination, and a larger measure of public support. The states interested had underwritten the enterprise to the extent of one-fifth of the estimated cost of slightly less than a quarter million dollars, the balance being covered by stock purchases of private citizens. Books were opened in Richmond, Alexandria, Winchester, Annapolis, Georgetown, and Frederick for the sale of shares, and by the late spring of 1785 nearly all had been subscribed—a sufficient measure of the enthusiasm with which the purse-bound and land-poor gentry of the Potomac after the war greeted the project. Construction of the canal at Great Falls, and the removal of obstructions to navigation between Harpers Ferry and Cumberland, were the two principal projects immediately launched by the company under the supervision of that pioneer in steam navigation, James Rumsey, of Shepherdstown.

At that time there was hardly a person in the United States who had seen a canal lock, and no canal had yet been built. Even in England only one canal was in operation, and its success was conjectural. Rumsey's design for the canal locks at Great Falls deserves to rank with his steamboat as an achievement of the first importance, and later English engineers who were consulted by the company, when it became apprehensive of the high costs involved, found little to change in Rumsey's original plans. News of the Great Falls canal was widely published in the United States and in Europe, and it was regarded as the principal engineering work of the eighteenth century in America. During the period of construction it was visited by many American promoters of other canals, as well as by a number of curious visitors both domestic and foreign. When successfully completed, it became an indispensable point on the serious tourist's itinerary through the New World democracy. But navigation on the upper river presented greater difficulties than had been expected.

From the outset it had been assumed that by "removing the rocks" the river itself could be sufficiently improved to make it navigable; but little was known of fundamental engineering, the rate of the river's fall, or the practical difficulties of clearing the obstructions. Underwater blasting—itself a considerable technical achievement—was undertaken to open channels. In rapid waters a system of chains anchored to rocks under the surface, and attached to floating buoys that could be removed during the winter, was highly valued as a means of working the boats upstream. The long, narrow riverboats drew hardly eighteen inches, and in some places a sufficient channel was provided by sinking saplings weighted with stones to form a veritable corduroy foundation to the stream.

Rumsey commenced his work with assurance, using his crew of workers on the upper river projects during low water, and moving them to the site of the Great Falls canal at other seasons. Almost from the beginning he experienced difficulties with the unskilled indentured workers who were employed, and especially with the rebellious Irish immigrants.

Often they ran away, and although punished by shaving their heads and eyebrows, it was difficult to keep them at work. They were disorderly, and by supplementing the company's ration of three gills of rum a day, managed to be drunk much of the time. In the course of blasting, Rumsey even lost a few workmen who, he calmly reported, "used the powder Rather too Extravagantly." Eventually the company gave up the use of indentured immigrants and fell back wholly on Negro slaves.

High waters during a series of unusually severe winters impeded even the work on the Great Falls canal. As the work dragged on, enthusiasm cooled among the shareholders. Fresh issues of company stock were offered, but they were not taken up, and serious delinquencies appeared when stockholders were assessed for their contributions. Some temporary relief was secured when amendments to the company's charter permitted sterner treatment of delinquents and the sale of stock to aliens, subscribers being found—a Potomac canal may have seemed logical to the Dutch—among the merchants of Amsterdam.

To demonstrate progress and reassure investors, the company determined to hasten the opening of limited stretches of the river. A town was laid out at Great Falls on land Washington conceived to be "very valuable," called Mathildaville, where a gristmill, sawmill, and forge, and later a market house, were erected by the company. Washington's secretary, Tobias Lear, wrote a helpful puff for the Patowmack Company in the *New*

York Magazine, or *Literary Repository*, under the title, "Observations in the River Patowmack, and the Country Contiguous, &c." The Maryland businessman, Thomas Johnson of Frederick, a boyhood friend of George Washington and his only compeer in a belief in the Potomac's destiny, succeeded him as president of the company. Matters steadied somewhat after these changes.

While the worst financial storms still raged, and even before the canal and its six locks at Great Falls were completed in 1802, boats began to accumulate above the falls, where their freight was carted across the stretch of still-unnavigable water to reach the lower river, where the little city of Washington now stood. With greater accessibility to the falls, the stream of curious visitors increased, and Mrs. Meyers's tavern at Mathildaville proved more valuable than the gold later discovered near the falls. The company secured an official seal, showing one of the narrow river batteaux, loaded with flour barrels, being poled into one of the locks; and behind one saw a perspective of the river, and another boat.

With the opening of the Great Falls canal, the cost of freight by river was reduced to less than half the equivalent cost by wagon. The year the canal was opened, traffic increased threefold. A severe shortage of boats developed. The astonishing total of 45,000 barrels of flour, together with much whisky, iron, and tobacco, was locked through, and the company received in tolls almost $10,000. In a burst of enthusiasm, the company declared its first dividend. Unhappily, it was to prove its last as well.

The passage of the Great Falls assured, the company embarked upon the improvement of the tributary streams—the Shenandoah, Monocacy, Antietam, and Conococheague—where it became mired in difficulties so great they could lead only to the eventual dissolution of the enterprise. Faith in the project as a whole declined. Particularism intensified. Along the Antietam a fund was raised for the improvement of navigation, but it was deposited with the company under strict injunction that it was to be used for no other purpose than the improvement of that stream. Along the Shenandoah dissatisfaction with the company reached such a pitch that eventually a separate corporation was set up to develop that tributary. In the financial doldrums, the company found it could not move forward to new undertakings, and could barely retire a little of its great debt each year, without regard to the dividend claims of its shareholders. Even lotteries failed to produce the revenue needed to improve tributary routes.

If further work to develop the Potomac route to the west was to be undertaken, a fresh stimulus would have to come from outside the company. An extraordinary request from the state of Maryland, in 1819, for a full statement of the company's condition reflected its growing concern with the failure of the Potomac navigation scheme. The undertaking of the 365-mile Erie Canal had clearly shown that nothing could be expected if the sleepy Patowmack Company was allowed to pursue its leisurely course of action. It also revealed the flaw in the basic assumption of the Patowmack Company that navigation in the river itself could be successful. Indeed, the error of slack-water navigation had already been recognized by the company. It had flirted with the idea of an all-canal route from the head of Tidewater to Cumberland, where it could join the National Road. But such an enterprise was beyond the resources the company could hope to command. When the joint investigating committee set up by Maryland and Virginia had completed its examination of the affairs of the Patowmack Company, it was obliged to report that the terms of its charter had not been complied with, and there seemed little doubt the company ever could comply. The most careful scrutiny failed to show that the river had been substantially improved at all. During the whole year, there were scarcely forty-five days of good navigation; as in the days before the improvement of the river, floods and freshets alone made it possible to navigate.

The new impulse in Potomac development was not long in arriving. The Chesapeake and Ohio Canal Company, chartered in 1824, began work in 1826, and received from the Patowmack Company all its property in 1828. But alas, the Potomac had no brilliant DeWitt Clinton to rally popular support and press matters to a speedy conclusion. It was to be another twenty-two years before the canal reached Cumberland.

3

While in Berkeley Springs—then called Bath—Washington had made the acquaintance of James Rumsey, whom he engaged to build a house for him on some lots he owned in that town. The Rumsey family had lived in Maryland for nearly a hundred years and it may be conjectured that James Rumsey moved from Cecil County to the Potomac to improve his fortunes. At Berkeley Springs he ran a mill where he ground flour and sawed lumber, and like George Washington and many other Revolutionary soldiers was impressed with the future possibilities of the place—enough to open a tavern "at the sign of the Liberty Pole and Flag," and to build there, so tradition runs, the first bathhouses among the maples and elms at the warm springs Lord Fairfax had earlier dedicated to public enjoyment.

In Rumsey's correspondence with Washington, matters of housebuilding are intertwined with reports on the progress of Rumsey's experimental riverboat. At the mouth of Sir Johns Run in 1784 the inventor had shown Washington a working model of his vessel and secured from him a certificate of endorsement, which he later used in obtaining a franchise from the state of Virginia, and to enlist the support he needed to continue his experiments. An air of secrecy surrounded his work, but in the inventor's search to develop some form of mechanical operation that would permit upstream navigation he did not turn at first to steam as the solution. Yet the importance of his "steamboat," as it was called, was instantly perceived by Washington, who was already preoccupied with the affairs of the Patowmack Company; and it is clear that he was sufficiently impressed with Rumsey's mechanical genius and business ability to nominate him as superintendent of the Patowmack Company.

During Rumsey's year with the Patowmack Company, which ended with his resignation in July, 1786, he was not able to make much progress on his invention, but at this time he did turn decisively to the use of steam propulsion. As in his earlier mechanical steamboat, he was still engrossed with the problem of mechanizing the hand operation of poling the boat upstream. His first crude effort in using steam was on a boat propelled by paddles and setting poles. The device worked, but it was so cumbersome that he soon abandoned it.

Hydraulic jet propulsion was Rumsey's ultimate solution, and with this means of power he was able to build a riverboat of three tons and demonstrate it successfully on the Potomac at Shepardstown.

Skeptics and hopeful people by the hundreds lined the banks on a mild December 3, 1787, the appointed day of Rumsey's public demonstration. The whole town and a large part of the county turned out. Among the audience were mechanics who had worked on the steamboat, dignitaries of the neighborhood, Revolutionary war veterans, businessmen well aware of the possible importance of Rumsey's invention for the river port town, and curious onlookers. Grouped at the ferry landing and occupying vantage points on the bluffs overlooking the river, they examined Rumsey's craft. It seemed conventional enough, save for the smokestack and boiler that protruded above the gunwale. Absent were the sails, the oars, the setting poles of ordinary river craft. Beneath the surface of the water a trough occupied the space where the keelson should have been, running three-quarters the length of the vessel. As steam power worked a cylinder, water was expelled through this trough to move the boat. Only a slight turbulence at the stern suggested an explanation of the mysterious movement.

Taking aboard a crew composed of his brother-in-law, Charles Morrow, and Dr. James McMechen, and a party of half a dozen ladies, Rumsey's unnamed craft was pushed out from the ferry landing into the current, where the engine was started. The astounded spectators saw it steady into the stream and begin to move against the current at a rate of four miles an hour. Between the steep banks it advanced up the river for half a mile, and then turned to make the return trip. As it came abreast of the spectators, enthusiastic cheers and applause rang out over the water. For two hours the vessel maneuvered up and down the river before steaming back to the ferry landing.

Then the successful Mr. Rumsey collected testimonials and certifications from General Horatio Gates and other notables who had witnessed the performance of the first steamboat.

With hardly more education than his mechanics, Rumsey had made the first practical demonstration of steam power on the water. Separated from his scientific colleagues and the skilled workmen of the port cities, he persevered. Yet, supported as he was by powerful friends and associates, he was not able to bring his steamboat to the point of its commercial application, or even to secure for himself protection in the use of his invention. He became involved in a long and acrimonious controversy with John Fitch, whose experiments with steamboats on the Delaware attracted the attention of delegates to the Constitutional Convention, a battle of invective continued by historians long after the death of both men. Indeed, the age of steam was beginning to fade before partisans ceased referring to Rumsey as "a would-be inventor" and to Fitch as "an inventive steam pirate." The successful steamboat awaited Robert Fulton, a long twenty years later, who established the first regular steamboat service in the world. But Rumsey was acclaimed by Franklin and Jefferson, and was honored in England and France for his work. In Birmingham, the steam pioneer, James Watt, offered to take him into partnership.

Rumsey's scientific efforts encompassed more than his steamboat. His design for a water-jacketed boiler was an important forward step. He developed a new type of pump, and incorporated it into his design for high-speed mills to secure greater efficiency from water power. Together with his canal locks for the Patowmack Company, this body of inventions bulked impressively in the brief decade of their origin. Without the protection later afforded by the Constitution, Rumsey had to fight unsuccessfully against favorite sons in the legislatures of individual states to obtain the recognition and reward rightfully due him. The knowledge of these difficulties, which had in part caused his departure for England, played a share in the Convention's desire to protect inventions by patent. More significantly, the successful demonstration of the feasibility of steam navigation caused men once again to reappraise the importance of inland waterways like the Potomac.

4

Checkered with discord and strife, the progress of the Chesapeake and Ohio Canal was hardly better than its ill-fated predecessor, the Patowmack Company. It commenced in fear rather than belief, fear of what the Erie Canal would do to capture the western trade rather than belief in the Potomac route to the west. The Canal was also born at a unique moment. When the merchants of Baltimore were finally persuaded to lend their indispensable support to the project, they still nourished the belief that a branch canal could be built, perhaps taking off from the Potomac at Point of Rocks or Seneca, and making Baltimore, in effect, the largest city on the Potomac River. When that dream was blasted by hard engineering facts that later came to light, the Baltimoreans promptly withdrew their support. From that point on the canal not only had to make its way alone as a Potomac valley enterprise; it soon had to compete with the new means of transportation the Baltimoreans had settled upon —the railroad. Indeed, it was a much-remarked coincidence that on that same July 4, 1828, the venerable Charles Carroll of Carrollton was laying the first stone of the railroad in Baltimore, while President John Quincy Adams in Georgetown had removed his coat to turn over the first spadeful of soil for the canal.

Had it not been for the original interest of the federal government, it is doubtful if the Canal would have been begun at all. Although limited to its initial subscription of one million dollars of canal stock, Congress authorized the three cities of Alexandria, Georgetown, and Washington, within the federal district, to subscribe a total of another million and a half dollars. Of the original capital stock, hardly an eighth was held in Maryland and less in Virginia, busy with her own canal scheme. And so badly had the cost of the work been estimated that the funds initially provided were scarcely enough to carry the canal as far as Point of Rocks.

Disunity went further. Even the three towns on the Potomac were unable to agree on the location of the canal. Georgetown pointed out that there was no reason for the canal to go any farther than Rock Creek. Washington City, as they all called it in those days, which had a municipal canal system designed by L'Enfant leading to the principal markets, felt that the Chesapeake and Ohio Canal should connect with this system. In Alexandria the citizens turned up their noses at the possibility of using the broad Potomac itself, and eventually organized a separate canal company which threw a viaduct across the river at Georgetown, and brought the canal down the Virginia side of the river to Alexandria. Thus all three cities were satisfied, but all of them nearly went bankrupt in consequence, and the feeder canals below Georgetown soon fell into disuse. The total volume of traffic produced by the canal, especially in its early days, was insufficient to support them.

In Georgetown the canal began at the Rock Creek basin, and climbed steadily up the Maryland side of the river, taking its water from the small tributary creeks and crossing the larger streams, like Seneca Creek and the Monocacy, on masonry viaducts built for the ages. In seven years construction had reached the first mountains at Point of Rocks. There it found the railroad, which had followed the line of the old wheat wagon road to Frederick, and across Carrollton Manor to the same gateway to the interior valleys the canal proposed to use.

Where the southern edge of the Catoctins nuzzles the Potomac at Point of Rocks, a precarious shelf ran at the base of sheer cliffs of bare stone. Scarcely forty feet wide, it was for this strip of right of way that the canal and railroad contended for four bitter years.

The controversy over the right of way at Point of Rocks revolved legally around the canal's claim to prior rights in the Potomac, and the courts eventually upheld this claim. Yet the years of political maneuvering showed only a bankruptcy of all constructive will in Maryland, the narrow views of both the canal and the railroad, and a hopeless compromise by which the state eventually shepherded the two routes along the narrow strip between Point of Rocks and Harpers Ferry and there specified that the railroad should stop until 1840. Another satisfactory route existed—but in Virginia! Extraneous matters kept intruding themselves into an issue that might better have been settled by concord between the states, or at least a few thousand dollars' worth of blasting powder. In the end the railroad submitted to the obligation to purchase 2,500 shares of stock in the canal, and the state imposed an array of conditions to meet the objections of the canalers. It specified fences to be built separating the steam engines from the horses on the towpath, and at one point the railroad even agreed to return to the use of horse-drawn trains over the narrow stretches where the railroad closely paralleled the canal. Even the flour milling industry, which had long demonstrated its ability to control rates on the railroad, was able to obtain a specification that no additional gristmills should be established along the contested route of the canal between Point of Rocks and Harpers Ferry.

The difficulties of the two systems were postponed by the compromise, but at best only until they reached Harpers Ferry, and by then, in 1836, there was no more readiness to deal forthrightly with the issue than before. In the internal improvements bill of that year, equal treatment was accorded the railroad and the canal, the state of Maryland subscribed three million dollars' worth of stock in each, as part of a pork barrel that included funds for a railroad on the Eastern Shore, plus a sop to Baltimore in the form of a crosscut canal from the Potomac to the Chesapeake Bay port city.

If the state of Maryland alone, to say nothing of Virginia and Pennsylvania, had been able to determine upon one route and one method of transportation, it would have undoubtedly recaptured a large part of the Ohio trade even at this late hour. By now, however, the disintegration was becoming so complete that it almost amounted to paralysis. No measure of statecraft would have been adequate to bring the opposing forces together.

The issue was hardly limited to Maryland. The federal government was largely interested in the canal, not only because of its importance to the national capital on the Potomac but because of its interest in the National Road. The Baltimore and Ohio Railroad found this out when it sought a federal subscription to its stock, only to learn after committees of both houses of Congress had reported the bill favorably that it was blocked by the chairman of the Committee on Roads and Canals of the House —who happened also to be president of the Chesapeake and Ohio Canal.

With all its obvious difficulties and limitations, and they were readily seen by shrewd contemporary observers, the canal had the strength of an established enterprise against the newer form of travel. This was the golden age of the canal-building era. By comparison with the Erie Canal, people could readily see what a Potomac canal would accomplish. Wild speculation in lands and trade accompanied its halting march to Cumberland, and each new spurt of construction was the signal for another boom.

But by the time the canal reached Cumberland it was already a stale issue. Men were living whose fathers had long since lost their fortunes building warehouses on speculation, or buying lots along the canal's route in anticipation of a rapid rise in prices. The principal speaker at the canal opening ceremony in Cumberland put it fairly, admitting that "Many of us were young when this great work was commenced, and we have lived to see its completion, only because Providence has prolonged our lives until our heads are gray. During this interval of four and twenty years, we have looked with eager anxiety to the progress of the work up the valley of the Potomac. That progress has been slow—often interrupted and full of vicissitudes. At times the spectacle of thousands of busy workmen animated the line of work, when to all human calculation, no cause was likely to intervene to prevent its early completion. But when we have turned to look at the scene again, it was all changed; contractors and laborers had departed and the stillness of desolation reigned in their place. Thousands have been ruined by their connection with the work, and few in this region have had cause to bless it."

The hopes of a brilliant transportation route, with packet service and "two-storied canal boats," had so dwindled by the time Cumberland had been reached that the canal officials looked principally to the coal traffic as their means of support. Its slowness, and the fact that its use was limited to an open season of about nine months, made it a hopeless competitor to the railroad in almost everything save the bulky and imperishable commodities. Yet the Potomac was rich in these, and the hauling of coal, iron, flour, and such limestone products as cement, plaster, and building stone gave the canal a prosperity of a sort that lasted almost through the century. At one time its advantage in this traffic was so great that hopes revived that the canal could be extended westward into the heart of the coal region; even the old plan for the Maryland canal to Baltimore was brought up again.

Measured by its promises, the canal was a failure. It did not become a great factor in the national transportation system, like the Erie Canal, linking the states of the Atlantic with the Mississippi. Like most canals, its importance was limited to the immediate region of the Potomac. Viewing the results, it would be difficult to say what national purpose was served that could justify the initial assistance rendered by the federal government. The fate of the canal also marked the tendency of the Potomac to recede from its position as one of the great American rivers, important in the life of the nation, and to become just another regional artery. The railroad alone prevented that from becoming altogether the case.

Limited as its impact was on the immediate region of the Potomac, the canal left there a lasting impression. Quiet agricultural villages along the canal, like Shepherdstown and Williamsport, which had already commenced to decline, took on new vitality as mills were built to grind wheat and crush limestone. Farms acquired greater value because of their proximity to the canal and the markets it reached; and many a farmer paid off his mortgage with the added income from the sale of hay, feed, and provisions to the boatmen. The relatively brief period of construction alone was notable, bringing a temporary prosperity to many new towns, and laying down a residue of new immigrants, like the silt of a river's flood plain, which remained as a permanent enrichment of the valley's population. Once cheaper transportation was made possible, new industries were born.

Like the turnpike and the railroad, in its modest way the canal also helped dissolve the barriers of localism in the agricultural compartments of the Piedmont and the valleys, bringing a cosmopolitan tone from port cities and larger markets high up the river. Not least, it contributed to the unified economy and the unified regional consciousness of the valley.

In the hard life of the canal for nearly a century lived many families, whose boat was their only home. The growing children caught a few fleeting months of school at some place where their boat was laid up for the winter, and a few days' work for the captain or his team of horses could be had. The survivors of this experience, children born and reared on canalboats, live today up and down the Potomac; yet so rounded are memory's corners that hardly one can be found who looks back on the experience with regret. To them it was an eventful life. They had their games, their songs, and their lore of boating. In the long delays at the locks, or during the days waiting in the basins for cargo there was plenty of time for these traditions to mature.

The horse's pace of two miles an hour spun extra verses to many a ballad, and the native gift for recitation found abundant expression. The murder of Captain Miller, a boatman, found its way into the lugubrious minstrelsy of the canal, and was later written down by Maurice Matteson from the singing of John Feldman of Eckhart. It begins:

> Himself a gay young fellow, as you can plainly see,
> His name was Johnny Howard, and a noble lad was he.
> He boated on the waters for many a night and day,
> Until he met a Negro, who swore his life away.

The singer then tells how Captain Miller, Johnny Howard, and the Negro boatman sailed the canal together, until one day

The Captain he got rageous and in an angry passion flew,
The hatchet in his hand young Howard for to slew.
Johnny being informed of this, and in no way slow,
Quickly picked up the spreaders stick and laid the Captain low.

The ballad continued to narrate the trial of the unfortunate Howard, and his conviction on the evidence supplied by the Negro boatman, giving a close version of the case recorded in the courthouse in Cumberland, each verse ending with the chorus:

My name is Johnny Howard, and that fact I'll ne'er deny.
For the murder of Captain Miller I'll hang on the gallows high.
The morning of my execution, heart grieving for to see,
My sister came from Jersey, to take farewell of me.

The native song was a literal art, minute in detail, and shrewd in characterization; it exhibited a legal erudition—of the sort gained loafing on the courthouse steps. When the choice lay between poetry and fact, the songster inclined toward fact. In his telling of tales, the solid mosaic of facts rather than freely wove fancies haunted him. As literature it resembled the canal-boats themselves, substantially loaded with cargo, slow-moving, perhaps not graceful but certainly characteristic.

Few episodes along the Potomac cut as deep as that of the Chesapeake and Ohio Canal, and few so spanned the unbroken years from the colonial dream of a western route to our own time. As it mirrored the crumbling regional unity, the uncertainty, the waverings, and the doubts that marked this land, so too its long history commemorated the fundamental stability of the valley with its wheat, and coal, and limestone, until the canal's very survival into the twenties of this century found it a beloved anachronism in a wholly mechanized world.

5

By 1802 the drive to the west had so filled Ohio that a new state was created. Congress recognized in its Enabling Act those fears of disunion which had long been expressed by everyone who knew the west, and took an important step to tie the bonds of unity. From the sale of public lands in Ohio, one-twentieth of the proceeds were to be used for building roads connecting Ohio with the eastern states. The "cement of interest," as the contemporary expression went, was to be applied. Nearly a decade would pass before the road fund was sufficient to undertake any construction, but the first National Road had been begun.

The years to 1811, when the National Road—also called the Cumberland Road—was finally approved, saw dramatic confirmation of the fears expressed in 1802. It saw the Burr conspiracy, and the Blennerhasset affair. It saw concrete steps to deal with them: the Louisiana Purchase and steadily increasing immigration. And it saw progress in planning the roads to be built.

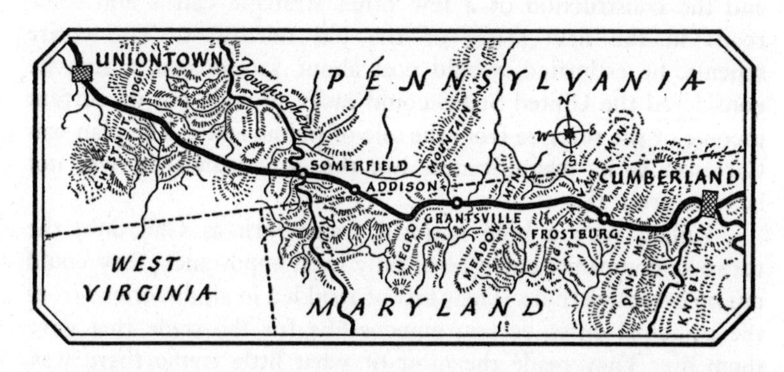

It did not require much reflection on the part of a special committee of Congress to report, in 1805, that the most practical route for the new road would be the one laid out by Nemacolin and Gist, which had been followed by Braddock's troops and later used as a post road and a deeply rutted summer road for migrants. Lying between the head of Potomac navigation at Cumberland and the Ohio River, this provided logical connections between the Maryland turnpikes reaching out from Baltimore and the Potomac port cities and the beginnings of practicable routes through the populous sections of southern Ohio. It took the shortest distance over the mountains, and the government's limited funds would here be spent most economically to link roads existing or proposed. Everyone who knew the west agreed the route chosen was inescapable.

Grasping for the rich trade of the Ohio were many seaboard cities, whose half-formed projects were concisely set down by Secretary of Treasury Albert Gallatin, in a classic *Report on Roads, Canals, Harbors and Rivers*—a document that might well be called our first national plan. Gallatin's scheme envisaged a protected coastal waterway from Massachusetts to Georgia, paralleled by a north-south national highway linking the principal Atlantic port cities; the development of four principal rivers flowing into the Atlantic, from the heads of their navigation; the construction of four turnpikes across the mountains to link the eastern rivers with four tributaries of the Mississippi; and the construction of a few other strategic canals and some roads in the new states of the old northwest. The entire scheme, he estimated, would cost about $20,000,000, a sum he considered the United States could easily provide out of current income or the revenue from the sale of public lands. The plan was Cartesian in its magnificence, but the American twig was not bent to grow that way.

Failing such a bold national framework as Gallatin's, the turnpike companies could be of only local importance; they could tap no extensive trade. Their bits of road led in all directions from the principal cities, feeling amoeba-like for the trade that gave them life. They made the most of what little traffic there was, building where the need was most obvious and the profits most sure. On the national map they showed as so many disconnected ganglia. In a few cases they were superbly built, but more frequently they were not, and barely a third of the pikes authorized were ever constructed.

Most important of all, the turnpike companies were chartered by the states, and they built only within state boundaries. Where the turnpikes approached the difficult mountain slopes, they petered out. They shied away from state boundaries. The turnpike companies had neither the taste nor the ability to pursue large-scale ventures. Into the tangled mountain wilderness of western Maryland, where three states stood between the Potomac and the Ohio, their leisurely penetration reached no farther than Cumberland, and then only when Baltimore banks came to the rescue.

The big covered wagons with their loads of wheat and flour, rolling down the Patapsco valley from the Monocacy, had quickly made Baltimore a city. People could scarcely remember the days of 1752 when John Moale's "Panorama of Baltimore" showed but twenty-five houses. The world trade in wheat had created a cosmopolitan Baltimore, a port city whose people were at home in Liverpool, Le Havre, and Bremen. The secrets of the Chesapeake Bay shipbuilders, the lessons worked out in constructing fast privateers during the Revolution, were now applied to the building of fast Baltimore clippers. With its ocean-carrying trade firmly established, and Europe clamoring for cheap wheat, Baltimore turned to the problem of getting the grain to its wharves. Her men returned from foreign lands with news of the latest advances in transportation. In Europe they had seen turnpikes, canals, railroads; now it remained to apply these lessons at home.

A flurry of turnpike construction had pushed out roads to the Potomac at Georgetown, then to the Susquehanna at Columbia and Havre de Grace. But the most important line led west to Frederick, to Hagerstown, and eventually ended at Cumberland. Over this route came the crops not only of Maryland valleys but of farms north and south in the trough of the mountains from the Genesee to North Carolina. Valley towns like Winchester were bound tightly to the port city of the Chesapeake by the road that crossed the Potomac at Williamsport. At other places along the Potomac, at Point of Rocks, Harpers Ferry, Shepherdstown, and Hancock, the "gundaloughs" ferried wheat across the river to reach the Fredericktown Pike, which led to the Baltimore markets. Rival ports from Head of Elk to Alexandria faltered in mid-career as the wheat was drained away to the booming metropolis on the Patapsco.

Yet the Baltimore merchants were not content. The slow wagon traffic along the turnpikes was scandalously expensive. You could bring a ton of merchandise 3,000 miles from Europe for $10, but that amount of money would hardly move the same goods 30 miles by wagon. Cheaper transportation was essential. Their vision traveled far beyond the plenteous wheat fields of Maryland and Virginia and contemplated the virgin farms of the Ohio where grain was selling for less than 50 cents a bushel. To get this limitless supply of cheap grain to Baltimore and sell it in the markets of the world was the challenge that stirred their blood.

Matters on the Cumberland Road moved with exasperating deliberation. Three years passed before the federal commissioners found they had $12,000 from the sale of Ohio lands that could be spent for road work. On their recommendation the road was laid out, and the states of Maryland, Pennsylvania, and Virginia were requested to consent to its construction. From the dragged-out political turmoil emerged a nicely balanced compromise. The route ultimately agreed upon followed generally the old road to Uniontown, and Washington, Pennsylvania. There it left the old Indian trail for Wheeling, where across the river in Ohio began Zane's Trace.

Congress took no chances in its detailed specifications for the first National Road. Four rods wide—60 feet—it was to be, and within these dimensions eight wagons were later to drive abreast. It specified the kind of road down to the kind and size of stone to be used by the road contractors—usually farmers along the route who commanded a quarry of some sort—and the types of permanent bridges to be erected. Finally, in 1811, work commenced. Once construction had started, considering the lapse of two years because of the war, it moved with rapidity, individual sections of the road being opened to traffic as they were finished. By 1818 the Ohio had been reached. A dense traffic immediately began to use the road.

From the time the Cumberland Road was completed in 1818 it received negligible maintenance, yet so well had the work been done that fifteen years later engineers found that while rough it was still a good wagon road. "The old Cumberland Road," Captain Delafield reported, "has worn sixteen years, and mile after mile has never been known to cut through at any season." Yet by the time the western sections of the road had been completed, the eastern sections were already wearing out under the heavy wagons.

When it became impossible to ignore the need for maintenance and repair in 1822, Congress agreed to provide the necessary funds. At this point it found a hardly unexpected obstruction in President Monroe. "Approving, as I do, the policy," the Virginia constitutionalist stated in his veto of this measure, "I am compelled to object to its passage." The president ignored the fact that the government had already built the road, and reminded Congress of his belief that the Constitution authorized no system of internal improvements. Although the Supreme Court, three years before, in the case of Mr. McCulloch, the cashier of the United States Bank in Baltimore, had outlined a somewhat broader view of the government's powers, he thought a constitutional amendment should be sought. However, in his "Views on the Subject of Internal Improvements," accompanying his veto, the president threw out a strong hint that so long as a national purpose was served, the power to raise funds carried with it the power to appropriate money for general purposes. Acting upon this suggestion, an appropriation the following year succeeded in gaining executive approval, and repair work commenced. When the repair program was completed, the road was turned over to the states, which had no constitutional qualms about collecting tolls. This episode, while negligible in the development of the Cumberland Road, was a significant step forward in the gradual evolution of public policy in matters of internal improvement.

Again the Potomac valley forced the development of national policy. But where the Mount Vernon Compact and the Annapolis Convention had emphasized the unity of the states, the Cumberland Road lance-headed a program of internal improvements over which the states were to become increasingly divided along the frontier of the Potomac River. The middle states of the Atlantic seaboard, and the states of the central west, were solidly behind the Cumberland Road, as they were behind the more comprehensive program. New England, save for eastern Massachusetts, and the south opposed it.

The road marked a decisive step in the progress toward a new kind of United States, a government dedicated more to supplying the services necessary to the nation's prosperity and welfare. As the people's idea of a federal government evolved, the road was a veritable bell, one that rang more soundly than the tinkling elegancies of strict constitutional construction. Its peals bespoke a rising commercial class as well as the burgeoning political authority of the growing west, whose Clays, Calhouns, and Jacksons, and lesser political figures came to Washington over the new road, their beaver hats chalked to evidence that they traveled free. In a way the road also stood for the power and accomplishment of the new government, although as the years dragged on and the road slowly made its way to Jefferson City, Missouri, a little at a time, as the revenue came in from the sale of public lands, it became in many minds an emblem of slowness and inefficiency, of governmental incompetence in handling construction work.

In rebuilding the road before turning it over to the states, which were to collect tolls and keep the road in repair, Congress had to meet the requirements of the states, which naturally wished the best construction possible in order to ensure the lowest eventual maintenance costs. They invoked the system of the Scotch engineer, John Loudon McAdam, whose English coaching roads commanded the world's admiration. McAdam had spent his youth in America at the dawn of roadbuilding, and this stimulus must have entered largely into his later engineering experiments. From well before the time he first systematized his conclusions in *A Practical Essay on the Scientific Repair and Preservation of Roads* (1819) his ideas had been widely influential in the United States; with this publication they acquired here, as in England and France, the authority of dogma. Fundamentally McAdam's system started with deep foundations, on which layers of rock were laid, beginning large at the base and growing smaller as the surface courses were reached. At either side were ditches, and the road had a distinct crown. The top of the typical McAdam road was tightly covered with gravel and dust, then rolled until bound together. Thus, much water was shed from the road, and none could percolate to the surface through the loose foundation courses. McAdam had struck upon perfect drainage as the central factor in good roads. Reliability was introduced in public works. Even more important to America, with its chronic shortage of skilled labor, his sytem permitted doing away with expensive and laborious methods of laid masonry courses with all the craftsmanship they implied. It made roadbuilding cheaper and quicker, permitted the use of less skilled workmen, and led directly to the use of machinery in roadbuilding, at first for crushing and sizing the stone and later for rolling and other steps in construction. To a nation of continental scale it was a capital discovery.

It was the McAdam system on which the army relied. The Corps of Engineers drew from McAdam's book the instructions to its officers in the field. When the unfortunate Lieutenant Mansfield attempted to secure the War Department's permission to deviate from these instructions, he was gently informed: "For these principles and rules you are referred to Mr. MacAdam's work on the construction and repair of roads, a copy of which is in your possession." Mansfield's more experienced successor, Captain Delafield, was able to secure the department's consent to similar changes he wished to make by citing extracts from *Macadam on Roads* to support his views.

Breaking stone and passing it through a 3-inch ring was well enough for farmers who had a quarry and wanted to earn a little extra money, but the later roadbuilders were of sterner stuff. They brought in gangs of Irish laborers, recruited at the wharf. One who saw them at work later recalled: "That great contractor, Mordecai Cochran, with his immortal Irish brigade—a thousand strong, with their carts, wheelbarrows, picks, shovels, and blasting tools, graded the commons and climbed the mountain sides, leaving behind them a roadway good enough for an emperor." In isolated regions, opened up by the road, the immigrants with their brogue, tight-fitting breeches, and stiff hats were stared at by the natives who had never seen such curiosities. The workmen drank and brawled, and men tried to keep their womenfolk inside of nights. The murder rate went up, and there were riots between the Far-downers and the Kilkennys. But the road went through.

By 1835 the observant John Pickell, who had been assigned to the National Road as a young engineer lieutenant, could find it "literally covered with horsemen, wagons, and other vehicles, forming an unbroken line, wending their way over its smooth but hilly and mountainous surface, to the far west. Thousands of immigrants, almost daily, were seen . . ."

From Cumberland up Will's Creek the road ran through the rugged Narrows that had enchanted President Taylor. Then it climbed up Braddock's Run into Frostburg, overlooking the valley of Georges Creek. Up, over Big Savage Mountain, and down again to cross the Savage River. Up again and again, over bald Keyser's Ridge, the backbone of the mountains at an altitude of nearly three thousand feet, and again the long western glide down to Somerfield, just above the Turkey's Foot of the three-forked Youghiogheny. So the road over the mountains unrolled, through compact and isolated valleys, separated by shaggy hills of ancient rock. Each of these places in time acquired a personality of its own in the minds of travelers. The very place names were vulgarized and overlaid with countless meanings.

Distance became dramatic, its eventfulness marked in the names along the pike: creeks became "Two Mile Run," taverns "The Six Mile House." Each weary mile was commemorated by a stone marker, giving the distance to Wheeling, or Cumberland, or to way towns.

One thought of the snow pockets of western Maryland around Frostburg, with drifts twenty feet deep and the drivers leading their teams off the road and along little used but more sheltered by-routes. Of drivers changing their loads to sleds for the trip over Negro Mountain, or shoveling out the wagons when nine miles in three days in the deep snow was counted an achievement. Or the windblown, icy glare of the steep descents on the eastward slopes of Laurel Hill, when it was not uncommon for a heavy wagon with chains on its wheels, braked by whole trees lashed to the axles, their branches dragging the road, to lurch out of control.

Around the 15-foot hearths such episodes grew with the telling into a salty tavern folklore. In one of the wagon stands James Murray said "he saw the wind blow so hard on Keyser's Ridge, that it took six men to hold down the hair on one man's head." A strong man was allowed to be "equal to a six horse team with a cross dog under the wagon." When a wagoner went east, he "left his religion on the Blue Ridge." With their 3-cent whisky, and their four-for-a-penny stogies, the drivers would talk the night out in the tavern bar and then curl up on the floor to "sleep like a mouse in a mill."

Long after the road was gone, the memory of the busy excitement of youthful days lingered fondly. Late in the century an old wagoner wrote: "I have stayed over night with William Sheets, on Nigger Mountain, when there would be thirty six horse teams in the wagon yard, one hundred Kentucky mares in an adjacent lot, one thousand hogs in other enclosures, and as many fat cattle from Illinois in adjoining fields. The music made by this number of hogs, in eating corn on a frosty night, I will never forget.

"After supper and attention to the teams, the wagoners would gather in the bar room and listen to the music of the violin furnished by one of their fellows, have a 'Virginia hoe-down', sing songs, tell anecdotes, and hear the experiences of drivers and drovers from all points on the road, and when it was all over, unroll their beds, lay them down on the floor before the bar room fire, side by side, and sleep, with their feet to the fire, as soundly as under the paternal roof."

The wagoners' life was hard but independent. Most men owned their horses and wagons and contracted independently with the shippers. Drawn by six horses, the commodious boat-shaped Conestogas were a peak of the wagonbuilder's art worth comparing with another contemporary marvel, the clipper ship. Red and blue the long, narrow bodies were painted, under arched white-canvas tops. When the road began they could haul 6,000 pounds, but this was speedily increased. The "eighty hundred load" represented a great advance. Toward the end of the traffic on the road, a maximum load of 12,000 pounds had been reached.

The heavy loads suggested favoring wagons with wide rims, and as loads increased the earlier narrow-rimmed wagon wheels were gradually replaced with standard 4-inch rims.

Over the National Road flowed an eastbound traffic of bacon and hams, whisky from the Overholt and other western distilleries, Ohio tobacco, lard, cheese, flour, corn, oats, and other farm produce. Westbound the traffic was lighter, and included paint, hardware, merchandise, and the great winter traffic in oysters to Pittsburgh that soon caused the establishment of a special express service for this article.

While the economic importance of the National Road was carried in its wagons, the glory and spectacle of the road was the stagecoaches with their gleaming side panels. In their heyday the coaches whisked travelers from Cumberland to Wheeling in twenty-four hours. Under pressure they could do better than that. The 222 miles from Frederick to Wheeling were once covered in less than twenty-four hours to carry a special presidential message, and over one stretch the stage reached an average of fifteen miles an hour. Perhaps the record run was that made by Redding Bunting, a six-foot-six Virginia stage driver on Stockton's "June Bug" line, who covered the 131 miles from Cumberland to Wheeling between two o'clock one morning and two the same afternoon, carrying the declaration of war with Mexico in 1846.

The stages were operated by regular lines, carrying such names as "Good Intent," "People's," "National Road," or Lucius Stockton's popular "June Bug" line. A Jerseyman, Stockton was colorful and imperious, and struck a patrician note as he raced over the road in his private carriage, the "Flying Dutchman." In the early days of the railroad, it was he who had challenged the locomotive with his horse and buggy—and had won. He was a superb driver, and commanded the respect of those who drove his stages. It became a legend of the pike that Stockton put whisky in his horses' water.

Another titan of the road was a giant six-foot-five Scotsman, James Reeside, who habitually wore a scarlet vest and tie. Called "the land Admiral," Reeside expanded his line to a total of four hundred men and a thousand horses, and branched out from the National Road to haul the mail from Philadelphia to New York and other points, becoming the largest mail contractor in the nation.

The stages were drawn by finely matched teams of four horses, changed at intervals of about twelve miles, and boosted by "postilions" of two horses more to climb exceptionally steep grades. When the stage drew up for a quick change at the end of a relay, announced by a blast from the driver's tin horn, the fresh team already harnessed would be waiting. The stages ran day and night on regular schedules, and at the height of the traffic fifteen were scheduled each way a day. The coaches usually carried nine persons, three to a seat, and the stage fare from Baltimore to Cumberland was $9. Tickets were purchased from tavern landlords whose taverns were the coaching stations.

Before the coming of the railroad a great lore had accumulated around coaching, part of which is presented in Thomas B. Searight's Homeric recollection of *The Old Pike*. On the National Road coaching never found the artistic expression it evoked in England, where Charles Dickens saw the nostalgia of "past coachfulness: pictures of coaches starting, arriving, changing horses, coaches in the sunshine, coaches in the snow, coaches in the wind, coaches in the mist and rain, coaches in all circumstances compatible with their triumph and victory, but never in the act of breaking down or overturning." Coaching on the National Road did create a small literature and larger folklore. A skillful driver was a joy to watch, and the young and confident stage drivers of the old pike were proud of their skill as they rounded the sharp turns of the mountain roads. "Old Red" Bunting could "turn team and coach on a silver dollar." Like most of the pike drivers, he employed the eastern system of "side rein" driving, but some others used the English "flat rein," or the Pennsylvania "top and bottom" system. Colorful and widely known by name, the stage drivers were as proud and independent as ships' captains and, while they could accept treats from passengers, were indignant when offered tips.

The many different kinds of vehicles and teams were well

recognized. There was a regular shipping line for fast freight and light packages, called the "Shake-Gut line." In addition to the "regulars," scores of wagoners called "militia" or "sharpshooters" were tempted onto the road with their narrow-tread four-horse farm wagons while the high winter rates prevailed. Bunglers and road butchers, as well as competitors, they were roundly detested by the professionals. A few proud wagoners had "bell teams" with open-mouthed bells hanging from an arch over the harness. Drovers were often seen, with their cattle, sheep, horses, and turkeys, making their way from Ohio to eastern markets. For a while the pony express galloped the U.S. mail in 6-mile relays over the road; but the boy riders and their saddlebags were shortly abandoned because the scheduled stagecoaches made almost as good time over the macadamized road. At one time the mail was carried in special coaches that could accommodate three passengers along with the locked leather pouches, but this, too, was abandoned and the mail reverted to the regular coaches.

Shoals of taverns gave a focus to life on the road. Each acquired its own reputation in the mind of the wagoner:

> Old Wheeler's sunfish,
> Bob Fowler's roast goose,
> Warfield's ham,
> Ain't that jam!

Taverns developed curious names, like the "Temple of Juno." Or hung out quaint signs, like:

> Out of this rock, runs water clear,
> 'Tis soon changed into good beer.
> Stop, traveller, stop, if you see fit,
> And quench your thirst for a fipenny bit.

The wagoners groused about Getzendanner's, where they put too much garlic in the sausage, and praised Middletown, where an obliging blacksmith replaced rims and horseshoes at any hour of the day or night.

The men along the National Road were young, hard, and gay. It was well they were, for it was a life that started when you shook out the kinks from sleeping on the tavern floor, threw the heavy gear on the horses, and climbed into the hard saddle. The wagoners danced with the girls all night and drove all day, testing themselves with bouts of hard drinking, plug tobacco, and the inevitable long, black stogies. They came from all parts of the nation, but especially from the valley regions at either side of the road itself. They included farmers and townsmen, the broken-down sons of Tidewater families, and the rawest Irish immigrants recruited from the gangs that had broken the stone to build the road. There were the Dutchmen and Scotch from the interior valleys, and Westley Strother, the well-liked Negro driver. They mixed and blended in the taverns, swapping their tall stories and jokes, and a veritable alumni of the road was forged in the hard experiences they shared. It lasted through the century, long after the stages and wagons had disappeared from the pike.

Not a life to be envied, most of the drivers soon left the road. Those who remained stuck with a fanatical loyalty, and when the railroad came they pushed back feeder routes to the iron railhead or went west in advance of the railroad to stake out new routes. Farming and tavernkeeping claimed most of the veterans of the road, but some struck it rich in Pennsylvania oil or coal, others rose to be politicians or labor leaders, roadbuilders or contractors, while others found themselves miners or laborers on the hated railroad. Many wound up in the county homes.

The Baltimore and Ohio Railroad reached Cumberland in 1842, and halted there a few years to gather its resources for the difficult penetration of the mountains. The men of the National Road easily saw what the railroad would do. As the earlier pack-horse drivers along the 2-foot forest paths had complained bitterly against the coming of wide roads and wagon traffic, the wagoners of the Cumberland Road resented the appearance of the railroad. They could recollect the once-busy wagon traffic between Baltimore and Cumberland. They remembered the Irish drivers at Frederick, a decade before, singing a sour jingle to the tune of "The Wearing of the Green," that commenced:

Come all you gallant wagoners,
That's for the railroad opposed.
It was once that I made money
By standing in the road.

Long before the first puff of steam was seen in the narrow valleys of the upper Potomac, it was clear to everyone that the railroad would extinguish the wagon traffic on the old pike. As the steam engines established themselves, the wagoners would intone the melody of resignation:

Now all you jolly wagoners, who have got good wives,
Go home to your farms, and there spend your lives.
When your corn is all cribbed, and your small grain is good.
You'll have nothing to do but to curse the railroad.

Opponents of the railroad were powerful enough to force significant deviations from the routes originally proposed, by organizing the opposition of farmers who sold hay and feed to the wagoners on the turnpike. They drummed up tales of the hazards of the steam trains until the farmers cried, "Compel them to stop at Cumberland. Then all the goods will be wagoned through our country, all the hogs will be fed with our corn, and all the horses with our oats. We don't want our wives and our children frightened to death . . . We don't want our hogs and cows run over and killed." The hunters were told that the railroad would frighten away the game. A piketown governor of Pennsylvania was found to oppose the extension of the Baltimore and Ohio west of Cumberland as "prejudicial to local prosperity." But by 1852 the railroad had bucked its way through the mountain ranges, and by 11 tunnels and 113 bridges, and such temporary contrivances as the zigzag "Shoo-fly Track" over Pettibone Mountain, had reached the Ohio at Wheeling.

The railroad, which had itself commenced with horse-drawn trains, and still spoke of "stalls" in engine houses for "iron horses," readily absorbed the life of the National Road, as it had that of the turnpike traffic in flour farther east. The stage-line agents in the old taverns were established behind wickets in railroad stations; locomotive engineers were made out of the stage drivers, and freight handlers out of the wagoners.

6

It remained for the railroad to complete what the National Road and the canal together had failed to accomplish: a through route to the west. Hardly the Potomac route of George Washington's vision, with its terminus solidly planted at salt water in Baltimore, the railroad nevertheless surrendered to geographic facts sufficiently to parallel the Potomac from Point of Rocks to Cumberland, before taking off across the mountains to Wheeling on the Ohio. More than another generation or two was still to pass, after this much had been accomplished, before the national capital would be linked directly to the west without the long detour via Baltimore.

Started by apprehensive Baltimore merchants as a way to secure their grip on western trade, the Baltimore and Ohio was for long uniquely a local enterprise. The first railroad in the nation to carry general traffic, the speculative chance it represented measured the anxiety in Baltimore over the prospects of the competing forms of transportation. Until it was undertaken, the predecessor railroads both in England and in the United States were hardly more than extensions of the plank railroads in mines and quarries, a bare few miles of track along which moved slow, horse-drawn trains of a few cars. To plan a railroad from Baltimore to the west over the mountains required a vision and boldness that has attracted much admiring attention.

The idea was stimulating, even when limited to horse-drawn trains and the stationary steam engines proposed to haul the trains over the high ridges up inclined planes. No less a figure than the architect Robert Mills was stirred by the plans of the Baltimore and Ohio to conceive a single-track wooden railroad from Washington to New Orleans, and supported his plan by an elaborate array of figures showing the economy of such construction, and an estimate of how much money the post office alone would save in transporting the mail. Already the American mind was at work transforming the new invention and adapting it to local conditions.

In the first ten or twenty years of the Baltimore and Ohio, that company was to work out the framework of national railroad practice, adapting it to the greater distances, superior natural obstacles, and fresh resources of this country, so much that its annual reports and technical progress were the principal object of study by railroad engineers everywhere; even in far-off Russia, experienced engineers from the Baltimore and Ohio were borrowed to plan that nation's first railroad system.

The technical advance of the railroad in its first formative years has been traced in detail by its official historians. Wood ties replaced the solid granite chairs originally used; the inclined planes over the ridges were discarded; steam engines took the place of horsepower; and solid iron rails gradually supplanted the strap iron on which the first cars had rolled.

The first trains to reach Frederick were still drawn by horses, and a considerable interval of experiment lay between 1828 and Peter Cooper's first American locomotive in 1830, during which it seemed worth while to try horse-drawn treadmills and even sails as sources of motive power. Even Cooper's locomotive, the Tom Thumb, built in a Baltimore carriage maker's shop, was in part accidental, for that ingenious and capable man would never have come to Baltimore had he not rashly speculated in city lots there and been temporarily drawn to the place by anxiety about his investment. Nor was the Tom Thumb, with its 3½-inch cylinder, of more than experimental importance. It was the watchmaker Phineas Davis (like that other watchmaker, Matthias Baldwin), who built the first practical steam engine for the Baltimore and Ohio, the *York*, opening the way for the long series of locomotives to the present.

The important difficulties encountered by the railroad were not technical; they were political. And in relation to this factor, they were financial as well. Supported by the savings of the Baltimore merchants alone, it would have been impossible for the Baltimore and Ohio ever to have reached its western objective. It suffered, as did the canal, from the limitations of local support, and with such resources was barely able to reach Harpers Ferry. It was idle to seek help in Philadelphia and New York, busy as those centers were with railroad projects of their own. Only British capital, recruited by President McLane in exchange for an agreement to buy British rails, eventually carried the railroads over the mountains.

In other ways political obstructions confronted the railroad. The long mileage of the Baltimore and Ohio that lay in Virginia, both along the Potomac and from Cumberland to Wheeling—a distance longer, in fact, than the length of the railroad through Maryland—was a source of constant hazard. The state of Maryland, unlike Virginia and Pennsylvania, did not reach to the Ohio, and the Baltimore railroad company had to seek favors in hostile states in opposition to native corporations busily promoting plans of their own. The railroad's position in Virginia was almost as precarious as it was in Pennsylvania, where the route to Pittsburgh, which it had originally proposed, was nabbed by the promoters of the Pennsylvania Railroad before the Baltimore and Ohio track was laid. In Virginia the Richmond legislators still inclined to Tidewater, and concentrated their attention on a railroad along the James that would run from Norfolk back to the mountains. Only the vociferous protests of the Virginia city of Wheeling salvaged for the Baltimore and Ohio at least the million-dollar stock subscription of that city, and obtained the necessary extension of the time limit of the road's original Virginia charter.

For years after the settlement of the right of way between Point of Rocks and Harpers Ferry, the railroad was embroiled in a political arena where crucial matters of routes and financial assistance were determined. By 1835 it was possible to leave Harpers Ferry by river steamboat for Point of Rocks. There one transferred to a canal packet for Washington City. One could journey thence by stagecoach over the turnpike to Baltimore, and return to Harpers Ferry by the steam trains of the Baltimore and Ohio. Each of these different forms of transportation had its own objectives, and they jostled each other in Congress and in the Assemblies as they competed for franchises and subventions. Gradually the railroad drew ahead. The river steamers soon disappeared. The turnpikes were deserted. The railroad and, for a time, the canal divided the river's traffic.

Not the least of the railroad's difficulties were self-made. In seeking to ingratiate itself with the western farmers and the milling interests, the railroad had inserted in its charter a statutory rate of four cents per barrel of flour. A dozen years later, when a quarter million barrels a year were being hauled, and flour was by far the principal commodity transported, it was claimed that the actual cost of such freight was six cents a barrel. The unfortunate railroad, when straitened circumstances caused it to appeal to the legislature for an increase in rates, found that all rate increases *except* those on flour were allowed. The legislative regulation of freight rates was becoming a Pandora's box.

Aside from the man-made difficulties of legislatures, the most important obstacles were those placed by geography between Harpers Ferry and Wheeling. Even in the valley of the Patapsco it had not been easy to secure favorable rights of way and lay track. In the increasingly rough country west of the Shenandoah River, the cost of the most carefully engineered routes, and the incessant requirement of tunnels and bridges, became so burdensome to the weakly capitalized road that its success was long in doubt. Matters were made more difficult when the possibility of running through Pennsylvania was snatched away by the legislature of that state and the Virginia legislature, on the other hand, specified that the railroad terminate at Wheeling and nowhere else.

In accordance with its usual practice of following the routes earlier explored by the turnpike, and thus to pirate the already developed traffic, the Baltimore and Ohio had crossed the Potomac at Harpers Ferry to the Virginia side, where it made connection with a new railroad, the Winchester and Potomac, that was beginning to run up the valley of the Shenandoah. Baltimore still had its eye on the wheat trade that centered at Winchester, as well as on the business stemming from the federal arsenal at Harpers Ferry, and the barge traffic in Cumberland coal that unloaded here.

From Harpers Ferry, it cut across the river's bend through the Shenandoah valley at a point opposite Hancock, and remained on the Virginia bank of the river, a twisting, difficult route with three tunnels and many bridges over tributaries, until Cumberland was nearly in sight. Below Cumberland it crossed to the Maryland shore to enter that city. Here the railroad was in a position to tap the heavy trade originating at the terminus of the National Road and, more important, the still richer and rapidly increasing coal traffic of the Georges Creek basin. With all its difficulties, the line the railroad had found was a water-level route, with a superior competitive position that has hardly changed to this day.

Unlike the canal, which reached Cumberland in triumph, the Baltimore and Ohio regarded Cumberland as hardly more than a milestone in its westward progress. It had already completed plans for the route to Wheeling, and its surveyors contemplated extensions into Ohio and even farther west. Before long, in fact, Cumberland was to find, like Frederick, that it was hardly on the main line of the railroad when a cutoff route was later built. Cumberland's usefulness was seen as an immediate source of traffic and the badly needed current revenue for the railroad; not, as by the canal, an end in itself.

For nearly a decade, however, the railroad hung with its western terminus at Cumberland, gathering its resources for the great leap out of the Potomac valley to the Ohio at Wheeling. Its difficulty was still financial. No longer could exhortations produce the needed capital in Baltimore, and even appeals to London were in vain. Already to the north, by the time Cumberland had been reached, the assemblage of small lines later to be gathered into the New York Central had reached the Great Lakes through the great natural river basin formed by the Hudson and Mohawk that forms New York State. The all-Pennsylvania route to Pittsburgh was rapidly pushing west. The Ohio country was beginning to be parceled out among the eastern lines that tapped various parts of it. But the Baltimore and Ohio could not budge from Cumberland.

What supplied the means for the expensive construction job that finally carried the road to the Ohio was the coal of Georges Creek; its traffic increasingly filled the tills of freight offices at Cumberland and Piedmont. Within two years the yearly earnings of the road had been almost doubled. Presently it was again possible to sell railroad bonds in Baltimore as they had not been sold for years. The policy of treating revenue as capital and issuing stock dividends, consented to by the patient and once again hopeful shareholders in Baltimore, ensured the funds for western operations.

West from Cumberland, Benjamin H. Latrobe, the chief engineer of the Baltimore and Ohio, laid out the route across the mountains and through the valleys of the Cheat and Tygart rivers. Four years of heavy construction work followed, when as many as 3,500 men and 700 horses were at work laying the track over difficult grades, boring the 11 tunnels, and building the necessary 113 bridges. From far and near the workers were drawn. Mostly Irish they were, and they left their mark. Keyser was originally called Paddytown. Green-coated bands of Irish workmen were found tramping across the countryside asking their way to the "big toonel" at Kingwood. They made an Irish city near Tunnelton, and the wars between the Irish factions here reached a climax. When Far-downers, moving east from Fairmont in search of work, after the completion of the railroad to that point, met the men of Cork and Connaught at the construction town near Tunnelton a pitched battle required a sheriff's posse of 130 men to establish order. This was the last great battle of the "Irish wars" that had marked the progress of heavy construction in the Potomac since the early days of the Patowmack Company. Hardly to be explained away by the fractiousness of the race or traditional enmities at home, the disturbances of "the railroad Irish" seem more likely the consequence of intermittent work, and in some cases may possibly have been fomented by contractors as a competitive device to keep wages low.

The effect of the railroad on the countryside through which it passed was phenomenal. It had helped boom Harpers Ferry from an arsenal town to an important junction point. The population of Cumberland had been trebled, and that mountain city became by far the most important center between salt water and the Ohio. The values of many farms on either side had skyrocketed, and the prices the farmer obtained for his crops made dramatic increases as the ease of getting them to market grew. Now, in western Virginia, a people who had lived such an isolated frontier life in the mountains for nearly a century that even agriculture had hardly been profitable were suddenly plummeted into close contact with eastern cities only hours distant. Abundant natural resources, heretofore worthless, became wealth.

After the last spike had been driven at Rosby's Rock on Christmas Eve, 1852, the feverish looting of forest and mine in the virgin territory beyond Cumberland made fortunes for fifty years. Sawmills were built and coal mines opened on lands which the railroad had made accessible. New towns sprang up in western Virginia along the railroad, from which turnpikes and branch railroads sprouted like suckers on a grapevine. County seats were reshuffled as the political balance changed. Steamboats commenced to reach high up the eastern tributaries of the Ohio, and new bridges linked the growing towns with wider trade areas. The National Road to the north and the Northwestern Turnpike to the south immediately began to decline, and towns along these routes faltered in their growth. A mountain empire, stretching deep into the south but inseparably linked to northern industrialism, was being created.

The railroad eclipsed the Potomac. Where the Tidewater planter had situated his house with a commanding view of the noble river that brought his goods and took away his tobacco, the railroad station on the riverbank turned its back to the stream, and the towns and industries into which the railroad had breathed life treated it as scarcely more than an open sewer. From the coal mines, the tanneries, the pulp mills, and the railroad towns oozed a bitter, brown stream that killed the fish, ended the lazy river recreations, and caused cities to seek elsewhere more costly but purer sources of water supply.

The Potomac had achieved its long-cherished route to the west. A century had passed since the Ohio Company had first envisioned it. With the years, the Potomac itself had changed. Vanished was the fur trade. Gone was tobacco as a major crop in the valley. Even the wheat boom had moved west over the mountains. The age of wood and water had passed, and one of coal and steam had come. Of the long procession of Potomac seers—Thomas Lee, George Washington, Thomas Cresap, George Mason, James Rumsey—hardly one would recognize what the valley had become.

Passing up the Hudson and the Mohawk valley to the Great Lakes and down into the Ohio, the New York railroads had reached the western capital of Cincinnati ahead of the Baltimore and Ohio. The railroads were knitting the country together more rapidly and efficiently than any efforts of statecraft. Industrial centers to serve the entire nation were arising in favored spots. The traditional commercial centers of the valley were declining like the old towns along the turnpikes, as the self-sufficient life they served rotted away. A new regional hegemony was born.

By the middle of the century the railroad reached nearly to the Ohio. The tide of emigration to the west had flooded again. To the wheatlands of the central states came increasing streams of immigrants over the new routes, not only from beyond the seas but from the Potomac valley itself. Along the turnpikes were seen westward-trending processions of immigrants from the worn-out Tidewater plantations, and again the wheat farmers of the Piedmont, once restrained by the Loudon system, had commenced to move.

An augury could be seen in the decision made by Cyrus McCormick to move his manufacturing of agricultural machinery west of the mountains. From his factory-farm in the upper Shenandoah, the McCormick reapers had been hauled by road to the head of the James River canal, then down to Tidewater and into coasting schooners; around Florida to New Orleans; into riverboats and up the Mississippi to Cincinnati, then reckoned as the center of the wheat country—in all a haul of 3,000 miles to reach a point barely 200 miles distant. Small wonder that McCormick located in advance of the wheat frontier, at Chicago, a mushrooming prairie city that a decade before McCormick's arrival had shipped but 38 bushels of wheat.

To the north and west agricultural and industrial expansion was to be seen, powerfully influenced by the spreading net of railroads. In this development Maryland increasingly shared. To the south the agrarian economy of staple crops, now overwhelmingly cotton in place of the earlier tobacco, rice, and indigo, prevailed. With all its agricultural decline, parts of Tidewater Maryland and Virginia shared in this economy. Between the two sections, with their growing differences, the Potomac marked a rough frontier.

In the lower river, the tobacco counties clung to the staple crop and the slave system. Not a bridge crossed the wide Potomac below Washington. No railroad penetrated the peninsula between the Potomac and Chesapeake Bay, but the plantations were tied as closely as ever by river ferries to the Virginia shore. Along the Piedmont the wheat farms, now enriched with Peruvian guano brought by railroad from Baltimore, were undergoing their final boom, but the landscape was already modified by the railroad and the small industrial centers along its route. In the mountains the coal empire was firmly knit to the railroad and the users of the fuel in northern cities.

Before the raid of John Brown's men on the federal arsenal at Harpers Ferry, the Potomac was seen as the ultimate frontier between the two rival systems.

TEN

THE WAR ALONG THE POTOMAC

THE NIGHT EXPRESS FROM WHEELING TO BALTIMORE ROLLED slowly into the Baltimore and Ohio station at Harpers Ferry early on the morning of October 17, 1859. As the conductor swung from the train, he found himself surrounded by armed men. Not a light showed in the valley town. In the darkness the train crew were being taken to one side when a disturbance occurred, in which the Negro station porter, Heywood Sheperd, was shot and mortally wounded. It was an ironic beginning for the great rebellion that the abolitionist zealot John Brown had planned to free the slaves.

John Brown and his raiding party had already occupied the federal arsenal and armory at Harpers Ferry, and there obtained the weapons they thought to use to arm the slaves. Long planned from an obscure retreat on the Kennedy farm in the Maryland hills across the river, the operation badly overestimated the readiness of the Negroes to revolt. By morning, when the night express was allowed to proceed across the bridge on its way to Baltimore, news of the uprising had already spread in the neighborhood, and the arsenal was ringed with excited farmers popping away with their squirrel rifles. All during the long day their forces were increased by companies of militia from far and near, and late that night a party of United States Marines, led by Colonel Robert E. Lee, finally arrived from Washington to take matters in hand. When daybreak came the next morning,

THE UPPER

POTOMAC

the Marines battered down the door of the firehouse, the last forlorn citadel of the raiding party, and the fighting was at an end. In six weeks John Brown was hanging from a gallows at Charles Town, convicted of treason, but the brand had been lighted that in less than two years fanned itself into the flames of civil war.

If John Brown had miscalculated the power of his cause to arouse the slaves, he also misunderstood the spirit of the farmers and townsmen in the vicinity of Harpers Ferry. Not a hand was raised to help him, not a voice was heard in his defense. The Potomac was not lacking in those who believed in the abolition of slavery, who had freed their own slaves, and supported such moderate measures as the colonization societies to give force to their convictions. But the tinderbox of a slave insurrection was real and horrifying in their imagination, as it was throughout the South. The effect of John Brown's raid in the Potomac country was to hasten the deterioration of the many movements attempting to deal with the slave question, to discredit steps to patch up the weakening bonds of national union, to erode the middle ground of compromise and moderation. Sympathy with the slave states greatly increased. The appeal to violence represented by Brown's raid, and the support of that policy by a large section of northern opinion, undermined belated efforts to find a means of compromising matters with the South.

Brown's raid gave food for thought to others. It stressed the weak geographic position of the national capital, and the absence of guards and fortifications. In less than a year the Richmond *Enquirer* was brazenly asking, "Can there not be found men bold enough and brave enough in Maryland to unite with Virginians in seizing the Capitol at Washington?" Easy as it seemed from a Richmond editorial swivel chair, Virginia had forgotten what she had done in the preceding half century to alienate ancient loyalties in Maryland, and she had forgotten how Maryland had changed.

Despite Lincoln's Cooper Union address, in which he denounced John Brown, northern opinion during the abolitionist's trial showed the readiness of large segments of the North to support the use of violence in coercing the southern states. The course of events showed a Union drifting apart, but the line of division was hard to draw on a map. Was it Mason and Dixon's line? Was it the Potomac? Even in Maryland a substantial sympathy with the southern states could be found, and it aroused false hopes and misunderstandings that later led the Confederate states into unreasonable political expectations that colored their military strategy to the point of disaster.

Yet in so far as a sudden change was indicated, much of this was illusion. The old bonds of unity between Virginia and Maryland had steadily deteriorated. Since the days when the Mount Vernon Compact symbolized the high point of homogeneity and unity, the common interest in staple crops, the plantation system, and the route to the west, Maryland had been drawing away from Virginia and the other southern states where the plantation system continued to dominate. Her interest in coal, industry, and railroads gave her more and more the character of a northern state. Northern capital participated in her development and controlled some of her most important enterprises. Maryland's number of Negro slaves had steadily diminished, the free Negroes increased; indeed, it became common to send slaves from the southern states to Maryland to obtain their legal freedom. The rise of a metropolitan center at Baltimore, even though that center had strong commercial ties with the southern states, further stressed the kinship to Philadelphia and New York rather than Richmond and Charleston.

Between the two states that bordered the Potomac, the river came more and more to be a dividing rather than a uniting force, to acquire the characteristics of a frontier. The Civil War itself, dividing the two states along the frontier of the Potomac, was to mark the tragic climax of that growing disunity. The sense of Potomac unity, regional as well as national, that had so animated the Revolutionary years had almost disappeared.

Where solid walls of belief were erected to the north of Maryland and south of Virginia, the two states that bordered on the Potomac long remained undecided. Virginia's vote on the secession ordinance, taken by the old *viva voce* procedure, "that old aristocratic thumbscrew which had kept a large part of the voters of Virginia virtually slaves," would probably never have been sustained had secret ballots been used. The commonwealth was shaken by sectionalism. With her 4 per cent slave population, western Virginia had the interests of a northern state. In laying the cornerstone for the addition to the Capitol in Washington in 1851, Daniel Webster had stated it: "And ye men of Western Virginia who occupy the slope from the Alleghenies to the Ohio and Kentucky, what benefits do you propose to yourself by disunion? . . . What man can suppose that you would remain a part and parcel of Virginia a month after Virginia had ceased to be part and parcel of the United States?" North of the Potomac, Maryland's vote to stay in the Union, taken only after the Assembly had moved its session from Annapolis, heavy with proslavery sentiment, west to Frederick, and its members of uncertain loyalty faced kidnaping by Federal troops, probably did not represent accurately the sentiment of that state either. Governor Hicks of Maryland, who owned slaves and was southern in his sympathies, was still a determined Unionist. He urged a policy of "neutrality," which the state legislature adopted.

To disentangle this complex skein, with its interplay of section and interest, is not the work of a chapter, or even a book; but to appreciate it is essential if the Potomac's part in the war is to be understood.

From John Brown's raid on the federal arsenal at Harpers Ferry to the final flight of Lincoln's assassin, John Wilkes Booth, down into Maryland and across the Potomac, the river was the principal theater of the war. It played the dominant strategic role. Potomac episodes of the war, important in themselves, epitomized the larger events in the national scene. And had the Potomac itself been better understood, the war would have run a different course.

When war came, the Potomac was the formal frontier. The issue was soon drawn on a front from the mouth of the Potomac to Harpers Ferry. From there west the line of the Baltimore and Ohio significantly became the effective line of battle. This road, a unique line of lateral communication, was a prize both sides struggled to possess, of an importance matched only by the broad Potomac below Washington—itself a veritable military highway.

At the outset of the war the Confederates were expelled from the Monongahela and the Kanawha, and after 1861 the southern army never attempted to regain the transmontane region. Union control of the Baltimore and Ohio and the western Potomac region, and West Virginia's loyalty, made troop and supply movement possible between the Potomac and Mississippi theaters of war.

The Potomac itself was only an approximate boundary. The Tidewater Maryland counties, where proslavery sentiment among the planters was strongest, were virtually in a state of military occupation from the beginning of the war. (Some thought they found here the initial pattern of occupation the federal government later applied to the southern states.) And from Harpers Ferry west, the old sectionalism of western Virginia, stimulated by the Baltimore and Ohio Railroad, which had good reasons for wanting to be out of the control of the Virginia legislature, soon resulted in the creation of the new state of West Virginia. That the eastern frontier of the new state arbitrarily included what is now the eastern panhandle of West Virginia, the Tidewater island around Charles Town, where ties with Virginia and sympathies with the South were pronounced, was due principally to the influence of the railroad, which entered the new state at Harpers Ferry. Indeed, the Potomac counties down to Alexandria and the counties of the Eastern Shore, barely escaped being included in "West (By God!) Virginia."

2

Over the wheatfields and green pastures of Loudoun, and down to the Potomac, in September of 1862 came the Confederate Army of Northern Virginia. At first a few gray-clad horsemen trotted down to the river's bank at White's Ford. Then the river was alive with them, "a magnificent sight as the long column of many thousand horsemen stretched across this beautiful Potomac," an observer recorded. Behind came the dusty files of southern infantry, in butternut homespun and ragged gray, those with shoes exceptions as the long columns passed over the dusty country roads. From the Union lookout atop Sugarloaf Mountain, flags signaled the news that soon reached Washington: the Confederate Army was invading the North.

For the first time, Lee was crossing the Potomac, carrying the offensive into enemy territory. Following the southern success at Second Manassas, even amateur strategists in the Richmond saloons could see the importance of an immediate thrust against the depleted Union forces, before their ranks again swelled with new infusions of recruits. The political demand in the South for an offensive stroke was compelling. But the strategy of the campaign that ended on the bloodstained banks of the Antietam was based upon a powerful misconception, one that had been cherished in the South since the beginning of the war. Once a Confederate army actually stood on Maryland soil, this theory ran, that state, with its many southern sympathizers, would rebel; at the least, Marylanders by the thousands, it was firmly believed, would rally to Lee's standard. With Maryland in the Confederacy, the argument continued, the encircled Union capital at Washington would of necessity capitulate. Once Washington had fallen, the martial spirit in the North would be dampened.

This popular view colored official thinking much more than a little, but added to it was a stronger and more realistic argument for the invasion of the North. Such a conclusive demonstration by southern arms, as would issue from a successful campaign in enemy territory, would hasten diplomatic recognition of the new republic and speed that British intervention which—spurred by the idle cotton workers of Lancashire—hung so clearly in the balance. Possessed of the matchless lateral communications afforded by the Potomac valley and the Baltimore and Ohio Railroad, a new southern frontier—possibly as far north as the Susquehanna—could be strengthened to the point of stability.

Not least, it was a campaign the material needs of the army made necessary.

Followed at some distance to the east by McClellan's army, and screened by Stuart's cavalry, Lee's forces advanced quickly to Frederick, where they found a quiet reception but little of the expected enthusiasm. When the southern army called for Maryland volunteers, few came forward. No demonstration by Barbara Frietchie was required to illustrate the coolness of the population once the plantation fringe of Carrollton Manor had been passed. Sprinkled as the countryside was with southern sympathizers, it was difficult for Lee's commanders to interpret their reception. They considered it probable that enthusiasm for the southern cause would mount with a prolonged occupation. They promised success on Maryland soil to energize their sympathetic but reluctant followers.

Yet even before it climbed the Maryland shore of the Potomac, Lee's army had commenced to fade. Thousands of men lagged behind and vanished. Discipline was powerless to control the straggling that rapidly grew to massive proportions. The personnel of some units was reduced by one-third. The weak supply system of the Confederate Army was unable to provide the men with food, and they subsisted principally on green corn, raw vegetables, and fruit plundered along the line of march. Dysentery and diarrhea were the crippling result of "the green corn campaign." The barefooted troops managed with difficulty even in Virginia, but when they struck the hard limestone roads of Maryland they fell out by the hundreds. The forced marches of the later stages of the campaign depleted "regiments to battalions, to companies even." Little wonder that even sympathetic Marylanders hesitated to join such an army.

From his camp at Frederick, Lee took stock of the situation. He learned that McClellan was still uncertain as to his movements, and he found to his surprise that the large Federal garrison at Harpers Ferry, now cut off by his northward advance, had not been ordered to withdraw from that hopeless trap. From Frederick it seemed a relatively certain operation to advance westward on Hagerstown and the rich harvests of the Cumberland valley, and to sweep Harpers Ferry as an incident in the westward movement. With this plan, Lee divided his forces and the armies commenced to move west.

The day following Lee's departure, McClellan's forces arrived and encamped on the same fields south of Frederick. Here, by sheer luck, the Federal troops stumbled on a copy of Lee's field order, wrapped around a few cigars, giving his troop dispositions and objectives. No time was lost in dispatching some elements to pursue the Southerners over the mountains. A sharp engagement at the crest of the Catoctins marked the passage to the Middletown valley, and hard fighting in the passes of South Mountain separated the Northern Army of the Potomac from the Cumberland valley. Caught with their army divided between Harpers Ferry and Hagerstown. Lee's divisions fought furious and costly delaying actions to give time for his entire force to rendezvous at Sharpsburg.

By the time McClellan's pursuit was fully understood at Lee's headquarters, it was too late to recall the forces under Jackson, McLaws, and Walker, already deploying along the heights surrounding Harpers Ferry. The situation of that unfortunate Federal garrison was hopeless. Barely the cavalry escaped the Confederate net, and when the trap was sprung, a fine harvest of military stores and loot, as well as a veritable army of 11,000 prisoners, fell into Stonewall Jackson's hands. The episode was summarized by Captain Henry A. Binney: "Harpers Ferry is represented as an immense stronghold, a Gibraltar. . . . Instead it was a complete slaughter pen."

Yet in the larger perspective of the campaign, Jackson's organization of his three striking forces in unfamiliar and rugged terrain had taken time, and even the management of so much of the spoils of war taken at Harpers Ferry required longer than had been expected. With his small forces assembled on the ground of his choice at Sharpsburg, Lee might well have faced the spectre of a crushing defeat had McClellan's army promptly driven an iron wedge between the two separated Confederate forces. When the dark news reached him of heavy losses in the mountains and rapid Union advances, Lee even considered a precipitate retreat into Virginia. For once it seemed, too, that in its daring strokes the southern army had undertaken more than it could accomplish. The difficulties experienced in co-ordinating the three forces that were moving on Harpers Ferry suggested not only a difficult plan to execute but, ominously, a measure of exacting staff work that no individual performance could overcome, and for which little preparation had been made in the southern army.

With a line thinned by the absence of Jackson's attacking force, which had not yet arrived from Harpers Ferry, and weakened by the thousands of stragglers lost in the marches over the Maryland mountains, Lee faced the approaching Union Army. His center rested a little east of Sharpsburg, and the slightly bowed line to either side paralleled roughly the road from Hagerstown to Harpers Ferry that ran scarcely a half mile from Antietam Creek. On the north the approach to the Potomac was heavily defended, but to the south, on Lee's right, it thinned to a few depleted divisions hardly stronger than regiments.

This was the position the Federal troops assaulted during the day of September 17. Fresh and strong, their units fully manned, they had poured over the mountain in solid blue columns and spread over the valley floor during the preceding two days. The Antietam lay before them and the enemy, spanned by its three stone bridges. The battleground offered little real cover other than the undulations of the earth itself, save on the Confederate left where patches of wood and cornfields were found. In this northern sector McClellan's first attack was directed at dawn, on what proved the strongest part of the Confederate line.

Massed troops against canister opened the bloody day, the bloodiest single day of the war. Never had artillery been more deadly; never had musket fire at short range exacted a greater toll. Whole brigades were decimated. The dead lay in windrows where they had stood in ranks. Even the corn in the field was cut by musket fire as with a blade. In one division, with its three brigades, all but two of the regimental officers fell. When asked where his division was, the Confederate General Hood had to reply, "Dead on the field."

Halted at the whitewashed Dunker church by the Confederates reinforced from other positions in the line, McClellan's forces slackened and the main force of the attack by noon had shifted to the center of the line. Again carnage, the Confederate line crumpling, the desperate fight in the sunken road and among the haystacks, the successful counterattack of the regrouped Confederates supported by their highly mobile artillery.

After noon the heaviest force of the Union attack fell on Lee's right wing, over the most difficult terrain, up the steep slopes overlooking the most famous of the Antietam bridges. Two massed Union regiments poured across "Burnside's Bridge," heavily supported by other units fording the stream, and by artillery. By midafternoon this bridgehead had mushroomed into a formidable attacking force that steadily drove the frantically reinforced Confederates back into Sharpsburg. Within an hour the Union troops commanded nearly the whole of the high ground west of Antietam Creek, and were a scant thousand yards from Lee's line of retreat to the Potomac at Shepherdstown.

Yet as the hard-pressed Confederates divided to either side of the Union attack, through their center came the fresh troops of A. P. Hill, just arrived after their hard march from Harpers Ferry. They flung themselves into the battle, rallying units on either side, and as darkness mercifully came, had pressed the still unbroken Union line almost back to the Antietam.

As the opposing lines drew apart, and the cannon were silenced, the cries of the wounded filled the night. Twenty-eight thousand dead and wounded lay on the battlefield of Antietam. One out of every three Confederates in Lee's army of 40,000 was a casualty. Counting the prisoners taken at Harpers Ferry, the Federal Army had lost almost twice as many. But the wounds had been started from which the South was to bleed to death.

McClellan's conduct of the battle, and more especially his failure to press Lee on the following day, have found few apologists. Admittedly Antietam was the Confederates' best-fought battle. Yet had McClellan conducted matters differently, the war might have ended that day with the Confederate Army annihilated, smashed against the Potomac in a new Cannae.

From Frederick, McClellan had not moved out promptly in pursuit of Lee. The fighting in the passes was deliberate and sluggish. He lost the opportunity to strike the Confederate forces while they were divided. He even failed to withdraw or to reinforce the cutoff garrison at Harpers Ferry. Although he arrived in front of Sharpsburg on September 15, with substantially his entire command, he waited for minor reinforcements to arrive the following day, and did not give battle until September 17. Even the conduct of the battle itself showed little appreciation of the obvious Confederate weakness, and McClellan's failure to smash the southern army, which he outnumbered by more than two to one, was considered inexcusable. But it was the failure to renew the battle on the following day and prevent Lee's subsequent leisurely retreat into Virginia that was most sharply criticized and cost McClellan his command of the Army of the Potomac. Even Burnside's ill-advised winter campaign that followed, ending disastrously at Fredericksburg, was in good part forced by the Lincoln administration's humiliation after McClellan's failure to pursue Lee.

With the retreat of Lee's army across the Potomac at Boteler's Ford it might fairly be concluded that the fighting on Antietam Creek, if hardly a clear-cut military success for either side, marked a decisive turn in the progress of the war. The invader had been hurled back. But for all its military details, and they were picked up avidly by a nation that followed the war in the penny press, in its larger aspects the Maryland campaign must be judged a failure from the Confederate point of view. Lee's Army returned to Virginia with hardly half the men who originally composed it. More than one-quarter of his army were wounded or lay dead on the battlefield; the remaining losses were due to straggling and sickness. Maryland had not risen to join the seceded states, but on the contrary, seemed more firmly in the Union than ever before. Washington was secure.

Most important, across the ocean, the course of British policy was setting firmly against the southern states. In his astonishing blindness, Mr. Gladstone might say, "Jefferson Davis and the other leaders of the South have made an army; they are making, it appears, a navy; and they have made, what is more than either, they have made a nation. . . ." But only three members of the British Cabinet could be found who so read the signs after Antietam, and were prepared to intervene in the American conflict. The British foreign secretary found the American minister gaining strength each day, and the episode of the Confederate warship, the *Alabama*, which had been built in England and negligently allowed to "escape," became more and more an issue that the secretary wished to forget.

Finally, at Antietam one could discern the future course of the war. Union corps against Confederate divisions; cannon against muskets; railroads against horses; organization against individualism—an industrial civilization against an agrarian civilization, the future against the past. One could see that the slowness of the Union Army was due in good part to its preparation, its deliberate organization, and its still-tentative procedures, its heavier supply trains, its mass. But clumsy as it seemed at first, it was the way the war would be fought to the inevitable and crushing end. Modern warfare had begun.

3

Again the gray and butternut men of Lee's army swarmed across the Potomac in the following summer of 1863. A long, slow, and indecisive winter campaign had been fought in the sticky red Virginia clay, terminating in the smashing repulse at Fredericksburg, and the venture into Maryland and Pennsylvania was planned to relieve the southern quartermasters as much as to draw the Army of the Potomac out of Virginia; and to threaten Washington was an added objective. Most important of all, the Confederacy would seize the initiative and disrupt Federal plans for the deployment of troops to other theaters of the war, disorganize its supply lines, and cut its lateral communications to the west. The hopes of the summer of 1862 had not wholly died, but the Richmond Dispatch explained the main objective of the campaign: "The South is for a time relieved, and the North is bearing the whole burden of the war." It was a consoling thought to publish July 2, 1863, the day the ultimate wave of the Confederacy's highest tide lapped farthest up along the rough hills south of Gettsburg—and then commenced to recede.

It began promisingly enough. In the rich Cumberland valley, Ewell's foragers found the stores that were such an important objective of the campaign: horses, cattle, provender, and flour; shoes, hats, and barrels of sauerkraut; even abundant ripe cherries. Towns were levied on, under threat they would be burned. Farther and farther up the eastward curving valley Ewell's corps pressed. The Pennsylvania natives made hex signs against the invaders, but to no avail. At Chambersburg, Early was detached and struck off over the mountains to Gettysburg and on to York and the Susquehanna bridge at Wrightsville. The main body under Ewell continued on to Carlisle. They had hoisted the Confederate flag, and were poised for an assault on the capital of Pennsylvania when word reached them to turn back and join Lee's main body at Gettysburg. The Army of the Potomac, now under the command of George G. Meade, had crossed the Potomac at Edwards Ferry near Leesburg and was advancing up the Monocacy. Here Lee wished to challenge them.

The main body of the invading Confederate Army under Longstreet had waded across the river at Williamsport, their trousers, cartridge boxes—and, for those who had them, their shoes—in bundles on their backs. They followed Ewell's corps up the Cumberland valley to Chambersburg. Of the rest of the Confederate Army, Stuart's strong cavalry force was presumed to be following a parallel course east of the mountains, and Imboden's troopers were burning bridges in the mountainous country west of the Cumberland valley.

As it turned out, Stuart's cavalry detached themselves so fully from the invading army that they were worthless for their most essential mission: reconnaissance. For six critical days "the eyes of the Army of Northern Virginia" were blinded, while the proud and ambitious Stuart sought to vindicate himself for the recent humiliating loss of his headquarters in the raid on Brandy Station by Pleasonton's newly concentrated Federal cavalry. As he had brilliantly done before on the Virginia peninsula, Stuart "rode around" Meade's advancing Army of the Potomac, crossing its path at Rockville and aiming at Hanover and a juncture with the Confederate Army somewhere along the Susquehanna.

While Ewell's foragers found fat pickings in the Cumberland valley, Stuart's cavalry were hard put to it for fodder in the less fertile coastal plain east of the mountains. They accomplished little. Two days of skirmishing around Washington, and another feint at the capital the following day, occupied Stuart's force as he followed the Army of the Potomac northward to Frederick. At Rockville he struck off to the east and ripped up a few rails on the Baltimore and Ohio line from Baltimore to Frederick. Continuing north, he found forage in Westminster and skirmished briefly with the Federal cavalry at Hanover. He swung high around the Federal Army and, with a force exhausted by night marches and slim rations, he passed through Carlisle and so to a reunion with Lee's main body at Gettysburg on July 2. All Stuart really had to show for his exploit was 125 sutlers' wagons, captured in Rockville. He had learned he was fighting a new kind of enemy cavalry. And he arrived at Gettysburg only after the battle had been in progress two days.

It was shoes that lured the Confederates into Gettysburg. Lee had planned to fight with his back to the mountains at Cashtown. Meade had planned his battle along Pipe Creek below Taneytown. But the combination of luck and the symmetrical road network centered on Gettysburg led to other conclusions. Henry Heth's brigade, leading the advance through the pass in the Catoctins at Caledonia, had come out at Cashtown, where they found hats, and moved on toward Gettysburg, where a stock of shoes had been reported. As he neared the little Pennsylvania college town he encountered Union cavalry, but because of lack of cavalry of his own he was unable to push his reconnaissance farther and concluded that Gettysburg was held by only a small detachment of Federal troopers. At the moment Heth was correct. The next morning, July 1, Heth began to move toward Gettysburg and the shoes his men needed, and before long found his command heavily engaged. Severe losses were encountered while the advancing force was astride Willoughby Run in its approach to Seminary Ridge, and two more regiments were disastrously trapped in an unfinished railroad cut. Heth withdrew and put his command in line of battle.

Behind and to either side of this advanced position, Lee moved in the rest of his army and prepared for a major assault on the following day. By then he knew that he faced a strongly reinforced Federal Army of the Potomac under a new commander, and a situation that left him no alternative other than to fight. A retreat through the mountains in the face of a superior force would expose him to defeat in detail. Delay would only weaken his poorly supplied force and strengthen the defending Federal Army. He summed it up, "A battle, therefore, has become in a measure unavoidable;" and even more concisely, "The enemy is here, and if we do not whip him, he will whip us."

When Meade took command of the Army of the Potomac on June 28, he inherited from Hooker an army that had steadily gained in strength, experience, and resources. Badly damaged at Fredericksburg and Chancellorsville, it had been rebuilt. It still suffered, as it was always to suffer, from its proximity to Washington, the tendency of the chief of staff to govern its movements, and its large number of politician-generals, and it was often the victim of anxiety over the safety of the capital. "A headless army," it was called.

As early as Pleasonton's raid at Brandy Station, it had been established that Lee intended to move north, and measures were commenced to counter this thrust. Hooker's withdrawal from the Rappahannock had been cautious, not only because he wanted to make sure of Lee's movements but because he was, like the entire city of Washington, sensitive to the exposed position of the capital. But once he was sure that Lee was well on the march, the Federal Army moved quickly. It crossed the Potomac at Edwards Ferry near Leesburg, and marched up the Monocacy, interposing between Lee's army and the capital. Meade had only the most general information about the Confederate movements, and on June 29 had still the impression that all of Lee's army was along the Susquehanna. Not until the pickets posted by Buford's cavalry division, the advance party of the Army of the Potomac, encountered Heth's men north of Gettysburg did the situation begin to unfold. Until then neither army had much in the way of specific information about the composition or movements of the other.

The long, bloody day of July 1, with its 7,000 casualties, which saw the fighting along the Cashtown Road, was mainly occupied in concentrating the armies and ordering the battle lines. Such other engagements were struggles for position and the possession of favorable terrain—preliminaries to the major battle still to be fought. After Hancock's careful estimate of the battleground, the Federal Army withdrew from the town to the commanding terrain feature, Cemetery Ridge, and determined to hold it at all costs and make it the backbone of their defense line. During the night the ridge was strongly occupied, and by noon of July 2 Meade's army was substantially up and in the line.

As the two lines faced each other, Lee's army was disposed along Seminary Ridge; Meade's along Cemetery Ridge. More than three miles long, the front was shaped like a fishhook. Both sides held strong defensive positions, but it was Lee who was obliged to attack. While Meade looked for an assault on the northern end of his line, Longstreet headed a wide flanking attack on the Federal left that aimed to hit the line where it was weakest and roll it up. Advancing up the Emmitsburg Road, Longstreet threw his forces against the Round Tops, Devil's Den, and the Peach Orchard, all difficult terrain features that had been occupied—by plan or by chance. Against these obstacles the Confederate assault beat in vain. It had come too late. Sickles had moved his troops from their initial position, and even incurred the risk of leaving both flanks "in the air" in order to take positions on stronger ground. Lee's generals delayed in reaching their decision, and Longstreet sulked. The attack was not made until late afternoon. The grand design of the battle, formed in Lee's mind in the early hours of the morning, and based on facts that were subsequently altered, ground to its inevitable conclusion; and Longstreet, who alone could have changed the Confederate attack, stubbornly and relentlessly carried out Lee's plan, with which he had never agreed. Division after division was thrown piecemeal into an ill-co-ordinated attack that in places became slaughter. At other points in the line similar difficulties were seen. As Douglas Southall Freeman's brilliant study in command concluded, Lee's army was not fighting as an army but as divisions, even brigades. As an organization it was disintegrating. The Army of the Potomac, "headless" as it traditionally was, was showing the fruits of organization and experience. These were factors as clear as superior Federal artillery, which at Gettysburg proved crushing, superior supply services, and a growing number of veteran fighters and commanders, with a maturing skill in leadership and staff work. Months were to pass before Colonel Theodore Lyman, a Massachusetts naturalist who became attached to Meade's staff, could see there in outline the organization of modern staff work, the selection of commanders, the careful planning of all aspects of the military job; at Gettysburg it must be deduced. The necessities of organizing the military resources of the North were making it imperative; it stood out sharply against the more informal, unspecialized habits of the Confederate headquarters, as they were reported, say, by Kyd Douglas of Shepherdstown.

Lee's attack on the Federal center the following day, July 3, only confirmed what could be seen on the first day of the battle. The attack was late in starting, supported by unco-ordinated artillery whose guns were short of ammunition, and directed by a listless commander carrying out a scheme in which he had little faith. The assault that culminated in Pickett's charge was an act of desperation that marked the substantial end of the battle.

But again the Federal failure to counterattack and press home their advantage that afternoon or the following day was an equal token of irresolution and, possibly, of inability to organize an attack as brilliant as Meade's defense. One more silent day the armies faced each other; hardly a gun was fired. As the southern army with its long supply trains begun its withdrawal through the mountains, rain began to fall; but the retreat was unmolested. When Lee reached the Potomac at Williamsport there was still no evidence of Federal pursuit in force. A vigorous cavalry attack was beaten off. The river was high and could not be forded. Five long days were spent building pontoon bridges, and on July 14 the army passed over them at Falling Waters to return to Virginia.

Like Antietam, Gettysburg was not a clear-cut decision. The Confederates did not consider themselves beaten; their only sense of failure was in not having marched on to Washington, Baltimore, and Philadelphia. "You will ask why we were not defeated then at Gettysburg?" Charles Francis Adams explained to his father, American ambassador in London: "We just escaped it by the skin of our teeth, and the strength of our position."

The bloodletting had been far greater than the one-day battle on the Antietam the previous year. Killed, wounded, or missing the Confederate Army counted 20,000 casualties, about one in four of those who had marched into Pennsylvania. The Union

Army counted 23,000. But, as usual, it was the Southerners who could ill afford to swap queens and who found the greater difficulty in filling up their ranks. Too, one-third of the general officers had been lost, and this was an even more serious affair. In command there was no substitute for experience, and no training place but the battleground itself.

The many critiques of Gettysburg have made it clear that no single factor on either side could have been altered to affect materially the balance of the scales. Chief importance must be given the terrain itself, superbly adapted for the defense, and the growing superiority of defensive weapons, especially the artillery. As Lyman later wrote: "Put a man in a hole and a good battery on a hill behind him, and he will beat off three times his number, even if he is not a very good soldier."

More clearly at Gettysburg than at Antietam could be seen the groping of the Federal Army for a means of organizing and employing its great potential strength; and more clearly at Gettysburg, too, could be seen the erosion of command in the Army of Northern Virginia. Territorially speaking, Gettysburg was the Confederate high-water mark; militarily speaking, it was, too; and its pattern foreshadowed the end of the war, and showed how it would be brought to an end.

4

Once again the Army of Northern Virginia crossed the Potomac on a major expedition into Maryland. Its Second Corps, under Jubal Early, which contained a bare 14,000 men, slashed its way down the valley (in the Shenandoah valley, north is customarily described as "down"), crossed the Potomac at Shepherdstown, and threatened Washington as the capital had not been threatened in the four years of the war. At Bull Run, when the war began, Confederates were within sight of the distant capital, but in 1864 Early's men were fighting in the very suburbs of the city. With a force so small, Early's audacious operation was hardly more than a raid. He could not have expected to do more than to panic the capital and secure the withdrawal of Union divisions from Grant, who was at that moment pressing Lee hard at the approach to Richmond. Yet that would be enough justification for the operation.

With Jackson's old "foot cavalry" Early moved swiftly. Over the South Mountain his column flowed, and down into Frederick (where they laid a $200,000 tribute; the city borrowed the cash from a local bank, and on this debt it is still making payments). The effects of the Confederate movements were perceptible as soon as they had provoked the evacuation of Harpers Ferry. Even the Pennsylvania militia were called to their posts. It was difficult to tell whether or not an invasion in force, such as that which preceded Antietam and Gettysburg, was taking place. In Washington the Federal chief of staff, Major General Henry Halleck, was scraping together a force to man the depleted fortifications that ringed the capital. From Petersburg the alarmed General Grant sent up a strong detachment of dismounted cavalry and an entire division, which landed at Baltimore on July 8, the day Early had occupied Frederick. By making good use of the railroad, General Lew Wallace had promptly moved 2,500 men under his command from Baltimore to the point where the Baltimore and Ohio crossed the Frederick-Washington road and the Monocacy River. Following him over the same road came Rickett's division of the Sixth Corps, just disembarked from Virginia. They were in place and ready to meet Early when he advanced south from Frederick the afternoon of July 9. The swift, decisive movement of Federal troops was a dramatic departure from anything seen in the earlier years of the war.

The battle of the Monocacy was the major engagement of Early's invasion. The greatly outnumbered Federal troops were readily dislodged from their positions by a flank attack, and retired down the road to Washington. Yet at a time when hours were precious, the battle itself had lost time for the invaders, and it indicated the loss of more. With what emotions Early discovered among his prisoners, veterans of Ricketts's division from the Petersburg-Richmond front, it is difficult to know. Here certainly was evidence that Early's operation was succeeding in its effort to draw men from Grant's force and thus relieve Lee; but it was also an indication that Washington itself might be strongly reinforced. Some thought it meant the entire Sixth Corps of the Army of the Potomac was already in the capital. Early approached the city with more than usual caution, making

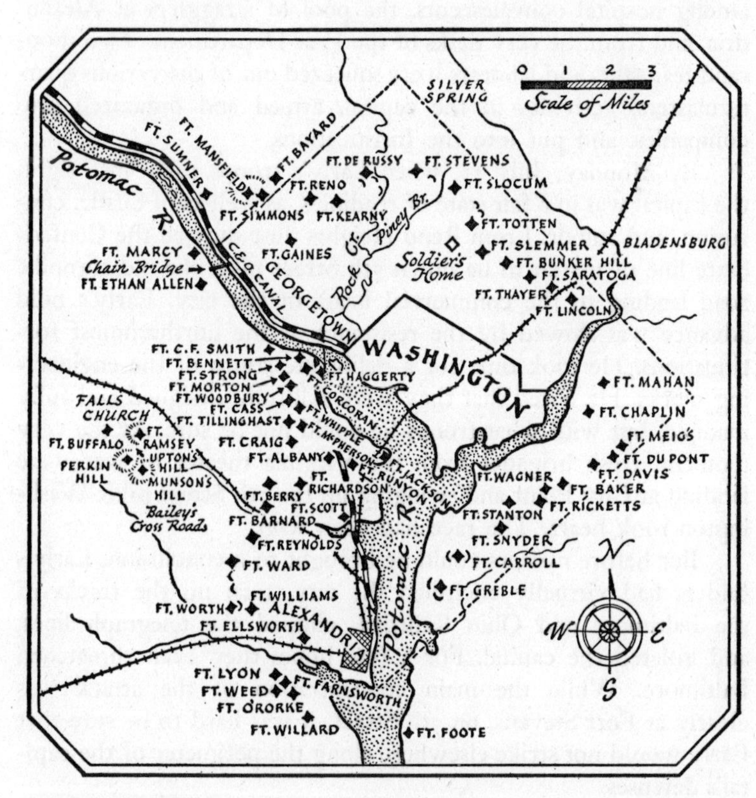

careful reconnaissance and throwing heavy skirmish lines in advance. On July 10 the Confederate leader moved his men over dusty roads under a hot sun to Rockville, and the following day continued on to Silver Spring.

Wallace's prompt defense of the road to Washington south of the Monocacy had given the apprehensive but not panicky capital another day to organize. Additional reinforcements sent by Grant—one entire corps and two divisions of another—were steaming up the Potomac, that highway to the Virginia peninsula, and could be expected to land July 11. Until then the city would have to make out with what forces it could pick up among hospital convalescents, the pool of stragglers at Alexandria, and from the very desks of the War Department. Two thousand teamsters and laborers were squeezed out of the various quartermasters' activities in the capital, armed and organized into companies, and put into the fortifications.

By Monday, July 12, when Early's troops came into sight, the capital was in a fair state of readiness, as well as of bustle, confusion, and rumor. From Reno Heights dust showed the Confederate line of march to be down 7th Street pike, the old turnpike road leading to the commercial heart of the city. Early's bold advance was slowed by the resistance at the northernmost fortifications. He took time for a deliberate study of the engineering works. He noted that they were skillfully designed and fully manned, but with what troops he could not be sure. At the very moment, fresh brigades from the Virginia theater of war were landing at City Point and swinging up the 7th Street pike. Washington took heart. The race had been won.

But before matters could be brought to a conclusion, Early's raiders had virtually encircled the city, torn up the tracks of the Baltimore and Ohio Railroad, clipped the telegraph lines, and isolated the capital. For a few hours they even threatened Baltimore. While the main concentration of the attack was clearly at Fort Stevens, on 7th Street, it was hard to be sure that Early would not strike elsewhere along the perimeter of the capital's defenses.

After two days of furious excitement, dawn came the morning of July 13 and showed an empty battleground before Fort Stevens. Early had disappeared. It was fully noon before troops started after him in pursuit. As became an audacious, if only partly successful maneuver with a relatively small force, Early had moved north at midnight, and on the next day had recrossed the Potomac at White's Ford and was back in Loudoun County.

5

The last act of the war along the Potomac was set in the decaying fens of southern Maryland. Here among the still-loyal Southerners, in the most northern outpost of the Confederacy, John Wilkes Booth, murderer of President Lincoln, was able to hide for a full week, while cavalry by the brigade and police by the thousands searched for him. As the strange tale of Booth's flight unfolded, the wonder grew. Dozens of people had seen the fleeing Booth and his accomplice Herold. The earlier planned abduction of Lincoln was common knowledge to hundreds of lower Maryland folk. But the $100,000 reward, the persistent combings of the country by cavalry, police, and detectives failed to produce either the criminal or a single clue of value. Much blame has been placed on the failure to co-ordinate the different search parties, and upon the fact that they were selfishly divided because of the large rewards offered, but the better answer to an escape that almost succeeded lies in the peculiar loyalties of the old Maryland plantation country.

Considered as "occupied enemy territory" since the beginning of the war, southern Maryland succeeded for nearly five years in performing valuable services to the Confederacy. Union troops never came to know it well. The long peninsula of Prince Georges, Charles, and St. Marys counties, lying between the Patuxent and the Potomac, was filled with ruined tobacco fields growing up in pines, and laced with tidal streams and swamps. The common use of the waterways had made it unnecessary to develop roads that were more than paths, and they threaded the sandy ridges between the watercourses in a fashion as obscure to outsiders as it was familiar to natives.

Through this territory all during the war a steady stream of goods from Baltimore and Philadelphia, cached in Maryland haystacks and barns against capture, had flowed to the southern states. The Confederate post office had a regular mail service here, and regularly employed Maryland agents, and prided itself that northern daily newspapers were read in Richmond within twenty-four hours after they had been published. The territory was also a familiar escape route for Confederate prisoners, for spies, and southern recruits. The most generally used road was the old trail that began at the slowly silting river crossing at Port Tobacco and followed the colonial post roads to Baltimore. The old ferry landings were all used, and the crossing of passengers in small boats was a nightly event.

As for interfering with this traffic, for most of the war the Federal gunboat blockade on the Potomac was a joke. Marylanders rowed across the river even when the moon was full, and ran the blockade in sloops and yawls. The occupying Federal troops and the force of wartime detectives were equally ridiculed: John Surratt, the Federal postmaster at Surrattsville on the road to Port Tobacco, who was also a Confederate dispatch runner, thought them "great boobies" who "seemed to have no idea whatever how to search me."

The population, too, was slowly decaying. Isolated from the life of the times, even from the war, in this region without railroads or industry, it looked backwards. There had been a steady exodus of the people, and of those who remained some thought they saw a human parallel to the declining quality of the Maryland and Virginia race horses because so many of them had been sent south and west. Old ways persisted. In this part of the Potomac one thing was plain: unswerving loyalty to the southern cause. They clutched it as the last familiar object in a vanishing world.

This region, which commenced just across the Anacostia River from Washington, contained the route John Wilkes Booth planned to use in his original plot to abduct President Lincoln and take him to Richmond as a hostage, and it was substantially the same route he followed in his escape after the assassination. It began at the Navy Yard bridge and proceeded directly south, closely following the river through Oxon Hill, Accokeek and Mason Springs, and then cut over to Port Tobacco and the heavily indented river area below the Big Bend. The alternate route that Booth actually used began at the same bridge and passed through Surrattsville (now Clinton), T.B., Beantown, and Bryantown, where Port Tobacco was again in close reach.

The much-told, much-embroidered tale of Booth's flight begins with his original plan to kidnap Lincoln and carry him off to the Confederate capital. This bold design the 26-year-old "Maryland Confederate" had formed in his mind nearly a year before the assassination, and to carry it through he had recruited the services of an ill-assorted group of men, most of whom were in his paid employ. The plot was preposterous and could never have been executed, but it was elaborately rehearsed and the conspirators had once attempted to carry it through when Lincoln was scheduled to attend a dramatic performance at the Soldiers Home in March, 1865. The president failed to appear and the plot was postponed. Within Booth's original scheme lay the notion that with the president of the United States a hostage, the South would be able to secure better peace terms (and Booth, who thought his earlier efforts to serve the South insufficiently appreciated, would emerge as its triumphant savior). But from this point on it faded. The conspirators began to drop away, and a mass assassination of high government officials rather than abduction became the governing idea. The rest of the plan was preserved.

The night of April 14, Booth made his escape from Ford's Theatre, and arrived at the Navy Yard bridge over the Anacostia well ahead of the news of the attack on the president, and hours before the assassin had been identified. To the guard at the bridge, Sergeant Silas Cobb, he gave his correct name and said he was going to Charles County, and was allowed to ride on. Wartime controls had been relaxed. Shortly afterward Herold appeared at the bridge, fresh from the attempt on the life of Secretary of State Seward, and was also passed. The two joined each other on the road to Surrattsville, where they had left field glasses, carbines, ammunition, and other equipment that might be needed. At midnight they reached Mrs. Surratt's tavern, then newly rented, but the tenant, John M. Lloyd, handed over the articles that had been left there and was informed of the assasination of the president and Secretary Seward. The conspirators rode on, Booth with his fractured leg, and near dawn reached the home of Dr. Samuel Mudd, three miles from Beantown. Mudd set Booth's leg, made a splint and crutches, and Booth shaved off his mustache. Mrs. Mudd and a household employee, Best, saw the injured man.

In the late afternoon of Saturday, April 15, Booth and Herold left Dr. Mudd's and rode toward Port Tobacco. They became lost in the marshes and were guided by a Negro, Oswald Swann, to the home of Samuel Cox near Bryantown. Cox put them in the hands of his overseer, Franklin Robey, and sent for his foster brother, Thomas E. Jones, a Confederate agent who knew the river crossings. For six days the fugitives lay in a little grove of pines while the cavalry search parties passed along the nearby roads and made inquiries at houses and crossroads taverns. Possibly an effort was made to cross the Potomac during those nights, because Booth scribbled an entry in his pocket diary under the date of Friday, April 21: "After being hunted like a dog through swamps, woods, and last night chased by gunboats till I was forced to return wet, cold, and starving, with every man's hand against me, I am here in dispair." On the dark and foggy night of Saturday, April 22, when the Federal searchers were concentrating farther down the river in St. Marys County, Jones put Booth and Herold in a rowboat and they started for Machadoc Creek on the Virginia shore. Through bad luck and adverse currents their boat was carried twelve miles up the river and around the bend to Nanjemoy Creek. Here the fugitives concealed themselves and their boat at daybreak, and Herold procured food from the home of Colonel Hughes near Nanjemoy Stores, and that night made a successful crossing of the river.

On the Virginia side they landed near Machadoc Creek, and Herold found the residence of Mrs. Quesenberry, to whom Jones had recommended them. She provided a meal which Herold carried back to Booth. Jones's brother-in-law, Thomas Harbin, and Joseph Badden guided them to the cabin of William L. Bryant, who took them on to Dr. Steward, who sent them on to the cabin of William Lucas, a free Negro in his employ. The following day Lucas drove them to Port Conway, where a ferry crossed the Rappahannock to Port Royal. Here they fell in with three horsemen returning home from service with Mosby's guerrillas, with whom they crossed the ferry. Because of an altercation the Negro ferryman later remembered them. So did a bystander, William Rollins. At Port Royal they went to the Peytons', who could not accommodate them but suggested the Garrett farm just up the river. Under the name of John W. Boyd, Booth spent the night at the Garrett farm and the next day learned that Union cavalry were in Port Royal searching for them. For greater security they went to a tobacco barn, now full of hay, and proposed to spend the night there. The barn door was carefully locked by the suspicious Garrett family, who feared their guests might try to steal their horses and escape.

There, at two o'clock the next morning, April 26, the searching party of Union cavalry found them.

Since the beginning of their flight nearly a fortnight had elapsed. In that time Booth and Herold had been seen by eight named persons and quite probably by a few others in Maryland; also by a dozen specifically identified persons and still more unknown, in Virginia. Despite the wave of revulsion that had swept both North and South at the president's assassination, and the huge rewards offered, no effort was made by any of these persons to turn Booth over to the Federal authorities who were searching for him on all sides.

The discovery of Booth appears to have been mainly the work of one man, Colonel Lafayette C. Baker, a tactless but highly competent detective who had been summoned from New York City by Secretary Stanton shortly after the murder of the president. When Baker arrived at the War Department on April 16, he was not put in charge of the search for Booth, and indeed was kept from some of the evidence that had been found; rather he ran a parallel course to others who were conducting their own parts of the unco-ordinated search.

Baker did it with a fine-tooth comb. He had a general familiarity with southern Maryland from earlier service in the war, and he knew the Northern Neck. Rather early in the search he seems to have concluded that Booth would have escaped in the direction of Richmond by the well-established underground routes. What clues his detectives found in Charles County to convince Baker he was really on the trail are not known. Baker's *History of the United States Secret Service* does not tell.

Something might have been learned from the other conspirators. Paine had been picked up as a suspicious character almost by chance at Mrs. Surratt's house in Washington, and Atzerodt, Lloyd, O'Laughlin, Spangler, Arnold, Dr. Mudd, John Surratt, and Weichmann were soon apprehended. It was somehow established that Booth and Herold had crossed the Potomac. Putting the pieces of this jigsaw puzzle together with his knowledge of the region, it is probable that Baker concluded the Rappahannock Ferry at Port Royal would be a reasonable point at which to search. There is no evidence that he did more than deduce this much, but it was enough.

By the afternoon of Monday, April 24, two of Baker's most trusted "detective officers," Colonel E. J. Conger and Lieutenant Luther B. Baker (a cousin of Colonel Baker), with Lieutenant Edward P. Doherty of the 16th New York Cavalry and twenty-five troopers—all these were disembarking at Belle Plain in King Georges County, Virginia, with instructions to search for Booth and not to come back until they had found him. Through the night they traveled the country roads toward the Rappahannock, stopping at farmhouses to ask about the river crossings, places where fugitives might be harbored, and doctors likely to give treatment to injured strangers. They showed photographs of Booth and Herold. By noon of April 25 they had found a man in Port Conway who identified Booth as the man he had seen crossing the Rappahannock ferry the day before. He identified Booth's four companions, and remembered that one of them had said he was going to Bowling Green, Va. Off went Conger and his sleepy cavalrymen to Bowling Green, where they arrived at midnight and found their man. With little hesitation he led them to the Garrett farm, and so to the tobacco barn where Booth and Herold lay asleep. In five minutes Booth had been shot, whether by his own hand, as some claimed, or by one of the troopers, as was probable, none could say. By morning he was dead.

The search for Booth showed the persistent loyalty of the lower Potomac, on both the Maryland and the Virginia side of the river, to the slave system, the Confederacy, and its ideals, even at the end of the war. In one man this loyalty was personified: Thomas A. Jones. Later Jones tried to explain what he had done and why he had done it in his book *J. Wilkes Booth. An Account of His Sojourn in Southern Maryland*, but nothing was plainer than the facts themselves.

When summoned by his foster brother to meet Booth and Herold, Jones already knew of the president's assassination, and he probably knew that Booth was the object of Federal searching parties. Yet into that isolated world of southern Maryland the news leaked slowly and obscurely, the bare facts untouched with comment or public reaction. There was little of horror in the first reports, especially to those who had hated Lincoln for five long years and whose world had been filled with little but hate. The old habits and attitudes of the war years lingered. The Union was the enemy, the southern states the friend. Some speculative writers have found an explanation in Jones as the "descendant of the Bayards and the Baltimores, whose life, if necessary, went with his fidelity to a cause once he espoused it," or in his "cavalier instincts." Others laid Jones's behavior to Booth's personal fascination and charm. Jones himself laid it to pity. All fell short of the mark: hundreds of others in the lower Potomac would have done exactly what Jones did.

In 1864, Jones said, it was general knowledge in Charles County that a plot was afoot to kidnap Lincoln, take him to Port Tobacco, and thence to Richmond. The name of Booth, who had traveled through Charles County on the pretext of looking at farm lands, had been linked to it. Jones showed neither surprise nor concern about the story. When he was asked by his best friend and foster brother to shelter Booth and help him across the Potomac, he took it as a matter of course that he should be expected to help. "I did not know Booth, but when Cox put him in my keeping nothing would have tempted me to betray him," Jones asserted. During the six days in the pine grove he brought Booth the daily papers, and listened to him talk about the assassination. "Murderer though I knew him to be," Jones explained a quarter century later, "his condition so enlisted my sympathy in his behalf that my horror of his deed was almost forgotten in my compassion for the man."

Across a table in a Port Tobacco tavern during the time that Booth lay in hiding, Jones had faced Captain William Williams, who had been told Jones had some knowledge of Booth's whereabouts. "I will give $100,000 to anybody who can tell me where Booth is," Williams said. "That is a large sum of money and ought to get him if money can do it," was Jones's reply. The man from Tidewater was not to be tempted. The Confederate states owed him $2,300 for his work as postal agent; he had $3,000 lost irretrievably in Confederate bonds; and the Confederate States of America itself was all but sunk—but Jones's loyalty was beyond reach.

If it were possible to argue from individual cases, it might be concluded that, much as the Potomac had changed the course of the war, the war itself had little changed the Potomac.

ELEVEN

THE SEAT OF GOVERNMENT

THE GREATLY EXPANDED CITY OF WASHINGTON THAT EMERGED from the Civil War, those who knew it agreed, still had a rustic flavor. It also wore a look of expectancy. On all sides could be seen evidence of an intention to grow. The half-completed dome of the Capitol that during the war commanded the eastern end of Pennsylvania Avenue was now finished; but it announced the seat of government of a not wholly formed nation. Down from the Capitol terrace, toward the west end of L'Enfant's mall, stood the half-finished shaft of the Washington Monument. Between the two stretched the principal built-up quarter of the city, from the poor shacks fronting the tidal marshes of the Potomac to more urbane houses as far north as M Street, with a scattering of country homes beyond to the old road that began in Georgetown and followed the edge of the hills eastward to meet the Baltimore Road. Midway in the half-grown Mall stood the Smithsonian building, red, naked, and alone. It was not an inspiring vista.

The city was commonly referred to by visitors as "a country town;" or sometimes "a sleepy country town," or "a southern country town." The descriptions were all perfectly correct, and held for nearly the whole of the city's first century. Most of the streets were still unpaved. The parks were unfenced, unimproved commons, or only too often dumps or borrow pits. Few shade trees had yet been planted, and the ungraded, muddy streets followed the roll of the land. Flocks of geese and herds of swine were encountered foraging and rooting in the streets. An outraged Vermont senator complained in 1871 of "the infinite, abominable nuisance of cows, horses, sheep, and goats running through the city," and thought something should be done to keep "the national representatives and national citizens who come here to attend business, from being obliged to share the sidewalks and streets with animals that ought to be kept enclosed." A little earlier the incorrigibly chatty Gail Hamilton had written of a walk with Senator Hale of New Hampshire in one of the principal residential sections when a pig ran against the senator and knocked him flat on his back. Describing the section north of Massachusetts Avenue, she continued, "you just go across the common and you are in the country."

Back from the American embassy in London, Henry Adams described the city as "the same rude colony camped in the same forest" he had earlier seen. European visitors (whose superior knowledge came generally from a study of steel engravings) thought they detected traces of Carthage, Palmyra, and other battered remnants of classical empire in its half-finished monuments. Dickens paraphrased the Abbé Correa de Serra's "city of magnificent distances" into "the city of magnificent intentions." Criticism was abundant, for the city was fair game for the satirist: often they were unfair. Horace Greeley told some complaining government clerks they should quit: "Washington is not a nice place to live in. The rents are high, the food is bad, the dust is disgusting, the mud is deep, and the morals are deplorable." More usually the satires were gentle. Mark Twain, in *The Gilded Age*, thought that if the mud were only diluted a little the city's streets would serve as canals; to meet the climate he recommended an overcoat, an umbrella, and a fan. A half century after Tom Moore wrote them, the poet's dry words were approvingly quoted:

> This fam'd metropolis, where fancy sees
> Squares in morasses, obelisks in trees:
> Which second sighted seers e'en now adorn
> With shrines unbuilt and heroes yet unborn.

What Moore had somehow gathered in his brief week spent in the British legation was that sense of the future which hung over the city, the sense of destiny, of becoming. It was more than the nineteenth century's belief in progress, and it tempered all but the rudest and most perfunctory criticism.

Ragged and incomplete the city most certainly looked. Abundant vacant lands lay within its central part, but a steady drift outward to new subdivisions and country properties was to be observed. By 1853 the Baltimore and Ohio had been asked to stop its Washington-bound trains at way stations for the convenience of commuters. The high cost of living in Washington during the war accelerated this outward movement, and the railroad put on special trains morning and evening for the convenience of those who wished to live in Baltimore and work in Washington. Population flowed steadily toward the more salubrious upland country north of Washington, and Meridian Hill, Mount Pleasant, Kalorama, Homestead, Kendall Green, and even Silver Spring grew popular, while many people still thought of the heights of Georgetown as a desirable place of summer residence. But the compact center of the city proper was ringed with lands platted long in advance of the city's need for lots. Since the early land bubbles of Robert Greenleaf and Thomas Law such excess subdivision had plagued the city with land speculators and syndicates of promoters whose activities often spoiled entire districts, aborted their logical development, and caused a wasteful leapfrog pattern of building, with large vacant tracts held at high speculative prices between colonies of new houses.

The slow growth of the city within its huge framework was the leading reason for its rural appearance. In literal fact the Federal District had embraced three cities—Georgetown, Alexandria, and what was sometimes still called Washington City. Rolled together the three might have made a city, but separated, any one of them was hardly more than a village—and Washington the newest, emptiest, and most rural of the three. In 1846, so slowly had the capital grown that the part lying south of the Potomac was returned to Virginia.

The closely built-up pattern of Alexandria (1749) and Georgetown (1751)—called "the city of houses without streets" in contrast to Washington, "the city of streets without houses" —at least reflected the compact urban ideal of the eighteenth century, one that sharply differentiated the town from the countryside with row houses, squares, and carefully defined public places. The Potomac country was not to see a distinguished expression of this ideal, although nearby Annapolis provided one of its best examples. Washington itself showed but a wistful emptiness, a plan undeniably grand on paper, but still lacking in principal buildings, monuments, avenues, and any quality of true urbanity. Between small neighborhoods of good residences were colonies of cabins and shacks. Often squalid alley slums inhabited by slaves or domestics lay directly behind the finest houses, like the old plantation "quarters." The impression of a patchwork coverlet, with silk squares sewn next to homespun, was sometimes commented upon, but more generally taken for granted.

2

The form and character of the capital, as it spread across the flat lands of the river bottom and sent fingers of habitation up into the surrounding hills, reflected faithfully the dominant aspects of life in the Potomac region. Its varying roles and impulses found apt expressions in the reluctant city.

From the surrounding plantation country the city acquired its individuality, its topography, and the durable fabric of its society. Indeed, the land where the city stood had once been but a series of Potomac river-front plantations. Daniel Carroll's Abby Manor occupied the high ground of Capitol Hill; David Burnes held Beall's Levels, the lower land between the White House and the Capitol along Pennsylvania Avenue and the Mall. So it went throughout the 10-mile square, in the landed Potomac manner, with plantations named Duddington's Pasture, Port Royal, Jamaica, Mount Pleasant, Mill Tract, Isherwood, Hop Yard, Mexico, and Widow's Mite. In the official reports sent President Washington by Thomas Johnson and Dr. David Stuart, the land had been found an agricultural shambles like other parts of Tidewater Potomac before 1800; and Andrew Ellicott, the first surveyor of the city, gave it as his private opinion that "this country intended for the permanent residence of Congress, bears no more proportion to the country about Philadelphia and Germantown, for either wealth or fertility, than a crane does to a stall-fed ox." For nearly seven miles along the surveyor's route he reported, "there is not one house that has a floor except of earth." Nearly a century would have to pass before travelers in the outskirts of the city could report more than broom sedge in abandoned fields, the wreckage of barns and vine-grown outbuildings, unfenced pastures, scrubby forest, and all the other consequences of the old agricultural exploitation of the Potomac. The tobacco planters could count themselves lucky in having the new city, and there many of them went to seek new fortunes.

To the speculative habits of the original landowners may be traced many of the early difficulties of the town, and the hesitation with which development advanced in a ragged, sprawling fringe. It negated L'Enfant's plan for a commercial center between the Capitol and the Eastern Branch. Each owner tried to make his land more valuable as a prospective center of the new city, and this diffusing tendency remained the decisive force in a place that reminded most observers of a thickly settled countryside rather than a town. The lack of an urban standard and distinct urban traditions in the Potomac region for a long time led to a free and easy tolerance of what the national capital's leading historian refers to somewhat meekly as "village conditions."

The initial terms of acquisition allowed the original plantation owners to retain half the lots in the new city, and to cut the timber on all land they had owned. So promptly was this done that within a few years David Baillie Warden was writing: "It is deeply to be regretted, that the government or corporation did not employ some means for the preservation of the trees which grow on places designed for public walks. How agreeable would have been their shade along Pennsylvania Avenue where the dust so often annoys, and the summer sun, reflected from the sandy soil, is so oppressive. The Lombardy poplar, which now supplies their place, serves more for ornament than shelter." The poplars had been planted by Jefferson and lasted a scant thirty years. They were soon replaced by elms, lindens, sycamores, and maples—the common trees of the plantation gardens and driveways—and a small but regular planting of street trees gradually became an accepted municipal policy. High post and rail fences around some public grounds further enhanced the impression of the countryside.

Some other values of plantation life also perpetuated themselves in the federal city. Its race tracks and its gaming houses owed something to it. In a larger sense the atmosphere of public service mixed with politics and place-holding, the emphasis on social values and caste, the prestige accorded the legal profession, the cliques and the art of the caucus that Virginians and Marylanders had brought to the national political scene, all had become characteristic of the capital. They stamped local government and some national government. All these characteristics had been seen in the microcosms of Alexandria and Georgetown; now they were transplanted to a larger theater. Even the social heritage of the isolated Potomac plantation and the decorous modes of the colonial capitals of Annapolis and Williamsburg were reflected in the gay hospitality and fashion of the capital, as indeed the very first arrivals there had discovered.

The city was literally built out of the natural resources of the Potomac. The architects of federal buildings, notably Latrobe, carefully surveyed its timber and its quarries. From the tan sandstone quarries at Aquia came the stone for the White House. To build the Smithsonian, temporary lateral canals were constructed down the Potomac and through the city, connecting the red sandstone quarries near Poolesville directly with the building site. Maryland marble built the Washington Monument. The city's streets were cobbled in mica schist from Virginia quarries above the Key Bridge.

The taste of the Potomac planter was seen, too, in the building of the city, in its houses and its municipal buildings, transmitted through such Potomac figures as Washington and Jefferson, and a host of lesser ones. It flavored the city's plan and even the buildings of the national government. Not only the older settlements of Alexandria and Georgetown, but the first houses on Greenleaf's Point, the Six Buildings, and the new residential sections showed it very well.

The self-contained mansion, free-standing on its own acres, surrounded by stables, carriage houses, kitchens, and other "offices," was transplanted to the town. Always the emphasis was on the social apartments, and what could be provided in the way of rich and impressive decoration was marshaled here, where lofty ceilings and fine proportions announced the purpose of the rooms. One of its most authentic surviving examples is Tudor Place in Georgetown; another is the Octagon House.

The White House itself perfectly mirrored this residential ideal, and fringing the park facing the President's House, which later became known as Lafayette Square, could be seen its further reflection. Here, around the old race track that first occupied the square, and to either side of Latrobe's freely adapted Palladian structure for St. John's Episcopal Church (whose parent congregation first met in a tobacco barn) arose a series of houses that stumbled piecemeal and by accident into a happily unified design. This was the exception. Elsewhere individualism was the rule.

One characteristic feature of all Washington houses was their gardens, a visible memento of the countryside brought to town, a repository of one's favorite plants and household herbs, a necessity in the humid summers, and a never-ending source of beauty, pride, health, and even scientific experiment. Cultivated gardens and lawns were regarded as a distinctive feature of the city. They were identified as "these rural accompaniments" of city life. The tradition of the private garden, usually behind one's house, took firm hold, and the later requirement that a parking strip be left in front of one's house and another between the sidewalk and the street fitted the ideal. Gardeners, horticulturists, and nurserymen appeared at the very birth of the city. Passed over as a federal architect, L'Enfant spent the remainder of his life as a designer of gardens. The earliest buildings with their experimental orders of tobacco, wheat and corn, constituted "a botanical architecture." The first gatherings of scientists were concerned with botany. One of the first books published in Washington was a treatise on gardening, and the city's first guidebook proudly listed 296 species of flowering plants found in the city. Public nurseries were early established to provide trees for the growing city. Led on by this activity, as well as to provide an agreeable spot for its more leisurely moments, Congress joined with the Botanic Society and established the Botanic Garden.

On the banks of the "river of traders," where the wealthy Scotch tobacco factors and merchants had come to form the ruling caste of the indigenous population, a commercial class existed even before the city became a seat of government. From this source, perhaps more than from the landowners, came much of the money expended in building the new city, and much of the appetite for the speculation in lands and buildings that marked its early ears. The grand design of this group of merchants was to make the city a major center of regional and national trade, and to this end they sponsored notable projects in canals, roads, bridges, and public buildings. Their aspirations were evident early enough for L'Enfant to reflect them in his plan for the federal city, and it was they who largely carried them out. The customhouses, wharves, piers, and harbor improvements were constructed to meet their needs. More especially so were the markets, around which grew up the city's first business districts, its bookstores and printeries, and its taverns and hotels.

Chief among these was the Center Market, where the turnpike down 7th Street met the turning basin of the City Canal at its lock, about midway between the Capitol and the White House, and on the old route from Georgetown to the crossing of the Eastern Branch, which early became the post road through the city. A few blocks west the main road to Washington City from Montgomery County came down 14th Street. Between these two locations, along Pennsylvania Avenue and the Mall, the core of the commercial city was established.

This spot on the Mall became the scene of the early fairs "for the sale of all kinds of cattle, goods, wares, and merchandise," each May and November—following an old Maryland custom. The fair became the first of the modern agricultural fairs where new animals and crops were exhibited for educational purposes. Here the first stores, banks, and a theater were built. The first appearance of a government office in this area came quite as if by accident from the location of the post office in a converted hotel and the use of the surplus space by other departments. The new city of national government was rising, literally, within the old river town of the Potomac traders.

Along the Eastern Branch, then navigable to Bladensburg, wholesale trade and manufacturing were located. Here a tobacco-inspection house, a sugar house, and a brewery were established; the navy yard, gun factory, and ropewalks were found; an arsenal here stored and distributed the guns made at the Harpers Ferry armory; and here some of the manufacturing that commenced under the spur of national independence and contracted world trade during the Napoleonic Wars found a foothold. Along the lower reaches of Rock Creek, separating Georgetown and Washington, where water power was available close to the river landings, the canal terminal and the highways, mills were built to grind the Potomac wheat to flour, to make paper, and window glass. The development of the canals stimulated the expansion of such mills, and the building of lime kilns, yards for lumber and coal, and other characteristic regional enterprises.

These early efforts, in the belief that Washington could be made a commercial metropolis, had a decisive effect on its future form. They helped in determining the principal quarters of the city, especially its concentration in the flat, unpromising area between the White House and the Capitol, the location of roads and turnpikes, canals, and railroad routes and terminals. In their way they were as formative as the drawings of Major L'Enfant; indeed, the French engineer's sketch was utterly lacking in specifications for a broader city plan, and concerned itself only with streets, parks, and the location of some public buildings and monuments. The forces of Potomac life added flesh and blood to the skeleton provided by L'Enfant.

The Potomac route to the west by river, canal, turnpike,

and railroad was reflected in the planning and growth of the city. As one means of transportation succeeded another in popularity, as the relative balance among the three cities of the Federal District was altered by the growth of the city of Washington, the transportation routes and terminals stamped its future pattern. The railroads running across the Mall, the abandonment of the City Canal, the decline of the turnpikes, the neglect of the river itself—all evidenced what was happening everywhere in the Potomac region.

Of chief importance in the early days had been the turnpikes. These dominated the approaches to the city and helped dictate the location of the markets and business districts. Even many years after the railroad had triumphantly threaded its way through the city, great droves of cattle, flocks of sheep and turkeys, and quantities of farm produce came from the surrounding farms of Maryland and Virginia to the city's markets over the turnpikes. Many of their routes were powerfully affected by bridges over the river and the Eastern Branch. As the work of private companies, in which businessmen were the principal directors, the turnpikes were naturally an important tool in establishing the trading position of the city in competition with the earlier river towns of Alexandria and Georgetown. To counter the completed turnpike from Rockville to Georgetown, which diverted the upcountry trade, the Washington merchants had an improved road constructed from Washington north along 7th Street to establish a more direct connection. They also altered the trading position of the Potomac region in competition with Baltimore by connecting it with producing areas and markets. Thus the turnpikes that began at the Virginia end of the Long Bridge and led out to Alexandria and Fairfax reached ultimately to the tobacco regions of the lower river and the wheat country of the Piedmont and the Shenandoah valley. The road to Bladensburg continued on to Baltimore.

The results of these efforts failed to give the capital city the commanding national position its trading class sought. Every approach to the city had its tollgate, but the good roads led only to Baltimore or to the south. In its national importance the city remained but another town on the great coastal route north and south. The western route proved impossible to develop in the face of undisguised hostility from Baltimore, and was not extended beyond Rockville to connect with the Maryland turnpike and the Cumberland Road, despite even Congressional attempts to do so.

A similar effect came from the development of water transport. Regular steamboat service was established in 1815 between Washington and Aquia Creek, at the Big Bend in the Potomac, not far from Fredericksburg, where the principal road to Richmond and the southern states commenced. Riverboat service in the Tidewater country was so satisfactory that the railroad hardly penetrated there. But despite great efforts to develop the City Canal and exploit the anticipated canal trade with the West, nothing of this sort transpired. The City Canal, today covered over by Constitution Avenue, a solitary lock house remaining at 17th Street, was a decisive element in the location of the commercial center of the city, but it failed to become of great national importance as its builders hoped. It fell into the slough of local competition with Georgetown and Alexandria, and the even greater competition between the canal and the railroad.

The city's easy connection with the South and with Baltimore was still further reinforced by the railroad. In 1835 the first railroad entered the city over the line of the Baltimore and Ohio, and found its terminus on Pennsylvania Avenue at the foot of Capitol Hill. Not for another twenty years was rail service established to the south, and then only from Alexandria on. Finally, on the eve of the Civil War, this line was extended to the Virginia end of the Long Bridge, and during the war the tracks were continued over the bridge and through city streets to a terminus at the Baltimore and Ohio depot. When the Pennsylvania Railroad broke into this territory from Baltimore in 1866, it took over the southern line as far as Quantico.

Thus by the end of the Civil War, except for the canal, the city was still connected only with Baltimore and the South. No direct rail connection with the West existed save for the roundabout route via Baltimore, 49 miles out of the way. Georgetown remained a canal and turnpike town. This condition continued until 1870, when the railroad link was finally completed between the main line of the Baltimore and Ohio at Point of Rocks and Washington.

To such factors the city owed in good part its southern Tidewater flavor. Under these conditions its commercial framework was established. Thus the city was built. By the time direct connections with the West had been made, it was too late to realise the city's commercial aspirations. By then the interest of the federal government had become supreme, and even the city government, long controlled by local merchants, had been extinguished. In moving to take the franchise from Washington citizens and inaugurate eventually the commission form of government that persists to this day, Senator Lot M. Morrill of Maine summed it up in 1866 when he said that the district was never intended to be a government: "It is a seat of the government of the United States."

3

Congress could not put up with it. The government of the United States could no longer stand the "village conditions" that prevailed in its capital and were so complacently regarded by its inhabitants. The inefficiencies of local government were intolerable and, whatever the cause, matters could no longer continue as they had. The steady growth of the federal establishment, and the dramatic events of the war, had thrown this into sharp relief. A growing sense of national pride, accentuated by the war, demanded that something be made of the city "worthy of the nation." Whatever else the city might wish to be, it must become a fitting seat of government.

"None of us are proud of this place," Senator Stewart of Nevada, a Washington property owner who took an active interest in the development of the capital, said in 1871 in turning down a proposal for a world's fair in Washington. "Let us have a city before we invite anybody to see it." Common experience told it was a reasonable view of the matter. Clouds of dust rose from the two-day Grand Review, when the victorious army of 200,000 men paraded up the cobblestones of Pennsylvania Avenue at the close of the war. Like other streets of the city, deeply rutted from heavy military traffic during the four years of war, the avenue was in such bad repair that business was leaving it. The confusion that followed Lincoln's assassination raised questions of the adequacy of the local police, and Early's 1864 raid had shown the lack of a local militia. The sickness in the National Hotel caused by poisonous gas from obstructed sewers provoked a sanitation scandal. The fire that destroyed the books of Congress in the room on the west front of the Capitol could not be extinguished because sufficient water could not be procured, and the fire department of the city was still weakly composed of volunteers. Members of Congress stumbled along the ungraded and unpaved streets, bumped into roving domestic animals, fell over bridges that lacked guard railings. Street lighting ceased at midnight, and to save money no lamps were lit on nights when the moon shone. Open brick conduits carrying sewage were encountered in all parts of the city. Floods on Pennsylvania Avenue —"the Grand National Broadway of the Metropolis"—were of common occurrence. Flooded basements and storm drains stopped by high river levels regularly announced the advent of spring. The Potomac River, increasingly clogged with silt and refuse, had become a malarial swamp. The Eastern Branch, where the entire navy, such as it was, had anchored in Jefferson's administration, was worse, stretches of it almost dry land—a classic example of coastal plain erosion and deposition. Conditions at the White House became so unhealthy that Presidents Van Buren, Pierce, and Buchanan moved habitually in the summer to the heights above Georgetown or the hills north of the city to escape the fevers. The water supply from springs and dug wells was contaminated by the universal privies, barns, and stables of the increasingly crowded city. Efforts by Congress to mend matters, so far as the public buildings were concerned, only proved it was hopeless to establish a virtual federal enclave within the district, with its own sewer and water system and municipal services. It ordered the army engineers to take over the city parks, but broader measures were clearly needed.

From a municipal point of view, the city was regarded as "a collection of neighborhoods," and any effort to treat it as a whole ran against the diffuse pattern that had developed in municipal management, in faithful reflection of the diffuse pattern of settlement. The collection of villages that was becoming the federal city expressed itself in the organization of the city by wards. Ward government ran the city. Many of the wards had their own municipal services. The fire companies were organized by wards. Street improvements were a ward affair. City policy dictated that ward improvements should be financed from ward revenue. Once the ward's share in the expenses of general government had been met, the balance—if any—was used for local improvements.

The wards themselves were arbitrary rectangular blocks laid out on the city map, reflecting little in the way of social and economic realities, much less topography. Matters were made worse by the archaic franchise, which almost until the Civil War denied the vote to all but owners of city real estate. Six per cent of the city's population voted. The citizen's tax bill was thought bewildering. Aside from the general property tax and the personal property tax, he paid a business license, special assessments for sidewalks, pavements, and street lamps, and a charge of one cent a bucket for removing garbage.

With all the continuous bickering between Congress and the city, between the wards, and among the citizens, the truth was that the wastefully sprawling city, with its overly spacious streets, its rapid growth, and its vast areas of cheap and tax-free land, was too expensive to manage without excessive drain on the local taxpayers. It was not inefficiency, nor the incompatibility of federal and local authorities, nor the disproportionate share of municipal services consumed by the federal government without full payment—although all of these and others were mentioned by way of futile explanation by a worried citizenry who felt control of the city slipping from their grasp. High municipal standards conflicted with the slender tax base of what was still almost wholly a residential city with neither industry nor trade to tax. The difficulty imposed by the spacious city was met also by the public utilities like the Washington Gas Light Company —although here large bills for services were paid by the federal government—and by the traction companies. By the end of the city's first half century it was beginning to be clear that its commercial ambitions, the ambitions of the Potomac traders, would not be realized; and, indeed, that they conflicted with the development of the city as the seat of national government.

Patchwork measures to remedy these conditions, including reform of the franchise, a new city charter, and a redistricting of the city, failed to bring about a strong and aggressive municipal administration capable of dealing with the problems of a growing city still in village clothes. Congress was impatient in its role of making extraordinary appropriations to maintain an appearance of decency in its capital. Irritations were frequent, and the Congressional response was often curious: at a time when Pennsylvania Avenue was the only street in the city illuminated at night, Congress stipulated that the oil lamps be lit only when it was in session.

For over a decade before the war a steady increase in Congressional expenditures in the city was noted. The government was growing as the nation grew, expanded, gained in wealth, and became more complex; and that growth was faithfully reflected in the federal activities in Washington. Committees of Congress became more numerous, and demanded places where they could work. Overcrowded government departments created scandalous conditions of employment, public property was lost, stolen, or destroyed by fire, and clerks were killed in the collapse of private houses rented for government offices. The extension of the Capitol, and the classical revival buildings of the Treasury, the new Post Office, and the enlargement of the Patent Office, were responses to these conditions in the decade preceding the war. Pressures of growth were further shown by office blocks like the Winder and Corcoran buildings constructed by private speculators for the purpose of renting them to the government. Congress began to spend increasing sums of money on street improvements, not according to plan or any financial formula, but in a way well calculated to demoralize any local efforts in this direction. Finally, there was the 14-mile aqueduct constructed by private contract under the supervision of the army engineers, which took the water from the Potomac at Great Falls, and carried it down through reservoirs and mains, crossing the deep indentation of Cabin John Run by the longest masonry span yet built, to bring the city its first reliable supply of pure water.

Senator Morrill stated a categorical fact that had become plain to Congress when he said, "it is a seat of the government of the United States." Had he needed to elaborate, he might have explained that the representatives of victorious northern and western states could no longer tolerate the "sleepy southern village" as their capital; that the commercial interests of the city had to give way before its role as a governmental center; that without industries or trade to tax, the city obviously could not support itself; and finally that Congress did not intend to foot the municipal bills without first controlling the city government. He could have added that the planters' and traders' city on the Potomac would have to become less a regional center and more a national one. At the end of sixty-six years' experience, the federal government, which had come here with its 131 employees in 1800, had grown to the point where it could say something like that.

4

After it became aware of its full responsibility for the federal city, the government was not long in taking its first step. It established the so-called "territorial government" and brought to power in 1871 a close friend of President Grant, the Washington contractor and housebuilder, Alexander Robey Shepherd. The spectacular career of this American Haussmann showed disinterested and far from predatory convictions about the development of the city as a metropolis, pursued with unequaled energy and resourcefulness. Shepherd was not inexperienced in the ways of the city: he had drafted the improvement resolution of 1863, when he was but twenty-eight years old. He had had a large share in formulating the territorial government plan that followed in its general outlines the earlier proposals of Senator Morrill. A self-made man, he had begun as an apprentice plumber and moved rapidly into the building business, where his activities in constructing row houses en masse became prodigious. He allied himself with the growing number of citizens who believed in improvements (and were willing to treble the tax rate to get them).

Shepherd was a recognizable type among his contemporaries. Physically a large man, his tastes were on the large side as well. He was a big eater and drinker, full of hearty, profane talk, and the kind of masculine virtues esteemed by his friend President Grant. At home in an age of politicos and robber barons, a contemporary of the great city bosses—and, mistakenly, himself called a boss—the wonder was his essential honesty. Repeated investigations found him clean. His political support came from radical Republicans, from the new Negro vote, and from the laborers of the city. It was a new alignment. An aggressive and capable politician, he was a disinterested and proficient administrator. Imperious, autocratic, even dictatorial, he was also independent. He could throw caution to the winds, plunge into risks that might cost him his last penny, and beyond all else, get the job done.

When Shepherd began his activities, the previous city administration had completed a large program of public improvements—but almost all of it had been located outside the then built-up area! Some streets had been opened and graded to the district line, but nothing had been done with the ticklish matter of grading the existing streets or levying upon the owners of abutting property the special assessments for their improvement and paving. The dirty work in reclaiming the "city of mud and dust" was still to be done.

Although only one member of the 5-man board appointed by the president in the new municipal government, Shepherd dominated the group from the beginning. He was elected the executive officer at its first meeting, and thereafter the group met infrequently, "the acts of Shepherd being recorded as the acts of the board."

There was no doubt in any Washingtonian's mind what Shepherd had set out to do. "The board of public works," he said, "when they entered upon their duties concluded that they had been created for something or nothing, and if for anything, it was to devise and carry out as rapidly as possible some system of improvements, in order that in this respect the capital of the nation might not remain a quarter of a century behind the times." The opposition—composed mainly of old Washington property owners, apprehensive over the tax rate—was already organizing to oppose Shepherd's program. Their weapons were feeble, but they made the most of them. They urged a moderate pay-as-you-go plan, and a slow approach to the work. They sought injunctions to stay the work. They put pressure on Congress, and secured three full-dress Congressional investigations in a space of three years. In the end more than half of them refused to pay assessments levied against their property.

On his side, Shepherd seized the initiative and never lost it. Within a month after the board met, a "comprehensive plan" had been drafted "to make the city worthy of the nation," as the favorite contemporary phrase ran. The plan was limited to street improvements, but it went far beyond anything the city had ever dreamed of, and reached levels of co-ordination that set new marks in engineering progress. In its execution it smacked more of a wartime operation than any engineering work the city had previously seen. These improvements still form the foundation of the modern city.

Shepherd's first brilliant stroke was to solve the problem of L'Enfant's excessively wide streets. This had made paving so prohibitively expensive that almost none of it had been done in three-quarters of a century. He used planted strips between the sidewalks and the curbs to narrow the street widths. These parking strips logically opened the way to the great program of tree planting that is the most distinctive and gratifying feature of Washington streets today. Since the city's birth some planting of street trees had been a regular expense of the city government. Now it became an integral and indispensable part of every city improvement. Under an expert parking commission composed of William R. Smith, superintendent of the Botanic Garden, John Saul, a Washington horticulturist, and William Saunders, who managed the gardens of the Department of Agriculture, 60,000 shade trees were planted in two years.

The street improvements themselves consisted of a carefully worked-out program of sewers and drains, grading and uniform street levels, and the best quality of paving known to the time. In all 118 miles of streets and 39 miles of outlying roads were improved in about eighteen months.

The onset of the improvement plan came with an intensity that seemed almost demoniac to the older Washingtonians. Suddenly the city was alive with laborers ripping up old sidewalks and streets. No time was spent letting contracts: day labor was employed directly by the city. Instead of doing a little of the work at a time, street by street, everything was started simultaneously. The work was compounded. Once the streets were opened the work of connecting houses and buildings had to be done. The street railways, the Washington and Georgetown Company, the Metropolitan Company, and the Columbia Railway Company, were all obliged to lay track before the paving was done, and to their surprise to pay for it, the paving between the tracks to be of the same quality as that employed in the streets themselves. Without regard to the traditional local building season, work commenced in the fall and continued throughout the winter. When Congress arrived, a few weeks after work had commenced, it was amazed at the dismantled condition of the city's streets. They heard President Grant commend the work. In his words, Washington "is rapidly becoming a city worthy of the Nation's capital."

Shepherd's attack was dashing and audacious. It carried him past obstacles that had long proved insurmountable to the old city government. The old Northern Liberties market, a jumble of shacks and sheds which stood where the Public Library now is, had long been scheduled for demolition as part of a city-wide plan of new public markets. No effort had succeeded in dislodging the marketmen who owned stands there. Two weeks after Shepherd put them on notice, and before the tenants had got around to applying to the courts for an injunction, a gang of workmen turned up at eight o'clock one night and summarily tore down the buildings and cleared the square.

The railroads, which had long ignored city specifications as to street grades, paid little attention when Shepherd served notice that they must rebuild their tracks to confirm to the proposed street level. In the course of one night, a force of two hundred of Shepherd's men tore up the entire line of the Alexandria and Washington Railroad along Maryland Avenue.

Slashing forays like this left the opposition bewildered, and even the city corporation was unable to halt the rapid march of improvements. Instead, Shepherd wheedled approvals for his bond issues and building certificates out of the district's territorial legislature. As the streets were graded, houses were left perched twenty feet above the finished level of the street or sunk in ravines with only their second stories peeking above the sidewalks. Senators and congressmen who owned houses, as well as outraged local property owners, were left wringing their hands as the unfinished business of generations was speedily transacted.

Shepherd's army of construction workers, Negroes and Irish for the most part, formed a solid political machine that at special elections rammed through additional bond issues by overwhelming majorities. He used every trick in the politician's book. Voters were imported from Maryland and Virginia to pack the ballot box in favor of the public improvements bond issue, and while printed ballots favoring the proposal were abundant, those opposing it were in short supply and most of those voting against it were obliged to write their own ballots. President Grant stood solidly behind Shepherd. So did the brother of the financier Jay Cooke, Henry D. Cooke, who was heavily interested in the city's traction system. Grant had appointed him the city's governor, and it was he who sold the improvement bonds at a spectacular increase in price over the city's ordinary securities. In the political sweep, contractors, builders, and owners of unimproved land swung behind Shepherd's program. So did many taxpayers, confident that Congress would appropriate funds to support the improvement program. Indeed, President Grant urged liberal district appropriations for this purpose. Many of the city's newspapers stood behind the program, and in his largehanded way Shepherd inserted public advertising in all sixteen of them rather than in a few, as had been customary. He was accused of subsidizing the press. The opposition started its own newspaper. Shepherd and Cooke organized in still wider circles, preaching their doctrine of One Big City. They organized the district militia, and founded the Washington Club.

The momentum of the program was irresistible. What had been intended as a long-range plan was completed in eighteen months. But what had started off as a six-million-dollar program came, in the end, to better than eighteen millions, far above even the debt limit imposed by Congress. The city wound up with a per capita debt about twice that of New York City after the worst depredations of the Tweed Ring. Even before the panic of 1873 the city was bankrupt, and could not pay the salaries of its employees.

What did it come to? Shepherd, with the full approval of the president and Congress, as well as the district's own territorial legislature, and with thumping majorities of the city's popular vote, had rebuilt the city along modern lines. He left it in bankruptcy, on the verge of repudiation, and Congress picked up the tab. Congress footed the bill for half the funded bond issue (it was finally paid off in 1922), and thereafter paid about half the cost of running the city. Congress completed its control of the municipal government, and to avoid the embarrassment of disfranchising the Negro voters while it was enfranchising them in the southern states took away the vote from the city. Most Washingtonians agreed it was the lesser evil. Not least, Congress's assumption of municipal responsibility finally put an end to the possibility, seriously entertained throughout the first three-quarters of a century of the capital's life, that the seat of government would be removed elsewhere. The decision to undertake the State, War, and Navy building was taken as evidence of that. The commission government that was made permanent in 1874 was the capstone of these new developments—rather for the worse than for the better, politically speaking. But the mold of metropolitan Washington had been formed.

The effect of the new improvements was immediately felt. The city plunged into its first real-estate boom. Property values rose, and trading increased as some owners sold their property to avoid paying the assessments, and others eagerly bought them in anticipation of the increase in value. Still more subdivisions were laid out, and building activity greatly increased. Many members of Congress and their secretaries, who had previously spent the shortest possible period of time in the city, living in hotels or boardinghouses, now purchased or rented homes.

5

The prewar government of lawyers and clerks expanded and made room for a new government of scientists and statisticians. For a while it seemed that every promising stripling in the Patent Office, the Smithsonian, or the nonfiscal services of the Treasury was destined to head a new bureau. Veterans were everywhere in the new civil list. A politician observed, privately, that the Grand Army of the Republic had saved the nation, and now wanted it. The wits were still at work on Washington, but as usual they were behind the times.

Their leading subject was the office seeker. Generally presented as an importunate wretch, his plea was put succinctly by Petroleum V. Nasby:

> 1st I want a offis.
> 2nd I need a offis.
> 3rd A offis would suit me. There 4
> 4th I shood like to hev a offis.

The twenty triumphant years of the spoils system, climaxed by the virtually clean sweep of Lincoln and the Republicans, had marred the prestige and respectability of the public service. Politics was in bad odor; the administrative services were equally in poor repute. The opening paragraph of *The Orpheus C. Kerr Papers* ran: "Though you find me in Washington, I was born of respectable parents, and gave every indication, in my satchel and apron days, of coming to something better than this." But by 1872 the Republican party platform was urging measures that "public station shall again become a post of honor." The war itself had shown disgraceful defects of maladministration, scandalous dishonesty, boodling, and inefficiency. It was summed up: "In the present struggle, as already seen and reviewed, probably three-fourths of the losses, men, lives, &c., have been sheer stupidity, extravagance, waste. The body and bulk came out more and more superb—the practical military system, directing power, crude, illegitimate." The writer was Walt Whitman, who had seen the war and eight years of service in government offices afterward; but the poet's view was one widely shared, and it was quoted approvingly by those of larger experience. A hard lesson had been learned.

The war that had revealed the need for competence in government had also in part shown the way forward. It was not by chance that even a presidential nonentity like Grant had espoused civil service reform. In a managerial sense the war had been the indispensable school for industrialists, financiers, railroad builders and executives, and the rising business class. It gave them experience in dealing with affairs large beyond anything civil life had had to offer. Nor was this less true in equipping men to handle the expanded work of government.

The need for change was beyond question. Even before the war Senator Marcy, who acknowledged himself as the source of "to the victors belong the spoils," had protested pointedly: "But I never said the victors should loot their own camp." One department after another had stirred restlessly under the anxieties of its secretaries, whose signatures had to be put to documents and reports of dubious accuracy and merit. The Treasury had experimented with pass examinations to qualify its candidates. The State Department commenced a systematic training program for consular aspirants. The Interior Department attempted to develop a career service for its far-flung administration. A regularized civil service was becoming an irresistible necessity.

The events following the war made it clearer and more urgent. In the expanding prosperity of those days a million demobilized soldiers had been painlessly absorbed into the economy. The war's physical devastation was quickly made good. But men of high ideals who had fought the war, because they had believed it necessary, cursed in despair at the opportunism and incompetence shown by the Freedman's Bureau and the military government in the southern states. The growing scandals in government were linked to incompetence as well as to political interference and graft. The Land Office was unable to investigate collusion and prevent illegal enclosures because its staff of inspectors was hopelessly inadequate to the job in hand; it had to be completely reorganized. The Star Route scandals showed a Post Office breaking down under the strain of meeting the needs of rapid westward expansion. The new Railway Mail Service, the nation's pride, which demanded men of prodigious memories and stamina to sort the mail en route, almost suspended because of inability to attract men to a specialized service that offered so little security of tenure. Bad as matters had become in the older and simpler branches of the government, they threatened worse in the newer and more complicated ones that demanded not merely honesty but technical competence, and where administrative breakdowns had large consequences.

The federal government was becoming directly involved in the expanding national economy. It gave land subsidies to the railroads. It had an expanding tariff system that eventually "protected" raw material producers as well as manufacturers, and gave subventions to native industries like growing sugar beets and sugar cane and the manufacturing of tin cans. The government's part in the financial life of the nation was enormously elaborated with the war. The war debt clinched it. The coinage, silver purchases, banking and credit controls were strengthened throughout the rest of the century.

An industrial and business civilization had to have efficient public services. Boiler explosions on steamboats, railroad accidents, a lax postal service, insecure banks, epidemics, even polluted drinking water from tin cups on trains—all brought demands from business and the public for regulation and control. With all the dissipation of natural resources more than a few could see the end in sight. The unintentional operation of the homestead system lost the nation its forests, oil, coal, and minerals—the larger part of it to the railroads. The auditing of land grants to reclaim lands granted but not earned could not take place without an adequate staff. Inaccurate accounts made enforcement almost impossible.

Dorman B. Eaton, chairman of the Civil Service Commission, thought the increasing size, difficulty, and importance of administration had become the most powerful argument for civil service reform. He quoted with approval de Tocqueville's observation that "A multitude of actions, which were formerly beyond the control of public administration, have been subjected to that control in our time, and the number of them is constantly increasing." Eaton specified "A larger and still larger number of officers are required, and their neglect and incompetency more and more tend to become evils of serious magnitude. The railroad, the steamship, and the telegraph; the system of national banks and the new departments of agriculture, education, and public health; the life-saving and the Marine Hospital Service; the money order system, and the light house, the internal revenue and the postal administration, greatly extended—are but illustrations of the growth of administrative functions created or enlarged during the present generation. . . . Year by year, the prosperity and morality of every enlightened people become, in still greater degree, dependent upon the character and capacity of those who fill their places of public trust. In no country is this more true than in the United States. . . ."

The change came, and it changed Washington. Far more stability from one administration to another accomplished what some have claimed for the cow; it turned nomads into a settled people, boardinghouse transients into residents and homeowners, who demanded schools, churches, and other civil facilities. To the life of the city, as great as the growth in size of the federal offices was the changed character of their personnel. Women appeared for the first time as employees. The sciences and the professions sanctioned by the age began to give their color to the federal service. A more cosmopolitan society emerged, with a greatly enlarged diplomatic corps of superior quality, a host of distinguished visitors, and a steady coming and going to far corners of the nation of army officers and civil officials. And (with the growing stress laid on civil service quotas) the federal establishment acquired a more even representation in the capital of all sections of the nation, making it a veritable microcosm. Not least, in numbers or impact, were the journalists, lawyers, scholars, and the swelling numbers of representatives of the national interests, who were drawn to the city as the government intervened more and more in the national life. A distinctively modern city was emerging.

Of all these changes, perhaps the most immediate and the most consequential was the development of the civil list on the scientific side. From what proved seedbeds of significance in the Treasury, the Patent Office, and the Smithsonian Institution emerged an attractive and competent group of public servants. A view of the government at the end of the war showed what was happening. In the Treasury the Coast Survey, Office of Standard Weights and Measures, Lighthouse Board, Steamboat Inspection Service, Revenue Marine Service, and Marine Hospitals were notable centers of scientific activity. So was the office of the secretary itself, as Hugh McCulloch's pleasant and discriminating memoir shows. The Patent Office examiners had long been a traditional and well-identified center of scientific personnel and activity. Among the detached agencies, the Smithsonian, the Library of Congress, the Botanic Garden, and even the Mint were places that could attract men of high caliber. In the War Department the Signal Corps took meteorological observations that laid the foundation for systematic weather forecasting, and army officers, carefully instructed by Smithsonian scientists like Spencer Fullerton Baird, went off to far corners of the nation to bring back geological specimens, bird skins, or wolf skulls, in addition to their more martial duties. Naval vessels called at Constantinople to pick up casks of pickled salamanders, and in consular reports a new fossil might be of importance equal to a new duty on wool. Long gone were the days of 1819 when John Quincy Adams wrote his classic report on weights and measures with the aid of a rusty bank scales he had procured from the Bank of the United States, and a chronometer he had obtained with difficulty.

When McCulloch arrived in Washington in 1863 to organize and direct the National Currency Bureau—the war-born agency that created the national banking system—he discovered the new temper. To his surprise and gratification he was allowed complete freedom in his choice of personnel, for without it his great task of persuading banks to join the national banking system would have been hopeless. He became aware of other individuals of striking competence in the department, like George B. Boutwell, who directed the Bureau of Internal Revenue (and who later prepared a report of fundamental significance on the Mint).

When he sought lodgings McCulloch found himself in the home of a Dr. Barnard, a clerk in the Coast Survey. Far from "a country doctor," as McCulloch at first surmised, Barnard, who later became president of Columbia University, was engaged in calculating the explosive power of gunpowder. He introduced McCulloch to the Scientific Club, and to Dr. Joseph Henry, secretary of the Smithsonian Institution, then the greatest contemporary force for the advancement of science. Henry had invented the magnetic telegraph—an honor he shared with Farady—but from a sense of public duty refused to take out a patent on it. McCulloch met the grandson of Benjamin Franklin, Alexander Dallas Bache, an outstanding practical mathematician and superintendent of the Coast Survey. He met the physician Peter Parker, whose deep and intimate knowledge of China was unequaled. Indeed, it formed the foundation for nearly all official relations between the two nations. The astronomer Simon Newcomb, who had tabulated the most distant planets and established the reputation of the Naval Observatory, was a member of the club. So was George C. Schaeffer, a former Patent Office examiner, and librarian of the Department of the Interior.

The topographical engineer, A. A. Humphreys, veteran of many Civil War campaigns, military historian and authority on the Mississippi River, was a member, as was a character fresh from the pages of Peacock, Jonathan H. Lane, whose hobby was the invention of means of creating extreme cold—for what purpose he could not say. William B. Taylor and Titian H. Peale, pioneer photographer and brother of the painter Rembrandt Peale, were both Patent Office examiners. And so it went through the group: astronomers of the Naval Observatory, scientists in the Army Medical Museum, army engineers, and the rest. Yet the entire membership of the club was hardly more than could sit around a single table.

The Scientific Club grew into the Cosmos Club, formed in 1878, and the Washington Academy of Sciences made its headquarters there. The National Academy of Sciences, formed in Washington during the Civil War, and the American Association for the Advancement of Science, with its headquarters at the Smithsonian, came to be major institutional expressions of Washington as a scientific center. But it is easiest to see the change in terms of biography.

The best representative was John Shaw Billings, one of that race of giants who were born about 1840. A young surgeon, and a born administrator, Billings came rapidly to the top in the Civil War. He organized and ran army hospitals, but by 1864 was happily writing: "I am to be what you might call the medical statistician of the Army of the Potomac . . . the sort of work [that] just suits me." After the war Billings's convictions about medical ignorance, as he had seen it in the war, led him to the Surgeon General's Library, a neglected storehouse of treasures, and in a few years he had established it as one of the world's greatest centers of medical information, and housed it in a new and appropriate building. His tools were those of the librarian, and with them he created a systematic library, an Index Catalogue of medical literature, and the periodical Index Medicus, an international index of current medical literature that became a worldwide reference. In 1873, against five entrants, he won a competition to design the new Johns Hopkins Hospital in Baltimore. The first modern hospital to reflect the advance in medical science since Pasteur, the first to show the lessons in hospital administration learned in the war, and the first to reflect ideals of modern nursing resulted. The Queen Anne architectural gingerbread tastefully hung on the building could not conceal its revolutionary quality. The new hospital was designed in conjunction with a medical school, but Billings swept away the old-fashioned clinical lecture that treated the hospital patient like a beetle on a pin, and took the step that did most to overcome popular antipathy to hospitalization. In the Census of 1880 and 1890 Billings was in charge of vital statistics. He was an original member and later president of the American Public Health Association, and did more to conquer yellow fever and open up the tropics than any man other than Walter Reed. In 1895 the consolidation of the Astor, Lenox, and Tilden libraries in New York City gave Billings another chance to show what he could do. He began by superintending the reclassification of the libraries according to his own system—one the library still prefers to use. He designed the fundamental library building, and then let Thomas Hastings hang his fashionable masonry garments of neoclassicism over the steelwork. And he created the first branch libraries the nation had seen, breaking open the way to bring books to the people in their own communities throughout the great metropolis.

Ideas and action flowed from men like Billings. It was he who selected the young English surgeon, William Osler, for the medical staff at Johns Hopkins. In a casual conversation during the Census of 1880 he suggested to a 20-year-old youth, Hermann Hollerith, that there ought to be some mechanical way to avoid the wasteful and often inaccurate tabulating of census returns by hand. Hollerith remembered it. Later he worked in the Patent Office from 1884 to 1890, experimenting after working hours with mechanical tabulation. By the time of the Census of 1890 Hollerith was ready to try his system. Electric current passed through holes punched in a nonconducting material was his basic idea. In competition with two rival systems, Hollerith's punch cards flew through the tabulating machine, and did the work in less than half the time of his competitors. Hollerith had revolutionized mass statistical methods, in Washington and throughout the world, not only in the work of government but in that of business and of research. His system was the foundation stone of what is now the International Business Machines Corporation.

The marked acceleration of the scientific bureaus following the Civil War took the form of offshoots from earlier beginnings. From Baird's studies in the Smithsonian came the United States Fish Commission, with its co-operative laboratory at Woods Hole, which laid the scientific foundations for fish conservation laws, successfully transferred shad to the Pacific Coast, rehabilitated the inshore cod fishery of New England, and even transported salmon eggs to New Zealand; not the least of the commission's work was the slowing of the decline of the once-prosperous Potomac fisheries. The national weather service was developed by Cleveland Abbe out of the earlier observations of the Army Signal Corps. The Geological Survey was formed under Clarence King and John Wesley Powell to carry through and co-ordinate the separate geological surveys the government had authorized. The Survey was rapidly expanded into the largest single scientific agency in the world. From its work evolved government activities in the reclamation of arid lands, mines, and forestry. From the Coast Survey's work in weights and measures came eventually the Bureau of Standards. Out of the older Marine Hospital Service grew the U. S. Public Health Service.

The prevailing accent appeared early: it was practical science that the government wanted. There was little patience with theories, and with pure research. Yet at the dawn of government scientific work it was almost impossible to draw the line; and, to be plain about it, there were few scientists who wished to, or few other places where scientists could find support for their work. Yet the predilections of Congress for scientific conclusions of practical value left most of the later scientific leaders battling hopelessly for their principles, and the rise of universities and research institutions left applied government science with the feeble solace that practical work might make some contribution to theory. Despite the increase in Washington scientists to the thousands and the tens of thousands, there was a perceptible change in tone. The dominant position of science itself began to alter.

With the Civil War had come a need for information, not only about the work of the government, but to show the state of the nation. Until then the statistical information gathered and reported dealt chiefly with the results of governmental activities; substantially all of it was included in the decennial census provided for in the Constitution, and, from 1821 on, in the import and export statistics of the Treasury. Later developments were largely outgrowths of this earlier work, which supplied both statistical methods and cadres of experienced men.

Some beginnings in the direction of increased statistical activity could be seen in the steady expansion of the Census, which by 1850 was reporting more fully on agriculture and livestock, and concerning itself with such details as religious bodies. But the forward surge was more marked in the first statistics on banking and currency, following the appointment of a comptroller of the currency in 1863, in the fuller agricultural data obtained following the establishment of the Department of Agriculture in the same year, and in the formidable array of statistics on education that commenced to issue from the Department of Education—as it was then called—in 1867. The production of minerals, and information on gold and silver, was reported from 1866 on. The next year information on marriage and divorce became available. In 1871 the first statistical series on wages began, and the same year accurate records on fisheries were first compiled. In 1878 regular reports on diseases were first issued by the Marine Hospital Service. By 1880 the movement was in full swing. The growing number of periodic statistical collections began to reflect the need for such information on the part of industry and business, and the same need was felt by regulatory agencies of government which were then beginning to appear. The change was evident in the Tenth Census in 1880, which showed an interest in retail prices, railroads and other forms of transportation, shipping, water power, and even newspapers and periodicals. By 1890 the outflow of statistical reports from Washington reached flood proportions, but the national demand had not been satisfied. Industrial statistics were still in their infancy; social and economic statistics in their modern sense had been hardly born. The statistical statement had become the ultimate authority; and neither government nor business could do without it.

The leadership in the new science was concentrated in the capital. Only there were yet to be found large masses of data that called for statistical techniques. The leadership was represented by Francis Amasa Walker, who superintended the Ninth (1870) and the Tenth (1880) Census. Walker had served in the war, and came out of it at the age of twenty-four, a mature man. In the Treasury, under the pioneer economist David A. Wells, a special commissioner of the revenue, Walker had become chief of the Statistics Bureau and there began to learn the new art. His chief stumbling block in public administration was the patronage system. It remained, in his judgment, the most serious difficulty in the 1870 Census, the cause of most of its inaccuracies and limitations. But by 1880 Walker had a new census law, and he was satisfied with the quality of work in the 22 volumes that appeared. Statistics led him to economics, where he spoke in the characteristic accents of contemporary business enterprise. Wages, said Walker, flying in the face of current theory, depend on current productivity and little else. He set himself to measure productivity. In an age of growing monopoly, he developed the doctrine of economic competition. In a period of contracting money and credit, he strongly advocated a wide monetary base, abundant currency and credit. He went on to become president of the Massachusetts Institute of Technology, and to raise it from a small college to a recognizable approximation of its present form. The superintendent of the Eleventh Census (1890), Carroll D. Wright, who in his government service also laid the foundations for the present Bureau of Labor Statistics, like Barnard and Walker, went on to direct an institution of higher education: he became the first president of Clark University. Closer links were forged between education and the public service, and the distinctively American foundations for a training system unlike that of any other modern nation were begun. The keynote was specialization.

If the character of the federal employees was changing, so too was the work they did. The South's fear of federal expansion crushed by the war, a system of federal land grants to the states for educational work was established. In 1887 the government moved for the first time to regulate the railroads. Presently there were moves to curb the trusts. The inspection of meat by agents of the Bureau of Animal Husbandry commenced and was soon greatly strengthened. Federal concern with pure food and drugs resulted in a vigorous administration headed by the picturesque chemist Harvey Wiley. On the heels of the 1862 Homestead law had come stricter controls over public lands. The president was authorized to withdraw forest lands from the public domain, and soon a professional corps of foresters in green uniforms were at work under Gifford Pinchot. With the closing of the frontier a movement to use federal funds for irrigation developed into the Reclamation Service, an engineering corps of high caliber. The inland waterways needed improvement and regulation beyond the intermittent attention Congress had given them; and were not long in receiving it.

With the turn of the century the pace had become faster. Theodore Roosevelt's impetus to conservation and industrial control, and Woodrow Wilson's measures to create a Federal Reserve Banking System, a Federal Trade Commission, separate Departments of Labor and Commerce, and important advances in agricultural and vocational education were major developments.

Yet the World War in 1917 showed the inability of the government to mobilize the nation. The Philadelphia *Public Ledger* commented: "The departments in Washington were never conceived or organized to meet the modern needs incident to mobilizing a nation." Citizens rushed to Washington to volunteer their services and see what they could do—or get—in the nation's hour of crisis. A war administration grew up alongside the civil organization of the government, and at the war's end much of it was incorporated in the permanent federal estab-

PAPER AND PULP MILLS,

LUKE, WEST VIRGINIA

lishment. Observers and participants of the war government saw more deeply. An organization problem of large magnitude had to be faced. The methods of training and civil service procedure that functioned adequately for specialized bureaus did not produce men and women who understood the government apparatus as a whole and knew how to direct it to carry through public measures beyond the scope of an individual bureau or department. Central administrative services were lacking, and the president was left powerless to co-ordinate the work of the growing number of agencies. The budget law of 1921 made some effort at overcoming this weakness, but it swam against the tide of postwar Congressional suspicion and hostility to the executive branch.

Steady growth of federal functions and activities throughout the period between the two wars was accelerated, of course, by the spectacular days of Franklin D. Roosevelt's first administration, when it was difficult to distinguish the temporary changes from the permanent, the measures to meet the emergency from the reforms, the final meeting of long-deferred problems from the anticipation of new public services. In the whole structure the capstone was provided by the measures required by World War II, and its outcome which left Washington the unprepared capital of considerably more than the United States of America. How much more could not yet be known.

6

Senator Stewart in 1871 could say, "None of us are proud of this place," but as early as 1877 the fastidious Henry Adams was writing: "This is the only place in America where society amuses me or where life offers variety."

The tide had been reversed. The Potomac had come far from Ebenezer Cook's description of "that shoar, where no good sense is found, but conversation's lost, and manners drowned." Washington had become physically a pleasant place in which to live, socially in many ways a fascinating one, and the federal service had recaptured its dignity and prestige. Washington stood out against the smoky monochrome of industrial cities. It was different from New York, where they talked only of finance; and it was different from the farming towns of America west of the Appalachians, the river city of St. Louis, the railroad city of Chicago. Even before the showy ranks of neoclassical buildings completed the skin treatment, the city's new character was evident. Contemplating the future in a less ironic vein than had Tom Moore three-quarters of a century earlier, Adams continued, in his letter to Charles Milnes Gaskell: "As I belong to the class of people who have great faith in this country, and who believe that in another century it will be saying in its turn the last word of civilization, I enjoy the expectation of the coming day, and try to imagine that I am myself, with my fellow gelehrte here, the first faint rays of that great light which is to dazzle and set the world on fire hereafter."

Washington had changed, but the times had changed still more. It would have been found incredible earlier than 1879 that in a popular novel like Adams's anonymous *Democracy*, Mrs. Lightfoot Lee would have chosen to live in Washington in preference to Boston, New York, or Philadelphia. And it was still incredible to most Americans that to increasing numbers the fascination of life in Washington lay not in politics or even in power, but in its society; that was the conclusion Adams, the anonymous novelist, put into the head of Mrs. Lee. The illusion persisted. "You will find New York speaking of itself as the capital," a shrewd young British diplomat advised a visiting friend. "After all, New York is not anything of the kind, and that is one of the commonest mistakes into which foreigners fall." Even Viviani fell into it forty years later when he told a French delegation landing in New York that Washington was "un Versailles négre."

It was the diplomatists who put the frosting on the cake, and gave Washington its special flavor. The increasing importance of the United States in a world of nations was reflected nowhere better than in the growth of the number and size of legations and embassies, and the steadily increasing stature of the ministers and ambassadors sent to Washington by their governments. By the close of the Civil War it was evident that the international position of the nation had changed. Its feats of military power, its wartime diplomatic achievements, and its economic strength were properly estimated in other capitals. The swift elimination of the French forces in Mexico heralded the new surge of power. The persistent drive for Canadian annexation worked itself out in another way. The firm, even ruthless hand with the remaining Indian tribes of the West saw their practical disappearance as a force for the first time in American life. Much of the remaining years of the nineteenth century was spent in thus securing the position of the United States in North America. It was not difficult; it was inevitable. Then began the Latin-American quest, led by the search for markets and that perpetual carrot, the Panama Canal. From small beginnings in the Pan American Sanitary Congress, developments led to the Pan American Union, a monument in the capital to the continuing interest in hemispheric affairs.

By 1900 this work was largely blocked out. Under the leadership of Theodore Roosevelt and John Hay the United States began to take its place as a world power. For this moment many nations had long prepared, and the strength of their diplomatic representatives in Washington was abundant proof of this prevision. A corollary—or, as some would have it, a contributing cause—of the rising tide in the State Department was the development of a stronger army under Elihu Root and the subsequently sterner necessities of World War I, and since the days of the "great white fleet," of a stronger navy. Army and navy officers thickened in Washington, making their own distinctive and colorful contribution to the city's life.

Much the same forces that were drawing men of competence to the federal service also attracted a leisure class that the accumulating wealth of the nation was making possible. The city was clean, quiet, and green; it had space and sunlight. The houses were seldom pretentious, and the easy standard of living was set by the low official salaries. Congressmen were paid a thousand dollars a year; Cabinet secretaries fifteen hundred. Army officers and civil officials were no better off. There was not much opportunity for elegance on that level, and above it elegance became pretension.

The cosmopolitan group of diplomats to be found in the legations was the new element in society. The world was their garden, and they broadened the range of values in the capital, bringing new ideas, new forms of recreation and entertainment, and a new social patter. Their objectivity and world-wide experience led to a fresh appreciation of the city, its values and its possibilities. They helped make it see what it could be. Cultivated and sincerely interested men like Bryce and Jusserand were listened to with respect and quoted with approval when they expressed opinions on American institutions or even the desirability of city parks. The diplomats also reinforced the genteel tradition: in manners, languages, literature, and even art they left their trace. They stressed "the idea side" of politics. Not least, in subtle ways, they helped a city trying to become a national and even a world capital break through the provincial forms of life and reach a more tolerant and catholic plane where different sections, races, nations, religions, philosophies could meet. And, in their turn, the diplomats agreed, far beyond the requirements of politeness, they took away something of permanent enrichment from their tours of duty in Washington; generally they called it friendship.

The genteel tradition was lovingly, laboriously, but undeniably cultivated. Whole pages of letters and memoirs describing Washington life seemed torn from the novels of Henry James. At odds with its time, the genteel tradition masked the vulgarity and brutality of much of contemporary industrial life, but it made possible a transition from the older forms of the eighteenth century, which had long persisted in the capital, to those of the twentieth. It offered a solution to a capital city itself devoid of industry. It perpetuated many lavender-scented ideas that were growing unreal and absurd, the idea of an elite or ruling class among them, but in trying to make them real it developed new values for a society that badly needed them. The interest in languages, literature, art, and culture were all to the good. The emphasis on values other than wealth, which fitted perfectly the requirements of the necessitous but increasingly intellectual population of the capital, were sharply opposed to those of the times, but led to a new appreciation of birth, breeding, manners, and intellect. Much as they felt it estranged from the life of the nation, intellectuals liked the capital because they were assured of appreciation there. Fewer people called them "high-brows." In the end it happened that, while the capital often seemed falling behind in its understanding of what was taking place in the nation as a whole, it was in advance in understanding what was to come; nowhere was this more true than in its appreciation of the shrinking size of the world and the importance of the growing international position of the nation.

Washington life was a simple life. The prevailing ostentation of the day found neither foothold nor appreciation in the capital. The houses were generally small and plain, and the life they framed was not elaborate. To a far greater degree than in most American cities, where flamboyance and display were common and to see and be seen the chief end of society, privacy was here counted desirable. Many factors conspired to produce this result. The old, durable nucleus of the city—always of decisive importance against the background of so many of its transient residents—was still predominantly drawn from the Potomac region. That is to say, it was still predominantly southern, and impoverished not only by the Civil War but by the great changes in wealth holding and wealth making that the war had accelerated. Industrialism had passed it by. Many residents of the city who were southern sympathizers—beginning with the mayor himself—had left it when the war began, and when they returned they felt themselves estranged from the new life of the capital and lived apart from it. They retreated to ancient strongholds in Alexandria and Georgetown. Public figures led even then a life of considerable exposure, and they wanted to relax away from prying and importunity. They also worked hard. A quiet evening with friends, a walk along Rock Creek, an occasional reception—

that was the normal thing. Social life was domestic in scale, most of it centered in private homes. Against the baronial standards of the day, servants were few. A secretary of state made much of preparing his own terrapin on the dining table in a chafing dish; a senator ceremoniously carved his Rhode Island turkey. The young Theodore Roosevelts, who lived in a small house off Connecticut Avenue when he became civil service commissioner, gave up champagne and served their guests California claret at twenty-five cents a bottle—boasting of their economy they wrote, "None of the guests have died yet"—and did most of their entertaining at Sunday high teas.

In such an intimate society good friends were readily made. Lord Novar, visiting Cecil Spring-Rice, one year after he had come to Washington, wrote in astonishment: "You might almost have thought they had been brought up together when I joined their circle." It reflected leisure, but also a knowledge of what to do with leisure. Entertainment in homes also magnified the importance of women, and helped shape the notable role they were then and later to play in the life of the capital, which is still one of its most deceptive and attractive features. Possibly the two most important persons in Washington for nearly fifty years were Elizabeth Cameron and Anna Cabot Lodge.

A tone of determined intellectuality prevailed, not so much because the city was filled with intellectuals but because it filled with specialists as the government itself became more specialized, and intellectuality became a conversational barrier against lecturing and boredom. The tendency for the city to take on the tiresome, uniform hue of the prevailing administration—an especial hazard during the long Republican years after the war—was also tempered by the presence of opposition party members, retired or relieved officers, many of whom hoped for subsequent appointments; and by factionalism. Many topics of current interest necessarily had to be avoided. The meeting ground of this varied people was literature, art, science, the broad issues of national life and world affairs rather than the narrow, acrimonious ground of political tactics or the discourse of specialties. It was the women who demanded that standard, and society was so small and of such a character that they could easily enforce their demand.

The delicate social balance could not long be maintained. As it grew larger the social life of the city tended to break up into groups, coteries, cliques. They touched and overlapped, and many people belonged to three or four such social circles; but the groups undeniably existed and left their mark. The very size of the city was making it impossible for everyone to know everyone else. Organizations like the old Literary Society that had once been representative became another clique. The diplomats retired from the world of Washington to the world of embassies and legations. Only New Yorkers like Theodore Roosevelt now called Washington a village. The days were fast passing when "society"—a handful of powerful and perennial senators, Cabinet figures, and diplomats—was either representative or dominant. Multiplication by cliques could go on endlessly through the entire city, like the courtyards of a Chinese palace, and it did, compartmenting the entire social life of the city.

Worse, power corrupted. "A friend in power is a friend lost," Henry Adams wrote sadly as the century was ending. "Absolute power corrupts absolutely." Between growing size and power, between cliques and corruption, the older Washington foundered, even though some of its ways and habits continued.

The small-town idea and many small-town ways persisted. Hardly 200,000 people lived in Washington in 1900; and of these one-third were Negroes, and the rest contained fewer and fewer whose impact on the city as a whole was very great. The atmosphere continued, however, but people began to compare it to a college campus. Those who came from the North continued to speak of its prevailing southern quality, but few Southerners did; to them it was the North. Much of the talk was the talk of small towns: it was gossip. The city was a whispering chamber where, as a later commentator said, the verb "to say" was conjugated in all its forms. Yet the fact remained that the city's shop talk was the world's business. And in addition to the gossip, which every observer then and since has found sufficiently remarkable to deserve comment, the characteristic life in private houses fitted the Potomac tradition of good talk. It developed raconteurs of mark. The art grew and it was said there were two kinds of people, the listeners and the storytellers. The stories were old and they were generally long ones, but the telling was appreciated for its own sake. If someone forgot a point, he was sure to be reminded of the omission. There were the old plantation stories, elaborated to half an hour and told as never before; the stories that had come to Washington with Lincoln; the embroidered tale, with acting and gestures, interesting by itself, that led to a thin point that was ironic or pathetic more often than humorous: The deaf gentleman of distinction who was mistaken for a clergyman and asked to say grace and replied, "I see you are talking to me, sir, but I am so God-damned deaf that if hell froze over, I couldn't hear the cracking of the ice." The Irishman who complained to city authorities that the pipes had burst and drowned the chickens he kept in the cellar, and was advised to keep ducks. Like a railroad train or a ship the city was a good place to swap tales.

This kind of life—a *Hauskultur*—projected itself into the city and affected its form. The city's response to its hundredth anniversary in 1900, the renewed appreciation of L'Enfant's plan and the well-meant determination to carry it through, was part of it. The acquisition of Rock Creek as a public park, agitated since the end of the Civil War, owed not a little to the taste for outdoor recreation and to Theodore Roosevelt's exemplary rambles, rides, and rock climbing there. But it remained essentially a city of homes. The interest in the city beyond the District of Columbia grew, and excursionists wandered farther and farther into Potomac country.

"We started, five of us, with horses and saddle bags and went along the Potomac through the wild country, with clearings among the woods and wheat and corn coming up," Cecil Spring-Rice was writing Ronald Ferguson in 1888. "We stayed our first night in a very rough inn where we were waited on by an old slave who had been butler in a southern family and had run off to the North to fight the enemy. He kept his society manners and rebuked us for any breach of etiquette. 'The napkin is to set over your knees, sirr,' he said to me. The next day we started early and rode along the Canal to a village for all the world like an English one, where we lunched with a farmer. He was six feet high—had eleven children, all of whom were in the neighborhood. He showed us a darkey whom he had 'raised' as a slave, who was passing and saluted him. Farming, he said, was bad, but one could live. The daughters told us a good deal about the life of the town. There were the Pooles, after whom the town had been named, and who lived in the big house. The old man Poole had restored the family fortune by making carpets in Baltimore. The people were divided into Episcopalians and dissenters. The former were the nicest to belong to. One could dance. The Methodists mightn't. On the table were the Annual Census of the U.S., a Bible, a Tennyson, and a German grammar.

"That night, after swimming the horses, we avoided the mists of the river and took possession of a house of which the owner was away. He came back while we were routing out his stable and joined heartily in the work. We had a splendid time with him and he told us how he had begun with nothing and now was owner of all the land about. The next day we passed Harper's Ferry where John Brown began the war. It is a beautiful place, deep down in the valley of the Potomac, where the Shenandoah runs in, but the glass was at 89 and riding is hot without proper hats. Then we entered the famous valley which supported both armies during the war till Sheridan burned every barn and killed every beast, so that as he wrote to headquarters, 'A crow couldn't fly through it without a haversack.'

"We stormed a house in the same way as before and were received royally. The house was an old plantation house. The Turners had owned it since the country was settled and all the land around. The last of them was a hard-drinking old fellow who, hearing of John Brown's raid, rode down to shoot him but got shot himself. A Pennsylvania Quaker took it and the stables and all the land—a fine fellow, 6 foot 4 inches, with a growing family. Then we went on through woods covered with blossoms red and white, and fields with very fine stock in them, till we reached Greenway Court where Lord Fairfax entertained Washington. The present owner's grandfather had been a steward to the family and had received a grant of the demesne. He was a great fighter in the war, and told us stories on stories about his cavalry raids into Maryland and Pennsylvania, and his final encounters with Sheridan. He had a 'Life of General Lee' and a Bible. He loved the Lees and respected Sheridan and his friends the enemy, but hated the Republican politicians. He was quite reconciled to his new position. The next person who entertained us was a German who didn't care about the fighting at all but had been impressed when every man in Virginia had to take arms. Then we came to the region where the English live. Loads of them are down there, hunting, shooting, growing vines and raising horses. Those that don't drink do well. The whole country is as fine as any you ever saw. High hills, splendid rivers, beautiful woods and smooth rich valleys."

The Potomac was being rediscovered.

7

It was to take Congress some time to learn how to build along the Potomac that seat of government which it thought would reflect the dignity and power of the great nation that emerged from the Civil War.

In its first building acts the nation's founders had shown themselves to be well aware of the importance of architectural values. Their stanch support of L'Enfant's unpopular plan was a cardinal instance of it. In the early buildings, the Capitol and the White House, their immediate desire and ultimate hope was for an appearance of permanence to strengthen a new and still tentative nation—to say nothing of its new and still tentative capital. They were aware of the importance of symbols. They knew that the concept of sovereignty was strengthened when associated with splendor, scale, and dignity. Marble and granite must take the place of brick and wood. In the years that had passed there was no retreat from these principles. But as the century wore on there were increasing differences of opinion on how they should be translated into reality.

"Hints on Public Architecture" had been given by the geologist Robert Dale Owen in a Smithsonian Institution publication in 1849, but the rising tide of the Gothic revival that was steadily pushing the classical forms into the background was itself swamped in the miscellaneous eclecticism of the mid-nineteenth century. Almost every one of the architectural vogues that swept the nation found some sort of expression in the public buildings of the growing city. The Georgian buildings of Hadfield, the Federal architecture of Latrobe, the classicism of Thornton and Mills, formed a substantial background. But to these had promptly been added James Renwick's romantic confection for the Smithsonian, Alexander's martial Gothic of the Soldiers Home, the sturdier Romanesque forms of the old city post office, the Italianate campanile of the Baltimore and Ohio Railroad depot (modeled after the station in Baltimore), the mansard-roofed brick château of the Pennsylvania Railroad depot (which Congress, in a moment of exhuberance, had stipulated should be copied from the Pennsylvania residence of Thaddeus Stevens), and sundry others whose charm was equally miscellaneous and equally fugitive. Carroll D. Wright thought Washington, when he first saw it in the sixties, was in the last year of the famous course of studies: Ambition, Distraction, Uglification, and Derision. The high point of this flamboyant array was the building to house the State, War, and Navy Departments, its hôtel-de-ville façades adorned with rich cascades of columns and much admired.

Less than a decade had passed before it was equally reviled. Charles Follen McKim called for an ax. He charged that the building was an architectural obscenity. The taste of the times was changing too fast for comfort. What was to become of permanence, splendor, scale, and dignity? In the end the government settled for the world's fair classicism of Mr. McKim, and

Daniel H. Burnham of Chicago and his associates—who spoke in the accents of those who alone have communed directly with the Almighty—energetically applied it to everything built in Washington from 1907 to the end of the Hoover administration.

The welter of styles might be excused. It was in the air. The national purpose that served as the pretext even exaggerated the tendency. But the location of the buildings had betrayed an even more fundamental uncertainty. In an axial city the axes were left undeveloped—or they were ignored. The extension of Mills's Treasury blocked the view of the White House from Pennsylvania Avenue; Mullet's State Department blocked it from New York Avenue. The Library of Congress blocked Pennsylvania Avenue's vista of the Capitol, as L'Enfant had planned it. The Washington Monument was off its axis, and never the type of structure the city's original designer had intended at this point. Public buildings were distributed throughout the central part of the city without thought of a co-ordinated plan. Planning and architecture must go hand in hand. The time was ripe for "the imperial façade," and in 1902 it was finally introduced.

That the McMillan Commission plan coincided with the great expansion of federal activities during the administration of Theodore Roosevelt, or even with the rising spirit of imperialism in the United States, was sheer coincidence. The plan itself had been conceived at the time of the centenary celebration of the removal of the seat of government to Washington, and was to a far greater extent the work of Congress than of the executive. To an extraordinary degree it was the accomplishment of Senator James McMillan of Michigan; one looks in vain for any indication that the president had anything to do with it, except to carry it out in part.

Simultaneously with the centenary the American Institute of Architects devoted the major part of its 1900 meeting in Washington to a discussion of "The Future Grouping of Public Buildings"; nearly all of the participants agreed with Glenn Brown, the architect who six years earlier had championed the idea of public buildings around the Mall and recommended that the city should go back to the original principles laid down by L'Enfant. The institute worked closely with the McMillan Commission.

The plan took its form from the ideas that had been so thoroughly tested at the Columbian Exposition in Chicago in 1893: the conception of "the Great White City," with its uniform ranks of plaster neoclassical buildings, marshaled by the great architect in chief Daniel Burnham. The designers were not so naïve as to imagine a perpetual world's fair on the banks of the Potomac, but references in the report to illuminated statuary, fountains, lagoons, and the other sure-fire effects that had been found so popular in Chicago disclosed an unmistakable ancestry. The experts appointed in 1901 to advise the McMillan Commission were the architects Burnham and McKim, the landscape architect Frederick Law Olmsted, and the sculptor Augustus Saint-Gaudens—the same men who had been chiefly responsible for the Columbian Exposition. All of them were professionally active in the capital—McKim as the architect in charge of remodeling the White House, Olmsted as the landscapist of the Capitol grounds, and Saint-Gaudens as the designer of the Adams Memorial. They served without compensation for nearly a year, and in the end presented a brilliantly persuasive report, and a complete set of plans that formed the foundation for a new capital city. For all its respectful allusions to L'Enfant, it was not a revival of the city's original plan but one wholly new.

By the first of 1902 the report of the experts was ready. The published document was handed to Congress by the Senate Committee on the District of Columbia, and the drawings and models that accompanied it were exhibited in three rooms of the Corcoran Gallery of Art. The commission had been asked to study and report on a park system for the capital, but before it had finished a comprehensive city plan was the result. The commission did not hesitate to point out that the plans "are the most comprehensive ever provided for the development of an American city." The plan faced realistically the needs of the growing city, based on an optimistic projection of its growth in size, and the expansion of the government agencies. It covered the entire District of Columbia, making recommendations for new parks, the development of each park, and outlining park belts and parkways connecting them all into a single system.

Beginning with the problems presented by Rock Creek Park, acquired piecemeal from 1889 on and still undeveloped; the malarial marshes of the Anacostia River; and river-front parks, Soldiers Home, and other pieces of park lands in the city, the commission offered a plan to bring them together into a single system. The model was Olmsted and Eliot's plan for the Boston metropolitan park system. To create such a system the commission had first to consider the vestiges of L'Enfant's 1793 plan (which was essentially a plan for streets and public buildings rather than open spaces or a comprehensive city plan) and that led it speedily to the question of sites for public buildings. Before its work had greatly advanced, furthermore, the not unwilling commission found it was being consulted by the executive departments on specific matters of building locations and design.

Much hostile criticism since has obscured the genuine greatness of the plan: its grasp of the city as a whole, and its concentration on what seemed to it the essentials. It rediscovered for Washington the importance of the region, with all its specific conditions of climate, geography, and history; it dealt boldly and with authority with the special problems of the city as the seat of government; and it found the Potomac River a factor of major consequence again. In many of its recommendations the plan was a tidying-up operation, a co-ordination of recommendations earlier and individually made, like those of the army engineers for the river improvements. It accepted most of the city as it was—even the blundering plan adopted in 1898 for the street system—and while it did not balk at clearing dozens of built-up blocks, it did not concern itself either with the over-all street plan, with traffic, or many other aspects of the city. In the final test, the amount of work recommended that had been actually carried out proved remarkable, and had more been done the city would be in advance of where it is today.

The commission commenced with the founding fathers and with L'Enfant. They visited "those historic towns and estates on the Potomac and James rivers and on Chesapeake Bay among which Washington passed his life." While disappointed with their scale, the commission found that the colonial towns of the Potomac region, Annapolis and Williamsburg particularly, "nevertheless possess a simple dignity and stateliness, and they evince an acquaintance on the part of their designers with the fundamental principles of art." They noted in the old colonial capitals, as in L'Enfant's design, the principle of radial streets, and traced it back to Lenôtre's seventeenth-century forest rides in the woods at Versailles. Examining the Potomac, they found that the army engineers, in the thirty years they had been in charge of the capital's parks, had done much to overcome the swampy condition of its banks; by dredging and filling they had created a tidal basin which daily scoured the harbor, and reclaimed hundreds of acres of land along the waterfront.

They studied the city's climate and found that "from the first of October till about the middle of May the climatic conditions of Washington are most salubrious," but "during the remaining four and a half months the city is subject to extended periods of intense heat, during which all public business is conducted at an undue expenditure of physical force." It particularly noted that Congress was frequently obliged to continue its sessions during the hot months. To meet these conditions the commission placed special emphasis on fountains, wooded parks, and the development of swimming, boating, and water sports along the river.

The commission took the view that their work, and the development of the capital, should not stop at the district line, and it noted: "Indeed, the best of the scenery lies beyond, especially in the neighborhood of Cabin John Creek and in the region just about and below the Great Falls." A romantic taste for the picturesque still lingered. The report recommended preservation and careful development of the entire river front from Great Falls to Mount Vernon, with a parkway to each of these key objectives.

To allow a new place for the river in the life of the city, the commission began with a proposal for quay development to extend from the site of the Lincoln Memorial to Rock Creek, and possibly through Georgetown, and cited a formidable number of instances in Paris, Budapest, and other European capitals it had recently visited in support of this recommendation. It recommended that Rock Creek valley become a parkway, with lateral parkways running up the Soapstone, Piney Branch, and other tributary streams. In passing, it pointed to the increasing use of the Chesapeake and Ohio Canal for boating and other forms of recreation and urged its preservation and development. Turning to the Eastern Branch of the Potomac, they noted "the present outrageous condition of the Anacostia River" because of the malarial swamps. They endorsed an earlier engineering proposal to deal with this tidal estuary, and then recommended an Anacostia River barrage at the head of the navigable channel, above which the land might be filled; and below, a water park with parkways along the embankments should be built.

The flats recently reclaimed from the Potomac, it recommended, should be treated in "the landscape of natural river bottoms," with great open meadows framed with earth dikes and masses of trees. This was in keeping with the recommendations throughout, which characterized in broad strokes the nature of the development proposed rather than attempted to make specific plans.

As to the city itself, the commission's major attention was given the redevelopment of the Mall, and the provision of sites for public buildings and memorials. It objected to Downing's informal landscape treatment as unsuitable either to L'Enfant's original plan or to the public buildings it proposed on either side of the Mall. Later development of these ideas, particularly their greater formality, and oppressive scale have obscured the more human touch of the McMillan Commission. To relieve from monotony "a meadow-like stretch a mile and a half long," it proposed "parterres of green and large basins of water, with frequent seats tempting the passer-by to linger for rest." The lavish plantings of trees proposed a more intimate scale than that we know now: the commission specified a 300-foot width, "walled with elms." Its enthusiastic recommendation of water fountains, now wholly forgotten, would have made lively what is now admittedly green but undeniably bleak. The plan recognized these needs of a *tapis vert* on a gigantic scale, and specified that streets crossing the Mall not be masked or depressed. "Indeed, the play of light and shade where the streets break through the columns of the trees, and the passage of street cars and teams gives needed life to the Mall," it said. One of its renderings showed a flock of sheep nibbling the sward.

The present sterile area south of the Washington Monument the commission proposed to treat as "The Washington Common"—a fully developed recreation ground for the city's population, where a stadium, tennis courts, ball grounds, open-air gymnasium, children's playgrounds, "all should be provided, as they are now furnished in the progressive cities of this country." The Tidal Basin, not yet ringed with Tokyo cherry trees, it proposed should be further developed for boating, swimming, and for skating in winter. "The positive dearth of means of innocent enjoyment for one's leisure hours is remarkable in Washington, the one city in this country where people have the most leisure," it slyly observed.

Impressive recommendations were made for a Lincoln Memorial, substantially the same as it was later built, and the related bridge across the Potomac to Arlington National Cemetery. The bridge had been urged since 1851, and scores of speakers from Daniel Webster on down had constructed it in oratorical imagery as the ever-enduring granite link between the North and the South. Republican to the core, the commission identified other memorial sites: a Pantheon to the makers of the Constitution south of the Washington Monument; a "Union Square" at the foot of Capitol Hill with statues of Grant, Sherman, and Sheridan; and minor statuary in public squares.

As to the parks themselves, the commission selected the hilltops and the deep, cool valleys and ravines as the best locations, and acknowledged that the summer heat of the city presented the most compelling arguments for such a plan. Most of the hilltops were sites of Civil War fortifications, and had thus also acquired historical associations. The plan proposed that they be linked together by a system of parkways. Minor parks provided in L'Enfant's original plan it planned to develop as neighborhood playgrounds.

In considering the Mall as a major element in the park system, as well as in the city plan, the commission faced the problem of removing the railroad tracks that crossed it, and so came to its first specific project: the Union Station. Putting together the necessary changes that had to be made by the Pennsylvania and the Baltimore and Ohio was made easier by the fact that Burnham had been commissioned to design the Pennsylvania's new station in Pittsburgh, and had drawn preliminary plans for its station in Washington. The key to the railroad puzzle was provided, however, by President Cassatt of the Pennsylvania, who offered to join with the Baltimore and Ohio in building a union terminal, and stated the terms on which he would move forward: that the railroads be compensated for extra expenses incurred in the move, and that the city provide suitable approaches to the new station. As it turned out, an additional sum was paid to secure what was considered a suitable architectural treatment of the station, considered as the gateway to the capital.

The result was the first union station of the modern type to be erected in any American city, a gateway building that was the first of the long series of neoclassical buildings in Washington that persisted for thirty years, and the first convincing demonstration that in the McMillan plan a proposal that had been advanced could really be carried out. It was a smashing beginning for a bold plan that had recommended the immediate purchase of all blocks fronting on the White House and the Capitol; all land lying north of Maryland and south of Pennsylvania and New York avenues for public building sites and the Mall improvement; thousands of acres of park lands and rights of way for parkways throughout the city and up and down the Potomac. The land acquisition program was the first significant reversal of a century-old melting away of the stock of public lands in the district, and the commission faced it without blinking an eye.

The fact had to be recognized that all this was but recommendation, and the McMillan Commission report fell increasingly short of meeting the requirements of a comprehensive city plan as urban problems increased during the first half of the twentieth century. Congress, in its role of city council for the District of Columbia, acted on parts of the plan, and as the years passed it took on even greater authority than at the time it was first prepared. The idea of "the Great White City," as it had been seen in the Columbian Exposition and the plans for the nation's capital, was translated into civic centers throughout the nation, and the high standard of park development proposed, like Olmsted's earlier gardens and lagoons in Chicago, were the foundation for the "city beautiful" movement. But so far as Washington was concerned, the 1902 report was carried forward by two permanent agencies: the Fine Arts Commission, established in 1910 to advise on the design of public buildings, and the National Capital Park Commission (later the National Capital Park and Planning Commission), established in 1924 to acquire park lands and later to serve as the city planning agency for the capital. The McMillan Commission's ideas were kept alive by the American Institute of Architects and numerous other professional and civic organizations, and in a somewhat watered-down fashion most of those which have still a current application have now been carried out.

The McMillan Commission had proposed both an architecture and a plan, and while they lived longer than any proposals that had preceded them, their time came to an end. The last stand of the old government, the government of absolute centralization, was the Triangle. Into a redeveloped area, the heart of the old commercial city, of some seventy acres from 1926 on were crammed the principal administrative departments. Between Pennsylvania Avenue and the Mall they covered the land. But the development was never completed, and between the District Building and the Labor Department, in what was planned as its single park area, the reason can be found; the park is a parking lot. Impressive offices, like the smaller conference rooms in the Departmental Auditorium, look out not upon living grass and trees but the shiny funereal surfaces of parked vehicles. The automobile halted the Triangle development even before it had been carried through, and the traffic problem it generated, still unsolved, is today Washington's greatest.

The Triangle was also the first wholesale application of the neoclassicism that had commenced with McKim and Burnham. Heretofore only individual buildings had been built, like the Union Station or the Department of Agriculture. Now came rank upon rank of columns, block after block of masonry façades masquerading as if structural steel had never been invented and used in their construction, uniform cornice lines and the rest of the eye-glazing spectacle. It was too much. The movement had exhausted itself.

Subsequent federal buildings avoided the area or any repetition of it, in spite of the urgings of the city's planners, and began to strike out for themselves. The new Census Buildings at Suitland, Maryland, ten miles from the Zero Milestone behind the White House, showed the trend. The Naval Hospital and the National Health Center at Bethesda, an equal distance northwest on the old road to Frederick, showed it. The 35 square miles of the Beltsville Agricultural Experiment Station, 20 miles out on the Baltimore road; the testing grounds of the Public Roads Administration, up the Potomac; the Carderock Model Testing Basin of the Navy Department, well beyond Cabin John Run; the mushrooming army and navy establishments on the lower river that stretched for nearly thirty miles—these were the reverse side of the Triangle, the developments as characteristic of the nineteen-thirties as the Triangle had been characteristic of the nineteen-twenties. All of them were begun when the Park and Planning Commission was energetically trying to promote buildings on the south side of the Mall, buildings west of the White House, and a new Mall east of the Capitol. The mammoth Pentagon Building, containing 40,000 employees—equal to the total federal employment in Washington at the peak of World War I—was located across the river in Virginia, wholly out of the jurisdiction of the Planning Commission, without a second thought for the sacrosanct ideas of the McMillan Plan or the objections of its contemporary guardians, the Fine Arts Commission and the Park and Planning Commission.

Neoclassicism disappeared as well. The new buildings of the Interior Department and the Social Security Board, the Pentagon, or even the lesser ones like that designed for the Public Health Service and later occupied by the Federal Reserve Board (and perhaps most strikingly, the so-called "Undesignated" buildings), stood for a simple, more utilitarian style, one that suited the automobile and the elevator, and did not conjure up images of toga-clad proconsuls getting into convertible coupés. They had movable partitions between offices, excellent mural paintings in the lobbies, escalators, and cafeterias. The windows could be easily cleaned, the floors scrubbed with mechanical equipment, and they had parking spaces in the basements.

The last stand of neoclassicism was John Russell Pope's Jefferson Memorial. Intended to occupy one of the key points for major memorials identified in the McMillan Plan, and taking its Pantheon theme from the same source, it encountered such violent opposition from Republicans and architectural modernists that when the outcry died it had been scaled down by one-third, reduced in price by half, and moved off the axis (in the center of the Tidal Basin, and requiring both a complete reconstruction of that celebrated water and new approaches to the Long Bridge across the Potomac). The National Gallery of Art, a later pink-marble variation on the Pantheon theme by the same architect, was not built by the government but by Andrew Mellon (who as secretary of the treasury had espoused the Triangle Plan of 1926) and presented to the United States as a gift. And with the death of Mr. Pope the last dogmatic exponent of the decorative ideas that began in Chicago in 1893 disappeared.

A more mature appreciation of the city's architectural heritage takes the Smithsonian and even the State Department buildings, which an earlier generation would have demolished, in its stride. It finds new values in the city's few remaining Georgian and Early Federal buildings. But it does not seek to imitate past modes of design.

What form the future buildings of Washington should take, like the form of the city itself, was never more open to question than today. The vague institutionalism of contemporary public architecture is a hesitant style, lost in the land between what was and what is to be. When the will to build is reborn, the city will probably take its theme from Eero Saarinen's unbuilt design for the Smithsonian Gallery of Art, or other buildings that give opportunity to the genius of our times.

TWELVE

A POTOMAC PROSPECT

FROM THE FAIRFAX STONE, WHERE THE POTOMAC RISES, TO POINT Lookout, where it flows at last into the Chesapeake, the diverse yet unified Potomac region has changed in the three hundred years since white men began to settle here. One cannot but conclude that it will change further. This we can learn from history: of the making of Potomacs there is no end. Each age and even each man makes his own. The plastic river is ever being shaped anew by a restless nature and a dynamic society. What it has been we have seen: fishery, granary, harbor, route, homesite, plantation, hunting ground, mine, power source, factory, capital, swamp, pleasure ground, and more. What it can be remains for us to discover. As men have shaped the river, so it has shaped them. It has made traders, trappers, planters, clerks, miners, farmers, drovers, merchants. What manner of men history will beat out on this anvil no man today can know. But the Potomac will be known by the men it makes.

The Potomac, what it had promised and what it has meant, is a stubbornly established fact in our national life. Today that fact is masked by the rôle of the region as the seat of national government. Washington has flooded the region. The most ancient Potomac cities, Alexandria and Georgetown, are engulfed in the sprawling metropolis, helpless in the face of federal authority. But they can say with the Scottish village that lies under Castle Rock in Edinburgh:

Musselburgh was a borough when Edinburgh was none,
And Musselburgh will be a borough when Edinburgh is gone.

For the true life of the Potomac is independent of the part played by Washington as the seat of government. That is but one, the most recent, today the most important, of the many roles it has played. Dominant as that seems today, it is the fact of a bare century, and in another century it may no longer be true.

As the city of Washington spreads out from its historic "ten mile square" it comes into collision with the older life of the Potomac. Absolute federal power and authority do not carry beyond the Federal District. The most desirable land for the future city is not the valueless, looted soil of the old tobacco plantations in the immediate suburbs (although the Quakers at Sandy Spring, and the Amish at Charlotte Hall, show what can be done with it). It lies to the west. There is no longer enough water in the river, or land in the valley, as they are now being used—or abused—for the region and the metropolis at Washington too. The Potomac country into which metropolitan Washington is growing is no more able to support by taxation a high standard of municipal services than was the city of Washington in 1860. The city and the region are finding themselves increasingly at odds with each other; and the conflict must grow.

Both must come to terms with the natural facts of the region. Its resources are not inexhaustible. Many of them are depleted. Others have substantially reached their limits of present use. Yet resources can be cultivated and developed. The valley is still rich, and it is full of promise for those who will trouble to understand it, discover it, and work with it. For those who will not, it is a fickle mistress.

Its history is one of failures and frustrations, of misunderstandings and broken promises. Its broad tidal bosom was not the arm of a sea that led on to China and the Spice Islands. No gold or silver was found by its early adventurers. Its plenteous furs proved not to come from the region but from the western country beyond. Its tobacco and its slaves became a dragging burden to its planters, and even its wheat but the prelude to a vaster farming area in the Mississippi valley. The brilliant prospect of the route to the west that so enticed the early capitalists dimmed when the turnpike and canal failed and the railroad found its most lucrative traffic pre-empted by northern routes. As a trading spot it proved a failure. Its early manufactures died stillborn. Its lands, in which so much had been ventured, turned out to be worth less than had been expected. Despite exhortation and persuasion, no great cities arose on its banks: even the seat of government, for half a century, was but a village, a national joke; for another half century it was a pitiless drain on local resources; and for another half century it was a disintegrating, sprawling metropolis whose growth was independent of the region and the will of its people but rested upon federal whim and caprice, a jelly escaping the planners' mold. Those who can read this record cannot easily misunderstand the region.

Today Washington has spilled over into the Potomac region and become its most important fact; but the city is only beginning to rediscover the region. What is to be done here in the future, when nation and region must live side by side, and national, regional, and metropolitan interests must be reconciled, must reflect the fact that the region is not only a hinterland of the city but that it is also the potential city of the future, the city that the vacuum tube and the internal-combustion engine, to say nothing of future atomic energy and the menace of the atomic bomb, are making possible, and even necessary. It must reflect the fact that the city is in the largest sense but part of the Potomac region, that it is here because of what the Potomac has promised and meant, and will remain here only so long as the region has meaning for it and can support it. The fine illusion of federal power and authority must reckon with stubborn facts of geography, and acquire the humility one learns from the history of the region. Not least, federal power must accept home rule and the franchise for Washington itself.

The seventy miles over which George Washington rode from the Eastern Branch to the Conococheague is the site of the green, fluid Washington of the future, and the Potomac will form its backbone. To make this prediction one does not have to assume an ever-growing bureaucracy, but an ever-spreading city, the city of the future that, like all our large cities, will be everywhere and nowhere. Indeed, it is likely that the number of federal employees in Washington will decrease, even as the federal government itself continues to grow. We know more today about administrative organization, and it is neither necessary nor desirable to have all bureaus or even all departments located in Washington. To permit the necessary expansion of the government during the war, over a dozen large bureaus were torn out by the roots and dispatched to Chicago, St. Louis, Richmond, Atlantic City, and other places where there was room for them. They have found their way back to Washington now, and throwing bureaus out of the capital is not the way government decentralization of the future will happen, any more than shuffling bureaus from one department to another will secure administrative efficiency and economy. Government activities will be decentralized to get them closer to the people they serve, to get more face-to-face contact and less paper work, to get more participation by citizens in the administrative process that has become characteristic and inevitable in our complicated world. In this process, as Professor John M. Gaus has pointed out, we can well afford subcapitals in Atlanta, Denver, and half a dozen other large regional centers. The movement in this direction is already well advanced, and when it reaches flood stage it will change Washington as it has not been changed for nearly a century.

As the city expands, the place of the river in its life will become greater. The cardinal fact in the Potomac today is the city of Washington. Now a metropolis of over 1,400,000 people—if we take the census estimate for the metropolitan area—it has long since spilled over the formal boundaries of the District of Columbia. The built-up area reaches 10 miles beyond the district line. Mass housing developments are being built still farther out. The commuting zone—one hour from work in the central city—now has a radius of 35 miles, a diameter of 75 miles.

The spreading of the city responds to dominant living values of our times, as well as to new technical resources. People want to spread out, they want more land. Today individually, tomorrow in groups and communities, they are on the march out of the old city. Like the residents of industrial cities since the Wage and Hour law, Washingtonians can jump into their cars at four-thirty in the afternoon, drive 30 miles to the well-stocked trout stream beyond Herndon, fish for an hour, and be home for dinner at eight; if they live near their favorite recreation spot, on salt water at Annapolis or in the mountain foothills, it is even easier. The five-day week, the two-day weekend, extends the range. You can find Washingtonians 150 miles out on a Saturday or Sunday, and in this radius the landscape is speckled with their vacation cabins. They fish at Solomons Island, hike on the Appalachian Trail at Snickers Gap, swim in the South Branch, and ski in the mountains above Frostburg among the sugar maples. In a driving range of three or four hours the whole valley becomes their playground. Every highway and parkway development extends that range. Increasingly more and more of them want to live where they play. As future pressures build up, more and more they will want to work and play where they live.

Francis Makemie was eternally right when he said that the best part of the Potomac region, the country above the falls, was yet to be settled. It still is. But without regional planning, conservation, and development the result is chaotic, a self-defeating despoilation of town and countryside, as both fight to live in the same place.

Beside the Potomac the Piedmont farmer still grows his wheat, but it is no longer his principal cash crop. His diversified farming allows it in the role of winter wheat, sown in drills between the shocked corn, its green ribbons gaily to remind us on the occasional springlike midwinter days of the true spring that is never far distant. Or, in a carefully planned rotation he may make a crop of wheat every four years or so. But his crop is harvested with the huge combines, and the grain is trucked to the railroad whence it goes to the milling centers. The old self-sufficient life is gone.

The number of traditional Potomac farms is diminishing gradually as the hungry metropolis reaches out for fluid milk and cream, and the routine of dairying supplants the old general-purpose farming of the Quakers and Germans in the steadily expanding Washington milkshed. The milkshed dominates farming, turning old wheat and tobacco fields into pasture, and rebuilding the old fieldstone bank barns of the German settlers from Pennsylvania into modern, sanitary dairy barns. In the old plantation country much tobacco is still grown, but it is yielding steadily to diversification. Truck growing, chicken raising, and the electrified egg factories hold its future here, as on the Eastern Shore. But even the truck gardens, whose produce is still sold in public markets in the city, are increasingly specialized. Up in the West Virginia panhandle an entire mountainside has been rebuilt in a series of terraces for the production of water cress. Fishponds along the Monocacy in Frederick County grow the bulk of the nation's goldfish, and a good many of its frog legs. At Hancock the development of applegrowing and the production of other fruits heralds an agricultural future in "tree crops" for the hill country long ago predicted by the geographer J. Russell Smith; another instance of it is the recent plantations of cork oaks in western Maryland. The old wheat-corn cycle persists, and the Loudoun system with its clover and lime can still be found, but greatly improved. The Potomac fisheries, subject of properly gloomy scientific reports since 1850, have not been restored, although the shad still run in small numbers every spring and are caught by pole fishermen where the river narrows below the falls at Chain Bridge. Downriver the oystermen's wars with the machine guns on Maryland conservation cutters chattering at the boats of Virginia poachers continue; and not long ago the governors of Maryland and Virginia, with a high sense of history, met at Mount Vernon to agree upon a new Potomac Compact.

The economy of the valley has long been closely integrated with the rest of the United States—and with the world. Bargeloads of coal from George's Creek, shipped to New England, become electricity that lights many a Connecticut salt-box. The Albermarle pippin has dominated the British fruit market since the days of Queen Victoria. The cement, plaster, rock wool, and other limestone products of the valley have built whole cities in Ohio and the western states. Its agricultural specialties are found everywhere in the nation.

The notable recent invasion of the countryside by city families involves more than "hobby farming" or "tax-loss farming." There can be few other agricultural regions where historically the line between town and country has been less distinct—and still is. Scratch the townsman and you find a farmer. Country life is in the back of everyone's mind when he talks about the good life, and work in the city is still for most a disagreeable penance until the time comes when the desk or shop can be locked and the landscape brought into view. Perhaps, like the language itself, it is part of the southern English heritage, this indifference to town life, this appetite for the land. But the consequences have been fabulous. Whole counties, especially in the Virginia horse country, are "fertilized with bank notes," latticed with whitewashed rail fences, and divided into estates. At its most spectacular the life of the horse show and the fox hunt over the springy bluegrass turf is at home in this perfect riding country. Between the Bull Run Mountains and the Blue Ridge, at Leesburg, Middleburg, Upperville, The Plains, and so through the horse country down to Warrenton, where you pass into the valley of the Rappahannock, there is scarcely a square mile that is not hunted over, not only by Virginians and Marylanders but by all who hear the hunter's horn, to the point where local customs have been diluted and the activity itself is lost in the social froth that now surrounds it. Elsewhere the city farmer is at work, building houses, barns, and fences that stagger the imagination of the Potomac natives. Yet a surprising number of such farms pay their way, and the advancement of scientific agriculture and the improvement of livestock is not a little due to such farmers who can afford to experiment.

Farming and country life along the Potomac, its ancient, conservative bulwark, is today in flux. Gone are the years when Potomac farming led the nation and men created new farming methods and invented basic agricultural machinery. The predatory hunter and the land speculator have been tamed, on the farm at least, and the men of Potomac have learned the patience that comes with milking twice a day, the importance of co-operation, and they have begun to learn the meaning of conservation. They have come of age and settled down.

On hard roads with autos and trucks, the farmer thinks little of going 15 or 20 miles for his seed, his implements, and his daily needs. He may make regular trips to New York or Wisconsin to buy his dairy stock. But even the dairy farmer may buy his milk at the store, as his wife buys her bread there. The radio near his dining-room table, like the movies he sees, the newspapers and magazines, and the good consolidated schools make him superficially indistinguishable from those who live in the towns and cities. But the life on the land is still not an occupation; it is a way of living, and the old ways live on in the new life.

When the first thunderstorm of spring comes the Potomac farmer says, "it wakes up the snakes." When his child brings home her first buttonhole from the school sewing class, his wife remarks, "It looks like a dead hog's eye with the lashes turned in." She criticizes the material offered at the city store saying, "It's woven so loose you could sift puppies through it." The simile drawn from the old days persists in the direct, humorous language.

The past lives, too, in country towns where old craftsmen do sums on a pocket ruler, and call five-eighths of an inch "half an inch and an eighth." To them a stepladder is a "stair trestle" and a helper a "servant" who "waits on" them.

By such tokens you can tell the Potomac farmer. You can tell him by his attitude: a confident and composed individualism, tempered with traditional grace and politeness and a new co-operation. His hospitality is still largehanded, and his table is still stocked with regional dishes. His hogs are fat, his cattle sleek. His house is large and his entertainment spontaneous, informal, and long.

With an individuality short of provincialism, he is one with the nation. Professor Barker has pointed to the remarkable political stability of the Potomac region, especially since the Civil War. In all those years the state of Maryland has but once voted differently from the nation as a whole in a presidential election. If the capital of the nation has acquired a national character, it is not less so than the region in which it stands.

One must admit, and regret, the fact that the people of the Potomac region are barely conscious of the Potomac as a whole, well as they may know it in part. And it is an encouraging sign that today the development of the river itself is at last in the hands of a Potomac River Basin Commission, an official interstate body created by compact by the states of Maryland, Virginia, West Virginia, and Pennsylvania, and the District of Columbia; the United States also has representatives serving on the commission. For the first time an agency of government is officially dedicated to the development of the Potomac. If an end has not been put to ancient rivalries, the machinery of compromise and conciliation has at least been established. The 14,500 square miles of the Potomac basin and the 390 miles of the river's length are the province within which the commission hopes to remedy pollution, restore the balance, and further the development of natural resources. Its powers are modest but comprehensive, and there is every reason to believe that the commission is the tool that must be strengthened and used in the future to remake the region. Yet its existence is still known to but a handful of people in the Potomac country, and the federal government has not yet put its shoulder to the wheel.

The army engineers, who have long had important responsibilities in the lower Potomac and a substantial record of activities in the river, have prepared plans for flood control and propose the construction of dams on the upper river, some of which will be used to generate water power and provide recreation facilities. Within its admitted limits, many of the recommendations in the army engineers' report are brilliant. The treatment of that historic and scenic spot, Great Falls, is an example. Above the falls, and out of sight around a bend in the river, the engineers propose a dam. They propose another at Chain Bridge, creating an 8-mile pool ending at the base of the falls. Thus the picturesque cataract, although becoming synthetic, would remain superficially undisturbed, its height undiminished, and by regulating the water released from the upper reservoir, the flows in the daylight hours and during the summer months would be about one-third greater than at the present time. The proposed Riverbend Dam, above the falls, would create a reservoir stretching back to Harpers Ferry, and up the Monocacy to Frederick and up the Shenandoah to Front Royal.

Although based on an earlier study published in 1934, the army engineers' report was completed just before the war, and that part of it relating to the Shenandoah was promptly shouted down by angry farmers, who may have known better how to object in principle than what to object to in practice. Some important objections can be raised against the work the army proposes to do in the river valley development, objections by no means satisfied by the provisions made for recreation, the preservation of picturesque views, and a few other concessions that are novel to the army's way of doing such work.

The objections come to this: the army hews to the strict line of the federal government's constitutional interest in inland waterways—and legally the Potomac is considered navigable to a point well up Will's Creek above Cumberland. The army believes it can correct flood conditions only with dams and dikes. Its plans are engineers' work plans rather than comprehensive, long-range development plans—which are bound to be controversial and to exceed the purely federal interest in the river. But if the proper view of the river's development be one that regards it as an organic natural unit, every part of which is related to every other part, and the whole integrated closely with the life of the Potomac's people, and if one is impressed with the magnitude of the Potomac's problems other than those that are of interest to the army, then certainly the engineers have offered a very small bone to a very large dog. Whatever the questions that may be raised on the engineering side, and they are substantial, the chief objection is that the limited plans proposed will obstruct more sweeping and more necessary measures of development in the future. The division engineer's report is now on the shelf, but little dust is allowed to gather there and the report is certain to be reconsidered. The demands for individual flood-control projects along the river are persistent, and none of them are worth undertaking unless part of such a larger plan. But above all, the other related needs of the Potomac cannot long be deferred. One look at them will show why.

The provinces of this river valley region, as we have seen them, are numerous and varied. Each has its own problems, and the problems are interrelated. In solving any one of them we help or hinder the wise and economical solution of the others. We cannot have clear water for swimming and recreation at Washington if we continue to allow deforestation in the upper reaches of the Potomac, the erosion of bare cornfields in the beating spring rains, the flooding of tributaries and of the main river itself as the melting snows and spring rains rush unhindered to the sea. And with afforestation, check dams, and other water-control measures, run-of-the-river dams are less necessary. Only a proper plan for the use of the land, like that now being developed by the Potomac counties with the aid of federal agricultural conservation funds, state forestry experts, and the teaching of the Friends of the Land and other conservation organizations, can form an adequate beginning. After three hundred years of agricultural plundering, the Potomac has at last come to that.

The chief problem of the river itself is pollution. As we have seen, it began early. The coal measures of the upper Potomac lay in great arches paralleling the mountain profiles, and it was possible to mine the coal from the hillside. After the coal was cut and loaded, the carts ran out of the mine almost by gravity. But so did the mine water, the beginnings of a stream pollution that has not ceased to this day, poisoning the green banks of the upper Potomac, killing its trout, and carrying its havoc downstream for a hundred miles.

In addition to the acid mine drainage, there are other sources of industrial pollution in the vicinity of Cumberland. The towns of the valley, Waynesboro, Hagerstown, Martinsburg, and Winchester, make their contribution. At the fall line, the Washington metropolitan region adds its lion's share. A careful survey by the Potomac Basin Commission shows: sewage-waste treatment has been provided for most of the important population centers, but almost nothing has been done about industrial

THE WASTELAND

wastes. If the Potomac is now to attract its share of the industries that are dispersing to smaller communities, if it is to have more plants like the American Viscose Corporation's plant at Front Royal, the Celanese Corporation's plant at Cumberland, and the Interwoven Stocking Company's plant at Martinsburg, the river must be cleaned up. Products manufactured in the Potomac basin had a value in 1943 of $285,000,000—as against one million dollars' worth of fish, two million dollars' worth of coal, and three million dollars' worth of lumber products. The value of manufactures was four times the total value of all farm products. To maintain its balance the Potomac must have still more manufacturing, but manufacturers will come—and stay and be useful—only if they can be fitted into the resources of the region.

Finally, the need for more comprehensive and more positive measures beyond pollution control and flood control is imperative if the river is to become the recreational center not only for the metropolitan area, the fifty towns and cities, and the rural population—a total of more than two million people—but for any number of visitors as well. Even the park-acquisition program, carried on co-operatively with the state of Maryland with federal funds authorized by the Capper-Crampton Act of 1930, has encountered qualitative as well as quantitative shortcomings. Afforestation plans in Maryland and Virginia are proceeding slowly but with great persistence, and constitute a major demand on the land that must be carefully fitted into the broad development of the region if forestry's greatest contribution is to be made.

If floods and pollution are to be eliminated, as we may expect, the job must be co-ordinated with the development of the river basin as a place to live, and work, and play. That work is of a larger order than has yet been attempted by the Potomac River Basin Commission, the army engineers, and the planning authorities of states and cities along the Potomac. It is the job that must now be done.

We must consider anew each part of the river and each part of the Potomac basin. Water supply, navigation, hydroelectric power, recreation, wild life, fishing, and a multitude of other needs must be met by the river. Their conflicting claims and requirements must be reconciled and satisfied if all the rich potentialities of the Potomac are to be realized. Whatever is done in the construction of dams or the design of pollution-abatement plants must fit these larger objectives as well. This is not the work of engineers alone: it is a job for everyone. In the end, considering the Potomac, Congress must decide. But before doing so it will take counsel with the Potomac River Basin Commission, the states affected, private organization, and all others concerned. Out of these deliberations, let us hope, a wise and farseeing plan worthy of a great and historic river will be evolved, and a river which many times has shown the nation the way forward will do so once again.

To give some more concrete indication of what such a plan ought to embrace is to outline a long-range master plan for the Potomac River basin. It should begin with the land itself. Great strides have been made in our knowledge of the soil, its proper management and cultivation. We know that much land now being tilled should be removed from agricultural use. Some of it, for example, should be reserved as flood plain or forested, or restricted to use as pasture or for certain types of crops. This work is now in progress under the leadership of county land-use committees and federal soil-conservation and agricultural programs. From our knowledge of what the soil can produce, we can establish reasonable estimates of the number and distribution of the agricultural population of the region. On this basis more reasonable and economical calculations can be made of the schools, roads, and other public services this population requires, and their desirable locations, and business and industry can be furnished a framework within which they can operate to advantage.

The location and size of the valley's cities and towns are closely related to the land use of the region, and its position as an interregional center of manufacturing and trade. Sound planning of towns and cities, much of which is now being undertaken by local governments assisted by state planning boards, can develop only in reference to broader regional plans. The chief shortcoming of local planning, in fact, comes from a legitimate hesitation of local authorities in facing such problems as the location of new industry and the amount of new employment that can be counted on. In the Potomac an additional uncertainty arises from the difficulty of predicting what activities of the federal government should be accommodated within a reasonable distance of the capital.

Through the valley must run express as well as local routes. The location and capacity of these highways and parkways can be established in terms of the valley's potential and desirable population, its industry and employment, and its other needs. Such calculations are now being made by highway engineers, but only in terms of the region's present population and needs, and often for the limited jurisdiction of a single state.

Taking a long look ahead—perhaps a century—the most probable schematic pattern such a development of the metropolitan Washington area could take would be "a rurban city," the larger portion of which would flow up the Potomac in thin ribbons of settlement, with lateral lines of development branching from the Potomac up the tributary valleys. The fertile soils of the valley floor should be zoned for agricultural use and thus protected from sterilization by the pressure of rising speculative land values. Above the valley floor, along the natural terraces and the foothills, the major residential development would be most logically located, most of it in small communities of a dozen to two or three hundred families; and above this the crests and ridges should be devoted mainly to forestry and—what comes to the same thing—parks.

The valley of the Potomac and its tributary valleys are also the logical routes for highways and parkways, and perhaps future landing strips. Most industry would be best situated near towns, the larger number of which already occupy points at the intersection of the major natural routes. These are also the logical major trading centers.

Such a decentralized pattern easily fits the geography of the region, the rural tastes of its population, its highly motorized transportation habits, and its decentralized pattern of industry and employment. It likewise fits the life-centered planning ideals of our time. It would allow, once again, reasonable standards of proximity to work, schools, and shopping, and a minimum of friction and waste in daily living. It would promote the most efficient and economical public services. Most of it would be channeling an inevitable pattern of regional development along the lines of least resistance, with a minimum framework of protection against unwise and preclusive land uses. As such it would be a natural development, and not one requiring large measures of coercive authority to maintain an arbitrary design.

In such a gradually developing future pattern we can assume sound plans for the Potomac and the measures of flood control, pollution abatement, power development, and water supply that are desirable and must be undertaken in any event. The river once again can be a place of fishing, swimming, and other forms of recreation, an ornament to the landscape and a fit token of its position in the life of the region.

Here, in such city and regional planning, is a device for relating the Potomac's past, as we have seen it, to its future, for putting our understanding of this historic river-valley region to work. If this can be done, the Potomac can once again make a unique practical contribution to a nation, two-thirds of which has now become urbanized. Despite great efforts, our cities have not yet solved the problems posed by the "slow explosion" of expanding metropolitan areas. Although our progress in the art of river-valley development has been rapid, we have fallen short in failing to relate city and open-country problems in one single regional plan. The whole world awaits a solution to these problems. Nowhere are they better posed—and with greater prospects for their solution—than in the Potomac.

The genius of the Potomac lies in its balance. It is a microcosm—it has been called a micro-chaos—of the nation. Lying between the ocean and the mountains, between the North and the South, its balance is epitomized in the profusion of plants, trees, birds, and wild life. On a foggy spring day the sea gulls are seen on the White House lawn; the mountain goldfinches rise in tiny yellow clouds along Rock Creek; and on a quiet evening in

Lafayette Square you may find a small white owl. The foxtail pine of New England and the southern sweet gum flourish equally. The balance of nature is faithfully reflected in the very climate, the landscape, and the activities of man. Within a hundred miles of Washington can be found in miniature nearly every aspect of the nation.

In its history the Potomac has been colored by the East and the West, the South and the North, and both the river valley and its principal city have become a cross section of the nation. Deliberately chosen as the seat of government because it possessed these qualities, it has mirrored faithfully through the years of its history the preoccupations of the nation. Balanced, it has adapted itself to changing times and changing needs, but its individuality has not been lost. The process of adaptation continues today. In its life the river has been the uniting force, first binding together the rival colonies of Virginia and Maryland, leading them to common interests in agriculture, trade, and transportation, and, when its promise was frustrated, to revolution and the building of a new government. It showed that it could be a dividing line, a frontier, when the nation itself was disunited and at war. The years following the Civil War that saw the rise of metropolitan Washington saw, too, growing unity once more and an increasing concern with the Potomac, until at last it came to hold the secret to the future development of the nation's capital, and perhaps more.

ACKNOWLEDGMENTS

I WISH to acknowledge the assistance of the following individuals and institutions: for constant advice and encouragement at all stages of this work, Hervey Allen, coeditor of the Rivers of America; for suggestions in handling material in Chapter XII, Julian E. Berla, A.I.A., Washington, D.C.; for early advice on bibliography, Luther Cornwall, Washington, D.C.; for detailed criticism and forbearance, Miss Jean Crawford, of Rinehart & Company; for the opportunity to inspect Mountvina and his collection of Amelung glass, Alden Fisher, Frederick, Md.; for data on Cumberland, Mrs. Joseph Harris, Rock Hall Manor, Dickerson, Md.; for information concerning coal mining in the vicinity of Cumberland, J. William Hunt, editor of the Cumberland *Times*; for friendly collaboration in the illustration of this book, Mitchell Jamieson, Linden, Va.; for information concerning her distinguished ancestors, Miss Emily Johnson, Frederick, Md; for recollections of the Chesapeake and Ohio Canal, the late Samuel Creighton Jones, Dickerson, Md.; for assistance in tracing historical documents, Miss Mae T. Keller, Cumberland, Md.; for sharing his collection of folk songs of Potomac miners and canallers, Professor Maurice Matteson, Frostburg State Teachers College; for troubling to read and criticize important sections of this book, and for suggestions on handling details of the early history of West Virginia, Darel McConkey; for suggestions on the treatment of the development and future growth of the city of Washington, Mrs. Chloethiel Woodard Smith, A.I.A., A.I.P., Washington, D.C.; for discussion of the planning of Washington reflected in Chapters I and XI, Elbert Peets; for the loan of books and suggestions concerning early explorers at the mouth of the Monocacy, Gordon Strong, Stronghold, Dickerson, Md.; for geographical insight and companionship on explorations of the Potomac valley, Leonard Unger, Rockville, Md.; for permission to examine her scrapbooks relating to Charles Town and Jefferson County, Miss Carrie B. Wilson, Charles Town, W. Va.

I am also indebted for the use of the facilities of the following libraries and historical collections, and to the individuals named: Library of Congress, Verner W. Clapp and Paul Vanderbilt; Norfolk Public Library, Miss Mary Pretlow; Maryland Historical Society; New York Public Library; C. Burr Artz Library, Frederick, Md., Miss Josephine Etchison; Frederick County Historical Society; the library of the Brookings Institution, Washington, D. C.; the Hagerstown Public Library; the Charles Town Free Library, Charles Town, W. Va.; the Jefferson County Historical Society; the Columbia Historical Society; the Washington Public Library, particularly the Georgetown branch and the Washingtoniana Division.

A particular debt of gratitude is due James W. Foster, editor of the *Maryland Historical Magazine*, for helpful criticism in the final stages of this work; and to Dr. Ernst Cloos, of Johns Hopkins University, for tempering a rash literary venture into Potomac geology. In the conventional formula, these gentlemen are discharged from any responsibility for errors or matters of interpretation.

The index has been compiled with her customary skill and insight by Mrs. Eric St. C. Purdon, of Washington, D. C.

Finally, I wish to thank King Gordon, who first encouraged me to undertake this book.

FREDERICK GUTHEIM

Mount Ephraim,
Dickerson Station, Maryland
1949

BIBLIOGRAPHICAL NOTES

The references that follow have been assembled to serve the general reader who wishes to pursue further some ideas or projects this book may have suggested. The notes organized by chapters will also serve as a rough substitute for the more conventional apparatus of scholarship, and together with references in the text will identify the major sources employed.

For the General Reader: The traveler, whether on the road or in the library, can make no better beginning than by consulting the relevant volumes of the Federal Writers' Project American Guide Series: *Maryland* (New York, 1940); *Virginia* (New York, 1941); *Pennsylvania* (New York, 1940); *West Virginia* (New York, 1941); and *Washington* (Washington, 1937; and later revised editions). All the guides cited have excellent introductory bibliographies. The several works of Paul Wilstach, *Potomac Landings* (New York, 1921), *Tidewater Virginia* (New York, 1929), and *Tidewater Maryland* (New York, 1931), are highly readable, anecdotal accounts of life along the lower river during the seventeenth and eighteenth centuries. They have been criticized for miscellaneous inaccuracies (for example, J. L. Kibler, "Numerous Errors in Wilstach's Tidewater Virginia," *William and Mary Quarterly,** XI, 2, pp. 152–156) and the point of view is often sympathetic to the point of sentimentality, but neither of these disabilities should deter most readers from enjoying them. They have the further merit of being available in cheap popular editions.

The more active wayfarer in the Potomac country who would travel beyond the highways should know of the Potomac Appalachian Trail Club's *Guide to Paths in the Blue Ridge* (Washington, 1934). Robert Shosteck's *Potomac Trail Book* (1935) has a similar value. An account prepared by Jean Stephenson and Marion Park, *Sketches of Places Visited on, and Itinerary of the Potomac Appalachian Trail Club Trip, June 8–10, 1934* (hectographed, Washington, 1934), is an excellently organized, annotated tour of the Potomac on the Virginia side from Alexandria to the headwaters.

A novel of small literary value may acquire a special interest because

* *William and Mary College Quarterly Journal of History*. The form of citation used here gives the volume in Roman numerals, the number in Arabic, followed by the page reference.

of its characters, place, or theme, and the Potomac and the city of Washington have drawn their share. For the Potomac valley, see Gath [George Alfred Townsend], *Katy of Catoctin* (1886), an exceptionally capable account of events, particularly in western Maryland, at the time of the Civil War; Anon. [John Pendleton Kennedy], *Swallow Barn* (1832) and *Rob of the Bowl* (1838); Minnie Hite Moody, *Long Meadows* (1941); Anon. [William A. Caruthers], *Knights of the Horseshoe* (1845) and *Cavaliers of Virginia* (1834–1835); and a few others mentioned in the text. John Jennings, *Gentleman Ranker* (1942), deals with the Braddock expedition, and his *Call the New World* (1941) with events along the Potomac in the war of 1812. Much of Hervey Allen's five-volume series *The Disinherited* deals with the mountain reaches of the Potomac, and his *Action at Aquila* (1938) is set in the valley during the Civil War. These are only a few from a long list that might be given. Novels of Washington are even more numerous, but still fewer are worth reading. Anon. [Henry Adams] *Democracy* (1880), with all its artificialities, is one of the few earlier novels that can be recommended. Mark Twain and Charles Dudley Warner, *The Gilded Age* (1873), is a celebrated satire of the postbellum capital. Of the modern novels, Samuel Hopkins Adams, *Revelry* (1926), and Harvey Fergusson, *Capitol Hill* (1923), have at least the qualities of good journalism.

Survivals: The guidebooks cited and the Virginia Conservation Commission's admirable directory, "State Historical Markers of Virginia" (Richmond, 1941), are of particular value in relating events and places. Considerable mention of mills, forges, routes, and other places in the text of this book may also be helpful. A representative gristmill (among many) is Powell's Mill, near Berryville, Va. George Washington's mill, several miles west of Mount Vernon, has been restored, as has Pierce's Mill in Rock Creek Park, Washington. A good survival of the iron furnaces will be found at the mouth of Antietam Creek, although the ruins at Catoctin Furnace and the furnace at Lonaconing are worth visiting. Survivals of this type are numerous still in the Potomac valley, but no systematic critical account of them has been written.

The local historical museums and historic houses should be investigated. Models and documents dealing with Rumsey's inventions will be found in the museum of naval history at Annapolis. The National Museum in Washington contains an uneven collection of material relating to some subjects treated in this volume. Applicable to Piedmont agriculture is much of the collection in the museum of the Bucks County Historical Society, Doylestown, Pa.

Pictorial Sources: No published collection of views of the Potomac country as such exists, but some of the plates in Edward Beyer, *The Album of Virginia, or Illustrations of the Old Dominion* (1858), showing Virginia on the eve of destruction, are often relevant. The pictorial resources of the Maryland Historical Society are exceptional, covering internal development, technology, and other useful subjects as well as portraits. For certain periods, especially the Civil War and the years immediately before and after, the illustrated magazines, such as *Harper's*, *Ballou's*, the *Century*, and others, are valuable. Two contemporary photographic surveys of the Potomac will be found in the *National Geographic Magazine*, and the same publication and *Life* magazine are convenient sources of numerous pictures of contemporary Washington.

A collection of more than 400 views of the city of Washington is in the possession of the Columbia Historical Society, but its public exhibition awaits the acquisition of a suitable home for the society. The Library of Congress, Division of Prints and Photographs, and the Washington Public Library, Washingtoniana Division, have rich collections of additional material of the same sort. I. N. Phelps Stokes and Daniel C. Haskell, *American Historical Prints* (New York, 1933), contains reproductions of the more important early historical views of the city with excellent notes. The Stokes collection itself is in the New York Public Library. H. Paul Caemmerer, *A Manual on the Origin and Development of Washington* (Washington, 1939), is a convenient place to consult some of the historical material relating to the city's plan and physical appearance, but otherwise is a most unsatisfactory book.

Portraits are abundant, perhaps the larger number already in public collections. Two useful catalogues are *Exhibition of Contemporary Portraits of Personages Associated With . . . Virginia, 1585–1830*, Virginia Historical Society, 1929; and Anna Wells Rutledge, "Portraits in Varied Media in the Collections of the Maryland Historical Society," *Maryland Historical Magazine*, XLI, 4.

Maps have been used extensively in this study, and they are of many sorts. Little cartographic advance, so far as the Potomac River is concerned, was made between John Smith's map, conveniently reproduced in facsimile in Arber's edition of Smith's *Works*, and the first complete survey of the river in 1736–1737 by Robert Brooke. The original tracing of Brooke's excellent map is in the Coast and Geodetic Survey. It is also the subject of two critical articles by James W. Foster, "Maps of the First Survey of the Potomac River, 1736–1737," *William and Mary Quarterly*, XVIII, 2 and 4. John Warner's map, "The Sources of the Potomac, Rappahannock, and Patowmack in Virginia, as Surveyed According to Order in the Years 1736 and 1737," was examined by Lawrence Martin, chief of the Division of Maps, Library of Congress, in *William and Mary Quarterly*, XIX, 1. On the same question, Fairfax Harrison, "The Northern Neck Maps of 1737–1747," *William and Mary Quarterly*, IV (2), pp. 1–16, is a useful discussion of a matter it has not seemed worth entering in this book: Who found the source of the Potomac?

Of working maps the author has found useful, first mention should be made of the National Geographic Society's "Reaches of the Nation's Capital" (Washington, 1938). The colonial development at its height is shown in Abraham Bradley, Jr., and W. Harrison, Jr., "Map of the United States, Exhibiting the Post Roads, the Situations, Connections, and Distances of the Post Offices, Stage Roads, Counties . . . and the Principal Rivers," a copy of which is in the Library of Congress; and Lieutenant Farley and W. Harrison, "Map of the Country Between Washington and Pittsburg" (Georgetown, D.C., 1826), shows the situation at the period of internal improvements, including the proposed route of the Chesapeake and Ohio Canal.

Sources for Chapters:

I. A Potomac Prelude: References to George Washington's view of the capital city are widely diffused, but they are pulled together succinctly by Brooks Adams, *The Degradation of the Democratic Dogma* (New York, 1919). The details of President Adams's trip from Philadelphia to Washington are given in H. T. Taggart, "The Presidential Journey in 1800," *Records of the Columbia Historical Society*, III, pp. 183 ff., and some further information will be found in John Clagett Proctor, *Washington Past and Present* (Washington, 1930), for which Edwin M. Williams, at I, pp. 67 ff., contributed a chapter on the removal of the seat of government. The general authority for the municipal history of Washington to 1878 is Wilhelmus B. Bryan, *A History of the National Capital* (2 vols., New York, 1914). For the anecdote concerning the bells in the White House, the *Letters* of Abigail Adams furnish a starting point, but the story is finished off by Mrs. Abby Gunn Baker, "The Erection of the White House," *Records of the Columbia Historical Society*, XVI, pp. 120 ff. Alfred J. Morrison, *The District in the XVIIIth Century* (Washington, 1909), is a useful compilation of travelers' accounts, and Margaret Bayard Smith, *The First Forty Years of Washington Society* (New York, 1908), is indispensable. Some other sources are noted in the text and elsewhere in these notes. Two extremely good views of Washington from the heights of Georgetown are the earliest known engraved view of Washington from Georgetown (1794) by G. I. Parkyns, impressions of which are in the New York Public Library and the Library of Congress, and a widely reproduced view of Georgetown and Washington in 1800, drawn by G. Beck and published in aquatint in Philadelphia by Atkins and Nightengale.

II. The Captains and the Kings: The available information concerning Smith, Spelman, and Argall will be conveniently found in Edward Arber's unexcelled edition of Captain John Smith's *Works* (Birmingham, 1884). Edward D. Neill, *Founders of Maryland* (Albany, 1876), contains the journal of Captain Henry Fleet. (While the text ignores the issue, the myth of early Spanish discovery of the Potomac is dissolved by Louis Dow Scisco, "Discovery of Chesapeake Bay," *Maryland Historical Magazine*, XL, 4, pp. 277–286.)

Mrs. Alice L. L. Ferguson has carefully excavated the large Indian settlement, Moyaone, and recorded her findings in *Moyaone and the Piscataway Indians* (Washington, 1937), making necessary a substantial revision of the data in the *Handbook of American Indians* (2 vols., Washington, 1912) and other standard works. William Wallace Tooker, *Algonquin Series*, Part VIII (1901), and "The Algonquin Terms Patawomeke and Massawomeke," *American Anthropologist*, April, 1894, gives the significance of the river's name.* William Henry Holmes, "Some Implements of the Potomac and Chesapeake Tidewater Province," *Fifteenth Annual Report*, Bureau of American Ethmology (1893–1894), contains some useful details, as does Gerard Fowke, *Archeological Investigations in the James and Potomac Val-*

* The following seventeenth-century forms of "Potomac" have been encountered, referring to the Indian tribe of that name, their settlement at the mouth of Potomac Creek, or the river: Patamack, Patawoenicke, Patawomeke, Patomac, Patomeck, Patomecke, Patomek, Patowamack, Patowmeck, Patowomack, Patowomeek, Patowomek, Pattawomeke, Petawomeek, Potomeack, Potomack, Potowmack, Satowomick, Satawomecke.

leys (Washington, Smithsonian Institution, 1894). John Leeds Bozman's delightful early *History of Maryland* (2 vols., Baltimore, 1837) has notes locating the various early Indian settlements, and some information of value will be found in the second part of Raphael Semmes, "Aboriginal Maryland," *Maryland Historical Magazine*, XXIV, pp. 157 ff. and 196 ff.

III. The Tidewater Frontier: The general histories of Maryland and Virginia become useful from this point. None are wholly satisfactory. For Maryland, John Thomas Scharf's *History of Maryland from the Earliest Period to the Present* (3 vols., Baltimore, 1879) and *History of Western Maryland* (2 vols., Philadelphia, 1882) most nearly cover the state's three centuries, and while diffuse and subject to some inaccuracy they contain the most information of relevance to this book. James McSherry, *History of Maryland from its First Settlement in 1634 to the year 1848* (Baltimore, 1848; revised edition by Dr. Bartlett B. James, Baltimore, 1904) and James V. L. McMahon, *An Historical View of the Government of Maryland from its Colonization to the Present Day* (Baltimore, 1831), are far more valuable than would be suspected from their date of publication. Matthew Page Andrews, *History of Maryland: Province and State* (Garden City, 1929), is the most recent effort to cover the entire history of Maryland in brief compass, but suffers from preoccupation with the earliest periods and an orientation to Tidewater—and other shortcomings. Another one-volume history that hardly got past the Revolutionary period is William Hand Browne, *Maryland: A History of a Palatinate* (Boston, 1884).

The richness and length of Virginia's three and one-half centuries have thus far defeated the efforts of any individual writer to discipline them in historical terms. The most considerable work is the first three volumes of Philip Alexander Bruce and Others, *History of Virginia* (6 vols., New York, 1924). The early years have been well explored, notably by Philip Alexander Bruce, *Economic History of Virginia in the Seventeenth Century* (2 vols., New York, 1896); *Institutional History of Virginia in the Seventeenth Century* (2 vols., New York 1910); and *Social Life in Virginia in the Seventeenth Century* (Richmond, 1907). Matthew Page Andrews, *Virginia, the Old Dominion* (Garden City, 1937), and John Esten Cooke, *Virginia, a History of the People* (Boston, 1883), are readable one-volume histories.

The various works of Charles McLean Andrews, especially *The Colonial Period in American History* (4 vols., New Haven, 1934) and

Our Earliest Colonial Settlements (New York, 1933), are of great value for this chapter. The complexities of land tenure are unraveled in part by Beverly W. Bond, *The Quit-Rent System in the American Colonies* (New Haven, 1919), and Clarence P. Gould, *The Land System of Maryland, 1720–1765* (Johns Hopkins Studies,* XXXI, 1, Baltimore, 1913). Archaeological studies of Henry Chandlee Forman, especially *Jamestown and St. Mary's* (Baltimore, 1838), are referred to in the text.

Of the large number of articles and monographs that have been consulted, only a few can be listed here. Hu Maxwell, "The Use and Abuse of Forests by the Virginia Indians," *William and Mary Quarterly*, XIX, pp. 73–103, concludes that "if the discovery of America had been postponed 500 years, Virginia would have been pasture or desert." One of the better accounts of a still shadowy figure is "Mistress Margaret Brent, Spinster," by Julia Cherry Spruille, *Maryland Historical Magazine*, XXIX, 4. Eugene Irving McCormac, *White Servitude in Maryland* (Baltimore, 1904), is enlightening, and Albert Emerson Smith, "Transportation of Convicts to the American Colonies in the Seventeenth Century," *American Historical Review*, XXXIX, pp. 232–249, leaves little doubt that "our jailbirds were criminals and their characters do not stand whitewashing." Of a number of references to the Northern Neck, the polished account of Fairfax Harrison, *Landmarks of Old Prince William; a Study of Origins in Northern Virginia* (2 vols., Richmond, 1924), has been of particular value in this and later chapters. An able summary of the Northern Neck proprietary is given in Douglas Southall Freeman, *George Washington* (2 vols., New York, 1948), I, pp. 447–513. Moncure D. Conway, *Barons of the Potomac and Rappahannock* (New York, 1902), is of uneven accuracy but valuable insight. C. W. Baird's *History of the Huguenot Immigration to America* (2 vols., New York, 1885) is an old but adequate account, with helpful references to the Potomac. Much information in this and other chapters is from William B. Marye's series in the *Maryland Historical Magazine*, "Patowmeck Above Ye Inhabitants," especially in Vols. XXX, XXXII, and XXXIV.

IV. The Tobacco Civilization: References on all aspects of tobacco are abundant and need not be repeated here. Much of the information on Chotank has been assembled by Fairfax Harrison in *Prince William*, cited above, but see also George Fitzhugh, writing in *DeBows Re-*

* *Johns Hopkins University Studies in History and Political Science.*

view, XXX, p. 78; "Recollections of Chotank," in *Southern Literary Messenger*, I, p. 43; and for the attempt to reproduce Chotank life in the Piedmont, Fairfax Harrison, "Brent Town, Ravensworth, and the Huguenots in Stafford," *Tyler's Quarterly Magazine*, V, pp. 164–185, and especially pp. 173–174.

Although much has been written on the golden age of the Potomac plantation, the subject still lacks the treatment it deserves. The information here has come chiefly from travelers' accounts, biographies, reminiscences, and similar sources generally indicated in the text, or from observation; and the orientation is largely that provided by Ulrich B. Phillips, *Life and Labor in the Old South* (New York, 1927); Francis Pendleton Gaines, *The Southern Plantation, a Study in the Development and the Accuracy of a Tradition* (New York, 1925); and those remarkable works of Avery O. Craven, especially "Soil Exhaustion as a Factor in the Agricultural History of Virginia and Maryland, 1606–1860," XIII, 1, *University of Illinois Studies in the Social Sciences* (Urbana, Ill., 1925). Lewis Cecil Gray, *History of Agriculture in the Southern States* (2 vols., Washington, 1933), is also of some value here, but more in Chapter VII and elsewhere.

The two principal volumes covering the indispensable observations of Philip Fithian are *Philip Vickers Fithian, Journal and Letters*, edited by John R. Williams (Princeton, 1900), and *Philip Vickers Fithian, Journal, 1775–1776*, edited by Robert G. Albion and Leonidas Dodson (Princeton, 1934). Louis Morton, *Robert Carter of Nomini Hall* (Williamsburg, Va., 1941), is an excellent economic analysis of the Carter plantations.

Thomas T. Waterman, *The Mansions of Virginia, 1706–1776* (Chapel Hill, 1945), is the leading architectural reference here and in Chapter VII, supplemented by Henry C. Forman, *Early Manor and Plantation Houses of Maryland* (Easton, Md., 1934); Fiske Kimball, *Domestic Architecture of the American Colonies* (New York, 1922); the earlier works of Robert A. Lancaster, *Historic Virginia Homes and Churches* (Philadelphia, 1915); Edith Tunis Sale, *Interiors of Virginia Houses of Colonial Times* (Richmond, 1927); and Henry C. Forman, *Jamestown and St. Mary's* (Baltimore, 1938).

V. Potomac Above the Falls: The most used references for the Ohio Company have been Kenneth P. Bailey, *The Ohio Company of Virginia and the Westward Movement, 1748–1792* (Glendale, Calif., 1939); Samuel M. Wilson, *The Ohio Company of Virginia*, 1748–1798 (Lexington, Ky., 1926); and Herbert T. Leyland, *The Ohio Company, a Colonial Corporation* (Cincinnati, 1921). The part of the Potomac planters is the chief concern of Kate Mason Rowland, "The Ohio Company," *William and Mary Quarterly*, I, pp. 197 ff. Kenneth P. Bailey, *Thomas Cresap, Maryland Frontiersman* (Boston, 1944); Lawrence C. Wroth, "Thomas Cresap, a Maryland Pioneer," *Maryland Historical Magazine*, IX, pp. 1–37; and further details on Cresap in *Maryland Geological Survey*, VI (1908), p. 160, are sufficient among numerous lesser references. *The Journal of Christopher Gist* (Pittsburgh, 1893) should also be mentioned.

The immigration of various national and religious groups is traced by a copious bibliography, the greater part of it biased as well as stimulated by pride and patriotism. Concerning the Pennsylvania Germans and their culture special mention should be made of Daniel Wunderlich Nead, *The Pennsylvania German in the Settlement of Maryland* (Lancaster, Pa., 1914), and Charles E. Kemper's many articles in the *Virginia Magazine*.* Immigrants referred to in this chapter may be traced further in William J. Hincke, editor, "The Journal of Francis Louis Michel," *Virginia Magazine*, XXIV, 3, pp. 275 ff.; Vincent H. Todd, *Christoph von Graffenreid's Account of the Founding of New Bern* (Raleigh, N.C., 1920); and, concerning Martin Chartier, Albert Cook Myers, *Papers Read Before the Lancaster (Pa.) County Historical Society*, XXIX, pp. 10 ff. Lyman Carrier, "The Veracity of John Lederer," *William and Mary Quarterly*, XIX, (2), 4, establishes it.

Substantially all references to George Washington and the West have been or will probably be superseded by Douglas Southall Freeman's *George Washington* (2 vols., New York, 1948), a work published too recently to affect this volume, but John W. Wayland, "Washington West of the Blue Ridge," *Virginia Magazine*, XLVIII, pp. 191–201; Louis Knott Koontz, "Washington on the Frontier," *Virginia Magazine*, XXXVI, pp. 305–327; and C. H. Ambler, *George Washington and the West* (Chapel Hill, 1936), might also be mentioned.

An adequate biography of Thomas Fairfax has long been needed. The references in Freeman's *Washington* revise our understanding of an important but largely misunderstood relationship. Most of the lore of Lord Fairfax at Greenway Court commonly encountered can be traced to the Rev. Andrew Burnaby, *Travels Through the Middle*

* *Virginia Magazine of History and Biography.*

Settlements of North America in the Years 1759 and 1760 (3d ed., 1798). Edward D. Neill, *The Fairfaxes of England and America* (Albany, 1868), offers some useful observations.

On the frontier movement itself, Frederick Jackson Turner, *The Frontier in American History* (New York, 1921), especially Chapters I and III, and the same author's *The Significance of Sections in American History* (New York, 1932), especially Chapters II and IV, are of obvious value. See also Alfred P. James, "Approaches to the Early History of Western Pennsylvania," in *Western Pennsylvania Historical Magazine*, XIX, pp. 203 ff.

VI. Cradle of the Republic: Much of the information and direction of this chapter, including the Bennet Allen episode, derives from Charles Albro Barker, *The Background of the Revolution in Maryland* (New Haven, 1940). Philip Crowl, *Maryland During and After the Revolution* (*Johns Hopkins University Studies*, LXI, 1, Baltimore, 1943), is a supplementary reference.

Potomac towns are described in Henry J. Berkley, "Extinct River Towns of the Chesapeake Bay Region," *Maryland Historical Magazine,* XIX, 2, pp. 125–134; Ethel Roby Hayden, "Port Tobacco, Lost Town of Maryland," *Maryland Historical Magazine*, XL, 4; Esther B. Stabler, "Triadelphia, Forgotten Maryland Town," *Maryland Historical Magazine*, XLIII, 2; Henry J. Berkley, "The Port of Dumfries," *William and Mary Quarterly*, IV (2), pp. 99–122; and L. W. Wilhelm, *Local Institutions in Maryland, Johns Hopkins Studies,* (3), Nos. V–VII, Baltimore, 1905). Francis Makemie's plea for urban development along the Potomac, *A Plain and Friendly Persuasive to the Inhabitants of Virginia and Maryland for Promoting Towns and Cohabitation* (London, 1905), is conveniently reprinted in the *Virginia Magazine*, IV, pp. 255–271.

Kate Mason Rowland, *The Life of George Mason, 1725–1792* (2 vols., New York, 1892), remains an indispensable source, although a fresh view is given by Helen Day Hill, *George Mason, Constitutionalist* (Cambridge, 1938). The letters of the Rev. Jonathan Boucher have been reprinted in the *Maryland Historical Magazine*, VII–X. Another eighteenth-century divine of some historical interest is presented in Elizabeth Hesselius Murray, *One Hundred Years Ago, or the Life and Times of the Rev. Walter Dulany Addison, 1769–1848* (Philadelphia, 1895).

Although of more general interest, John Chester Miller, *Origins of the American Revolution* (Boston, 1943), was of some value in orienting events along the Potomac within a larger framework. Some further suggestions were given in Clement Eaton, *Freedom of Thought in the Old South* (Durham, N.C., 1940).

VII. "Smooth Blue Horizons": A recent and localized survey of the geology of the Potomac region is Charles Butts, G. W. Stose, and Anna I. Jones, *Southern Appalachian Region* (Washington, 1933), a guidebook prepared under the auspices of the U.S. Geological Survey for the International Geological Congress in 1933. A geographical synopsis can be extracted from the larger work of J. Russell Smith, *North America* (New York, 1925), that leaves little to be desired. The observations of N. S. Shaler, *Nature and Man in America* (New York, 1891), however unsubstantiated, would comfort local patriots and enthusiasts for the Potomac Piedmont.

In addition to earlier references on agriculture in general, Clark Wissler, "Aboriginal Maize Culture as a Typical Culture Complex," *American Journal of Sociology*, XXI, 5, shows how the English settlers adopted the maize culture of the Indians, and provides a background to the absorption of the small-grain culture of the northern Europeans described in this chapter.

The first volume of James Morton Callahan, *History of West Virginia, Old and New* (New York, 1923), and the same author's *Semi-Centennial History of West Virginia* (1913) contain valuable accounts of primitive agriculture that survived in the mountainous region.

The introduction of lime and the Loudoun system can be traced in Rodney H. True, "John Binns of Loudoun," *William and Mary Quarterly*, January, 1922, pp. 20–39, and *A Treatise on Practical Farming* (Frederick, Md., 1803); an account of Israel Janney may be found in N. F. Cabell, "Some Fragments of an Intended Report on the Post-Revolutionary History of Agriculture in Virginia," *William and Mary Quarterly*, XXVI, 3, p. 164. References to mills and milling are widely diffused, but a convenient summary is Arthur G. Peterson, "Flour and Grist Milling in Virginia," *Virginia Magazine*, XLIII, pp. 97 ff. Much detail of Piedmont agricultural life is given in a reminiscence, William J. Grove, *History of Carollton Manor* (Frederick, Md., 1928).

Since most travelers' accounts are well known and generally identified in the text, there is no occasion to recapitulate here.

In addition to earlier references on architecture, W. F. Alexander, "The Bullskin Run," *Magazine of the Jefferson County Historical So-*

ciety, II, pp. 20–29; and virtually the whole of VII, December, 1941, the same publication, might be cited. The latter reference at p. 23 contains Thomas Tileson Waterman, "Early Architecture in Western Virginia."

VIII. The March of Industry: Dorothy Mackay Quynn, "Johann Friedrich Amelung at New Bremen," *Maryland Historical Magazine*, XLIII, pp. 155–179, is an account with illustrations that leaves little to be desired, although Mrs. Quynn does not appear to have seen a manuscript memoir of some value in the possession of the Frederick County Historical Society. The most important public collections of Amelung glass are in the Metropolitan Museum of Art.

In general, the state of material on early local industry is unsatisfactory. An entire local pottery industry has vanished without trace. Minor references to furnaces and forges are numerous in publications of local historical societies, but they have not received any systematic treatment for the region as a whole. A helpful parallel sketch is Arthur C. Bining, *Pennsylvania Iron Manufacture in the Eighteenth Century* (Harrisburg, Pa., 1938), the technical descriptions in which would doubtless apply equally to the Potomac works. Bernard C. Steiner, *Western Maryland in the Revolution* (Baltimore, 1902), and Edward S. Delaplaine, *Thomas Johnson* (New York, 1927), both support the view that the Potomac iron works were of importance to the Revolutionary war effort.

The early days of coal mining are in a similar historical state. One satisfactory account whose limitations are reasonably clear is the *History of the Consolidation Coal Company* (New York, 1934) compiled by Charles E. Beachley, secretary of the company on its 70th anniversary. Another sketch is Arthur Lovell, *Borden Mining Company, a Brief History* (Frostburg, Md., 1938). J. William Hunt, editor of the Cumberland *Times* and author of fugitive historical notes published there, has attacked numerous aspects of the topic and printed various recollections and documents which readers have contributed.

Henry T. McDonald, "The Armory and Arsenal at Harpers Ferry," *Magazine of the Jefferson County Historical Society*, II, pp. 4–19, blocks in the main lines of development.

Francis Thomas's pamphlet leaves little unsaid, and the text here does little more than summarize Meshach Browning's *Forty-four Years of the Life of a Hunter* (Philadelphia, 1859; reprinted with new illustrations, 1928).

IX. The Potomac Route to the West: The most recent and best general work on the Potomac route to the west is Walter S. Sanderlin, *The Great National Project: A History of the Chesapeake and Ohio Canal* (Johns Hopkins Studies, LXIV, 1, Baltimore, 1946). Cora Bacon-Foster, "Early Chapters in the Development of the Potomac Route to the West," *Proceedings of the Columbia Historical Society*, XV (1911), and George Washington Ward, *The Early Development of the Chesapeake and Ohio Canal Project* (*Johns Hopkins Studies*, XVII, 9, 10, and 11, Baltimore, 1899), are earlier works whose value has not been wholly superseded for the purposes of this book. On the question of competition between Washington and Baltimore, Walter S. Sanderlin, "The Maryland Canal Project—an Episode in the History of Maryland's Internal Improvements," *Maryland Historical Magazine*, XLV, 1, expands his more general earlier treatment.

The rich detail of life on the National Road is drawn largely from that fascinating recollection, Thomas Searight, *The Old Pike* (Uniontown, Pa., 1894). Archer B. Hulbert, *The Cumberland Road* (Historic Highways of America, X, Cleveland, 1904), supplements it. Edward Hungerford's pageantic *The Story of the Baltimore and Ohio Railroad* (2 vols., New York, 1928) must stand in place of the more critical account this subject requires, while Milton Reizenstein's more substantial *The Economic History of the Baltimore and Ohio Railroad, 1827–1853* (*Johns Hopkins Studies*, XV, 7 and 8, Baltimore, 1897) is specialized as to both time and interest.

X. The War Along the Potomac: Numerous standard references to the Civil War have been consulted, but those of particular relevance to this chapter are Douglas Southall Freeman, *Lee's Lieutenants* (3 vols., New York, 1944); Henry Kyd Douglas, *I Rode With Stonewall* (Chapel Hill, 1940); and Margaret Leech, *Reveille in Washington* (New York, 1941).

XI. The Seat of Government: To the establishment of the commission form of government, Bryan's *Washington* continues to be useful. A little new material on the regime of Alexander Shepherd is contained in Ulysses S. Grant, 3rd, "Territorial Government of Washington, D.C." (Washington, Women's City Club, 1929). Laurence F. Schmeckebier, *The District of Columbia, Its Government and Administration* (Washington, 1928) and other studies by the same author bear directly on the political problems of the city. The physical development of the city can be traced most readily in maps, models, views, and documents in the National Capital Park and Planning Commission, some of which have been published in numerous reports. To develop the changing character of the city, much use has been made of biographies, memoirs, letters, and other sources, nearly all of them published, to which reference is usually made in the text. Carl Russell Fish, *The Civil Service and the Patronage* (Cambridge, 1920), shows with particular clarity the evolution of a permanent class of public servants long before the Pendleton Act. The functional evolution of the administrative establishment by bureaus is best traced in the Service Monographs published by the Institute for Government Research, Brookings Institution; and an effort has been made to assemble much of this material in a more general form by Lloyd Milton Short, *The Development of National Administrative Organizations in the United States* (Baltimore, 1923). An adequate administrative history and biographies of representative administrators are generally lacking. Arthur W. MacMahon and John D. Millett's *Federal Administrators* (New York, 1939) contains some useful historical material, but its concern is chiefly with the present. Laurence F. Schmeckebier, *The Statistical Work of the National Government* (Baltimore, 1925), is still a useful historical reference on the growth of statistical series in response to the needs of the times, although in other respects it is badly dated. Helen Nicolay, *Sixty Years of the Literary Society* (Washington, 1934), contains some engaging glimpses of Washington society. To those who are intrigued by the relationships between public buildings and the activities they contain, the first edition of *Washington, City and Capital* (Washington, 1937) contains much specialized material wisely dropped from later editions of this guide. A manuscript written by the late Fred Wilbur Powell, in the library of the Federal Works Agency, in the course of examining the public buildings function, comes to some interesting conclusions on the planning of the Federal City.

XII. A Potomac Prospect: The growing consciousness of the river basin as an integrated unit can be traced before 1933, but its greatest impetus came from the regional planning work of the Tennessee Valley Authority commenced in that year, the publications of the National Resources Planning Board, and the indefatigable work of Morris Llewellyn Cooke. The annual reports and other publications of the Potomac River Basin Commission should be consulted. The basic current Army Engineers proposals for the river are contained in *Potomac River and Tributaries*, House Document 622, 79th Congress, 2nd Sess., Washington, 1946; but earlier and later reports are also illuminating. The State Planning Commission in Maryland and the State Planning Board in Virginia are sources of pertinent information and reports, and *Looking Forward* (Silver Spring, Md., 1942), a report of the Maryland-National Capital Park and Planning Commission, and the Maryland State Planning Commission's report on the *Baltimore-Washington-Annapolis Area* (Baltimore, 1937) are contributions toward metropolitan regional planning and the decentralization of the city. Major planning studies are now in preparation by the National Capital Park and Planning Commission, for some preliminary information concerning which the author is indebted by Mr. John Nolen. The decentralization of federal agencies is treated with breadth and wisdom by John M. Gaus, *Regional Factors in National Development* (Washington, 1938), a report of the National Resources Planning Board, and in numerous periodical references since, as well as in some unpublished studies of the Bureau of the Budget which the author has been able to examine.

INDEX